Algorithms for Intelligent Systems

This book series publishes research on the analysis and development of algorithms for intelligent systems with their applications to various real world problems. It covers research related to autonomous agents, multi-agent systems, behavioral modeling, reinforcement learning, game theory, mechanism design, machine learning, meta-heuristic search, optimization, planning and scheduling, artificial neural networks, evolutionary computation, swarm intelligence and other algorithms for intelligent systems.

The book series includes recent advancements, modification and applications of the artificial neural networks, evolutionary computation, swarm intelligence, artificial immune systems, fuzzy system, autonomous and multi agent systems, machine learning and other intelligent systems related areas. The material will be beneficial for the graduate students, post-graduate students as well as the researchers who want a broader view of advances in algorithms for intelligent systems. The contents will also be useful to the researchers from other fields who have no knowledge of the power of intelligent systems, e.g. the researchers in the field of bioinformatics, biochemists, mechanical and chemical engineers, economists, musicians and medical practitioners.

The series publishes monographs, edited volumes, advanced textbooks and selected proceedings.

More information about this series at http://www.springer.com/series/16171

B. Vinoth Kumar · P. Sivakumar ·
M. M. Rajan Singaravel · K. Vijayakumar
Editors

Intelligent Paradigms for Smart Grid and Renewable Energy Systems

Springer

Editors
B. Vinoth Kumar
Department of Information Technology
PSG College of Technology
Coimbatore, Tamil Nadu, India

P. Sivakumar
Department of Electrical and Electronics Engineering
PSG College of Technology
Coimbatore, Tamil Nadu, India

M. M. Rajan Singaravel
Department of Electrical and Electronics Engineering
National Institute of Technology Puducherry
Karaikal, Puducherry, India

K. Vijayakumar
Department of Electronics and Communication Engineering
Indian Institute of Information Technology, Design and Manufacturing, Kancheepuram
Chennai, Tamil Nadu, India

ISSN 2524-7565 ISSN 2524-7573 (electronic)
Algorithms for Intelligent Systems
ISBN 978-981-15-9967-5 ISBN 978-981-15-9968-2 (eBook)
https://doi.org/10.1007/978-981-15-9968-2

This Springer imprint is published by the registered company Springer Nature Singapore Pte Ltd.
The registered company address is: 152 Beach Road, #21-01/04 Gateway East, Singapore 189721, Singapore

Preface

The dependency on renewable energy resources is becoming vital to face the energy crisis that might show up shortly. It is a clean and green way to meet the ever-increasing energy demand. The revolution of the twenty-first century has begun with the advent of power electronic technologies to harvest energy from renewable sources. The rapid growth in the past decade even brings vehicular technologies to depend on power electronics. Thus, the integration of renewable energy sources with the grid plays an important role to regulate the energy generation and management. The renewable energy resources such as wind energy, solar energy, solar thermal energy, biomass energy, and hybrid energy always depend on the weather conditions. Thus, harvesting energy from such renewable sources needs not only the power electronic converters, but also a smart energy management system.

A separate field of study is devised to deal with smart energy management in the electric grid called the "smart grid". The smart grid is an electric grid equipped with communication devices, Internet, automation devices, and the like. It provides two-way communication between the utility and consumer. Thus, the electric grid becomes capable of knowing the varying energy demand, as well as generation from renewable energy sources and so smart energy management is made possible. This book is intended to provide knowledge on the integration of renewable energy sources with the smart grid. The smart grid must employ intelligent paradigms such as fuzzy systems, artificial intelligence, adaptive learning, and optimization techniques. Such optimization techniques yield a better result in terms of energy harvesting and energy management.

The chapters of this book will give a wide range of analysis on the application of intelligent paradigms for smart grid and renewable energy systems. A brief introduction to each chapter is as follows.

Chapter "Artificial Swarm Intelligence—A Paradigm Shift in Prediction, Decision-Making and Diagnosis" discusses the application of artificial swarm intelligence algorithms in smart grid and renewable energy systems.

Chapter "Robust Optimal Tuning of UPFC Using Single Objective Optimization Algorithm" depicts the tuning of the unified power flow controller (UPFC) using particle swarm optimization (PSO) to mitigate the electromechanical oscillations in power systems. The effectiveness of the proposed method is validated through MATLAB simulations and stability analysis.

Chapter "Hybrid Sensorless MPPT Algorithm for Detecting Fast Irradiation Changes and Partial Shading Conditions on PV Systems" illustrates a hybrid maximum power point tracking (MPPT) algorithm for detecting the rapid changes in solar radiation and the occurrence of partial shading on the photovoltaic system without employing sensors. The proposed method is validated and composed of conventional MPPT algorithms.

Chapter "A New Model of Demand Response in Smart Grid" presents a new model of demand response in smart grid which has an interactive approach to consider the interaction between utility and industrial customers.

Chapter "A Smart Bidirectional Power Interface Between Smart Grid and Electric Vehicle" provides the fundamental concepts of vehicle electrification from the smart grid. Also, it reviews the basics of an electric vehicle.

Chapter "Nature-Inspired Optimization Algorithms for Renewable Energy Generation, Distribution and Management—A Comprehensive Review" provides a detailed study of nature-inspired optimization algorithms for the optimization of renewable power generation systems with its recent progress. The classical paradigms of optimization and their deployment of the renewable power system to effectively manage and enhance their efficiency for power generation, distribution, and management are studied.

Chapter "Learning Automata and Soft Computing Techniques Based Maximum Power Point Tracking for Solar PV Systems" introduces the learning automata concept and its adaptability for the development of the maximum power point tracking (MPPT) algorithm. Also, it proposes a hybrid MPPT technique which is analysed by conducting extensive simulation studies for different input conditions.

Chapter "Fuzzy Logic Controller Based Plug-In EV Battery Charger" proposes a fuzzy logic controller-based bidirectional inverter for a plug-in electric vehicle battery charger to facilitate the G2V and V2G technology.

Chapter "Nature-Inspired Algorithms for Maximum Power Point Tracking in Photovoltaic Systems Under Partially Shaded Conditions" develops and analyses the performance of nature-inspired optimization techniques towards maximum power point tracking (MPPT).

Chapter "Soft Computing Techniques-Based Low Voltage Ride Through Control of Doubly Fed Induction Wind Generator" proposes a fault confrontation controller (FCC) design to augment the feature of a low-voltage ride through (LVRT) in doubly fed induction generator wind turbine. Considering the system's nonlinear nature, an attractive fault confrontation controller has been constructed using computational intelligence (CI) techniques, namely fuzzy logic, back propagation network (BPN), and an adaptive neuro-fuzzy inference system (ANFIS).

Chapter "Harmonic Current Estimation of a Non-linear Load Using Artificial Neural Network" proposes an artificial neural network (ANN) approach to compute harmonic current a nonlinear load. The proposed architecture is compared with the existing ones in terms of accuracy and complexity.

Chapter "Demand Response in Smart Residential Buildings" presents a demand response framework named as novel residential energy management system (NREMS) for prosumers to utilize the maximum in-house power generation from renewable energy resources.

We are grateful to the authors and reviewers for their excellent contributions to making this book possible. Our special thanks go to Prof. Dr. Jagdish Chand Bansal, Prof. Dr. Kusum Deep, and Prof. Dr. Atulya K. Nagar (Series Editor of Algorithms for Intelligent Systems) for the opportunity to organize this edited volume.

We are grateful to Springer, especially to Mr. Aninda Bose (Senior Editor), for the excellent collaboration.

This edited book covers the theory, case studies, and intelligent paradigms about the smart grid and renewable energy systems. We hope the chapters presented will inspire researchers and practitioners from academia and industry to spur further advances in the field.

Coimbatore, India — Dr. B. Vinoth Kumar
Coimbatore, India — Dr. P. Sivakumar
Karaikal, India — Dr. M. M. Rajan Singaravel
Chennai, India — Dr. K. Vijayakumar
August 2020

Contents

Artificial Swarm Intelligence—A Paradigm Shift in Prediction, Decision-Making and Diagnosis 1
V. J. K. Kishor Sonti and G. Sundari

Robust Optimal Tuning of UPFC Using Single Objective Optimization Algorithm 27
G. Kannayeram, P. S. Manoharan, N. B. Prakash, T. Sivakumar, and R. Saravanan

Hybrid Sensorless MPPT Algorithm for Detecting Fast Irradiation Changes and Partial Shading Conditions on PV Systems 39
Balaji Veerasamy, Takaharu Takeshita, Viswanath Anbazhagan, Angamuthu Ananth, and Selvagopinath Mayandi

A New Model of Demand Response in Smart Grid 71
Somayeh Siahchehre Kholerdi and Ali Ghasemi-Marzbali

A Smart Bidirectional Power Interface Between Smart Grid and Electric Vehicle 103
M. Nandhini Gayathri

Nature-Inspired Optimization Algorithms for Renewable Energy Generation, Distribution and Management—A Comprehensive Review 139
Vamsi Krishna Reddy Aala Kalananda and Venkata Lakshmi Narayana Komanapalli

Learning Automata and Soft Computing Techniques Based Maximum Power Point Tracking for Solar PV Systems 227
S. Sheik Mohammed, D. Devaraj, and T. P. Imthias Ahamed

Fuzzy Logic Controller Based Plug-In EV Battery Charger 263
N. Sujitha and S. Krithiga

Nature-Inspired Algorithms for Maximum Power Point Tracking in Photovoltaic Systems Under Partially Shaded Conditions 283
V. Vignesh Kumar and C. K. Aravind

Soft Computing Techniques-Based Low Voltage Ride Through Control of Doubly Fed Induction Wind Generator 305
M. Maheswari, S. K. Indumathi, and A. K. Parvathy

Harmonic Current Estimation of a Non-linear Load Using Artificial Neural Network 337
A. Venkadesan

Demand Response in Smart Residential Buildings 361
S. L. Arun and M. P. Selvan

About the Editors

B.Vinoth Kumar received the B.E. degree in Electronics and Communication Engineering from the Periyar University in 2003, and the M.E. and Ph.D. degrees in Computer Science and Engineering from the Anna University in 2009 and 2016, respectively. He is Associate Professor with 16 years of experience at PSG College of Technology. His current research interests include computational intelligence, memetic algorithms and image processing. He has established an Artificial Intelligence Research (AIR) Laboratory at PSG College of Technology. He is Life Member of the Institution of Engineers (India) (IEI), International Association of Engineers (IAENG) and Indian Society of Systems for Science and Engineering (ISSE). He is the author of more than 26 papers in refereed journals and international conferences. He has edited three books with reputed publishers such as Springer and CRC Press. He serves as Guest Editor/Reviewer of many journals with leading publishers such as Springer, Inderscience and De Gruyter.

P. Sivakumar received the B.E. degree in Electrical & Electronics Engineering from Anna University in 2006, and the M.E. and Ph.D. degrees in Embedded System from Anna University in 2009 and 2018, respectively. He is Assistant Professor with 10 years of experience at PSG College of Technology. His current research interests include model-based design of automotive software and computational intelligence. He is Life Member of the Institution of Engineers (India) (IEI) and International Association of Engineers (IAENG). He has published papers in peer-reviewed national/international journals and conferences and Reviewer of international journals. He has reviewed the 8th edition of a book titled "Understanding of Automotive Electronics" Elsevier.

M. M. Rajan Singaravel received his B.Tech. degree in Electrical and Electronics Engineering from SASTRA University, Thanjavur, India, in 2008, and the M.E. degree in Power Electronics and Drives from PSG College of Technology (Anna University), Coimbatore, India, in 2010. He was admitted as Ph.D. Research Scholar with MHRD (Govt. of India) fellowship in the Department of Electrical and Electronics Engineering, National Institute of Technology, Tiruchirappalli, and completed his dissertation in the year 2015. He is currently working as Faculty in the Department of Electrical & Electronics Engineering, National Institute of Technology Puducherry, Karaikal. His interests are battery storage systems, hybrid wind–solar systems and power electronics for renewable systems, micro-grids and distributed generator.

K. Vijayakumar obtained his Bachelor's degree in Electrical and Electronics Engineering (EEE) in the year 2006 from Coimbatore Institute of Technology, Coimbatore. He obtained his M.Tech. degree in Power Systems from National Institute of Technology (NIT), Tiruchirappalli, Tamil Nadu, India in the year 2009. Subsequently, he was admitted as Ph.D. Research Scholar with MHRD (Govt. of India) fellowship in the Department of EEE, NIT, Tiruchirappalli, and completed his dissertation in the year 2012. He was Postdoctoral Research Fellow in Nanyang Technological University, Singapore, from November 2012 to December 2013. He is currently working as Faculty in the Department of Electronics and Communication Engineering, Indian Institute of Information Technology, Design and Manufacturing, Kancheepuram, Chennai. He also holds the Adjunct Faculty position at National Centre for Disaster Mitigation and Management, MNIT Jaipur. He received Young Scientist Award for start-up research grant which was given by SERB – DST (Govt. of India) in the year 2015. His research interest includes design and development of DSP-based power electronics controllers for renewable energy, industrial electronics and control, smart grid, power electronics and electrical machines, FACTS controllers and SMPS for telecommunication systems.

Artificial Swarm Intelligence—A Paradigm Shift in Prediction, Decision-Making and Diagnosis

V. J. K. Kishor Sonti and G. Sundari

1 Introduction to Swarm Intelligence

"Thought", "Effort" and "Performance" are keys to Artificial Swarm Intelligence (ASI). The congregations of these key components from diverse sources roll into the making of the final product. It is an established fact in the history of science and technology that Nature nudges man into stumbling on unthought-of discoveries/inventions. The inspiration derived from nature always drives the human race, and Swarm Intelligence (SI) is no exception to this rule.

Swarm Intelligence encouraged the researchers to fine-tune the method of observing nature. Nature's role as a teacher at this juncture is unique. As an example, keen observation of ants, birds, and insects lifestyle polarized human thinking towards expanding the horizon of possibilities.

Swarm is a cluster of communicating agents. Bonabeau mentioned Swarm Intelligence as "The emergent collective intelligence of groups of simple agents" [1, 2]. Group behavior is the mark of SI. In simple terms, SI can be explained as identification of analogy from the examples available in nature, Understanding the possibility of application and Engineering according to the requirement. Some of the properties of SI are, simplicity, highly synchronized even though appears random, it also appears as a global behavior but this is a collective effort of independent agents. SI offers the creativity of other living organisms exhibited in their lifestyle, which shall add a new dimension to intelligence [3].

The thirst for knowledge never ends for enthusiastic learners. Alan Turing to Rosenberg Louis, researchers of mathematical, computing, and electronics fields

V. J. K. K. Sonti (✉) · G. Sundari
Department of ECE, Sathyabama Institute of Science and Technology, Chennai, India
e-mail: jayakrishna_adc@yahoo.com

G. Sundari
e-mail: sundariece16@gmail.com

B. Vinoth Kumar et al. (eds.), *Intelligent Paradigms for Smart Grid and Renewable Energy Systems*, Algorithms for Intelligent Systems,
https://doi.org/10.1007/978-981-15-9968-2_1

made notable contributions to the saga of Artificial Intelligence (AI) transforming into Artificial Swarm Intelligence (ASI). The last two decades witnessed this growth pattern, which is fast and fruitful. But this transformation is not complete and possible without Swarm Intelligence. In SI, direct and indirect interactions (Stigmergy) of agents (individuals) with the environment create a lot of information to observe, assimilate, simulate, and to realize into systems.

Swarm Intelligence is one of the precious gifts of nature to humans. Perception and Pragmatic approach offer necessary guidelines for the realization of this exciting knowledge into engineering systems. Earlier in history, human swarm-based predictions were observed to be more accurate in certain gaming outcome predictions. Later it was found that it has the potential to make such predictions in other fields. Swarm predictions outsmarting individual intelligence on number of occasions have become the new normal or a paradigm shift.

The evolution of any technology is exciting, especially in the process of unfolding new dimensions of science. AI is one such disruptive technology that is consistently engaging enthusiastic thinkers in terms of throwing lucrative challenges [4]. It is also important to understand that not every challenge can be addressed from the same perception. Perceptions changes with time in life so as in technology. Performance is also dependent on perception, the viewpoint from which the problem is examined also changes the way of problem-solving. Exactly, this is the freshness brought to the complex problem solving with the evolution of SI.

Majority of the aspects related to Swarm intelligence are centric towards decentralization and collectiveness. These two are diverse characteristics, yet are well-knitted in this concept of swarm intelligence, which is the inherent beauty and specialty. The flocks of birds and the swarm of insects teach the principles of courage, convergence, and collective effort. Swarm intelligence inherited these properties in its fabric of technical affluence.

Social Psychologists too reported the fruitfulness of "Altruistic" nature of individuals for a better society. Observation, replication, and re-organization are contributing factors to Cognitive thinking. Issac Newton's apple (Science), Albert Bandura's bobydall (Psychology), Beni and Wang's Swarm Intelligence (Empowering Technology) are the outcomes of observational skills in humans.

This kind of enthusiasm was possibly one of the reasons for proposing Ant Colony Optimization (ACO) by Dorigo and team in 1991 [5], Particle Swarm Optimization (PSO) by Kennedy and Eberhart in 1995 [6], Artificial Bee Colony algorithm by Karabago D. in 2005. The extension of this knowledge took place in 2016, in the form of Swarm Robotics by Bayindir [7].

Apart from the aforementioned developments, infusing swarm intelligence into wireless networks [8], use of swarm intelligence in financial market prediction [9], and Study of social interaction and Swarm algorithms as systems [10] are some of the recent developments.

Complex problem solving with Swarm Intelligent algorithms produces appropriate solutions. Less expensive, fast convergence, flexibility, and adaptability are the advantages of these algorithms. Modeling such thought-provoking behavior of swarms into systems is exactly the essence of Artificial Swarm Intelligence.

2 Artificial Swarm Intelligence

Artificial Intelligence in simplest terms shall be “borrow from humans and apply in algorithms”, whereas SI gives the knowledge of “inspire from swarms and apply in algorithms”. These parallels are coincidental here, but provoke innovative thoughts about the hidden interlacing nature of this creation.

The emergence of the ASI concept is exciting. Artificial Intelligence is the novel technology with strong computational background. Swarm Intelligence is the nature-inspired novelty offered to technology. Figure 1 represents fusion of ideas. Artificial Swarm Intelligence is the fusion of nature-inspired algorithms employed for technological advancement. In the history of science, any such fusion has changed the way the world looked at itself. ASI has been exactly contributing to paving ways for such avenues of innovation. This journey has been exciting technically, inspired biologically.

AI refers to Artificial Intelligence, SI is Swarm Intelligence, ASI is Artificial Swarm Intelligence, whereas BI refers to Bio-Informatics (A field that analyzes biological data using mathematics and other computing systems).

Artificial Swarm Intelligence is a dynamic concept, which is extending its horizon of applications. ASI helps in the amplification of natural intelligence of human groups. This will be carried out by interfacing human groups to real-time systems. ASI shall also be seen as the modeling of natural swarm behavior into real-time systems using intelligence algorithms [9].

Numerous algorithms equip technocrats and researchers to apply ASI in fields like Smart Grid Management, Renewable Energy Resources (RER) optimum utilization, medical diagnosis, and prediction analysis [11]. In this chapter, an attempt has been made to discuss the potential of intelligent algorithms and ASI concept to Smart Grid

Fig. 1 Fusion of disciplines leading to ASI

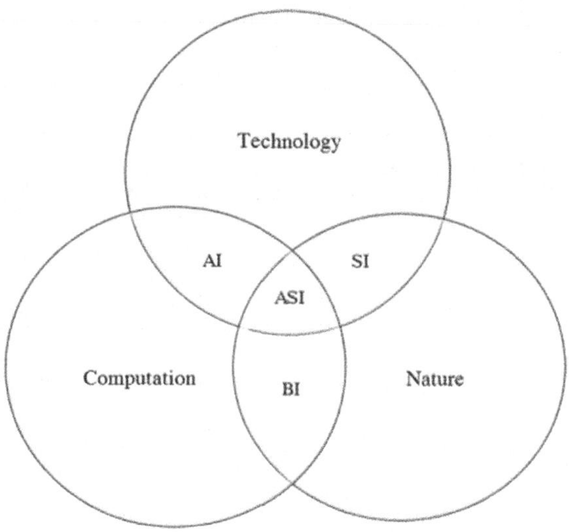

Management and Renewable Energy Resources Optimum Utilization. ASI possess such characteristics that provide scope for diverse applications.

2.1 Key Components

The key aspect of ASI is the "collective intelligence", where human swarms act as real-time systems that are guided by intelligent algorithms. The collection of data, processing, and convergence towards solution takes place with good amount of accuracy. This topic is elaborated in Sect. 3 of this book chapter (collection to convergence).

The interesting component of ASI is the decision-making process, which is the same as the neuron-based decision-making model. The excitation potential will reach the threshold and appears as action potential, the mechanism in both cases. In the former, it is neurological, whereas later this is swarm-based effort. Optimization, Decentralization, Collective Behavior, and Self-Organization is the key components of the typical ASI system adopted from SI, as depicted in Fig. 2.

Optimization is the basic outcome offered by SI in most of the intelligent algorithms. The utilization of available resources, adaptability, and reliability leads to the final anticipated global outcome. This is perhaps possible with the decentralization feature of SI inherited by natural swarms. Agents or individual units arrive at local best solutions, which contributed to the global best outcome. Therefore, decentralization in other ways assists in reaching out to optimal solutions. In the context of energy sector, the functioning of "microgrid" is a good illustration.

Self- Organization is a typical feature. The overall system appearing at a global level has numerous individual entities' interactions as the backbone at lower level. In other words, Collective behavior is a superficial expression of Self-Organization exhibited at independent or autonomous level. These key components adopted from SI in developing systems or methods used in ASI are well tuned to the technical transformation of applying to real-time environment.

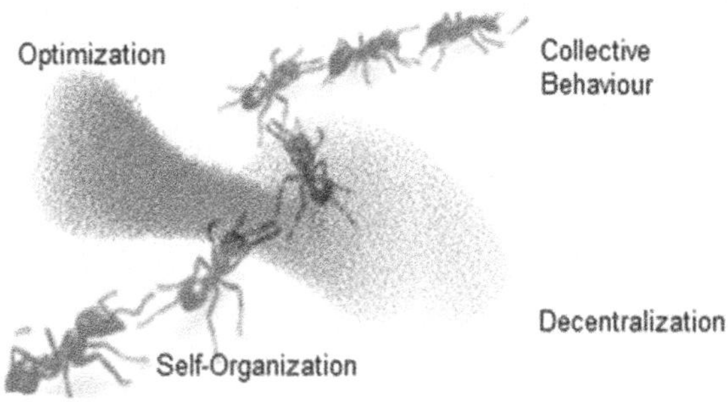

Fig. 2 Key components of ASI

2.2 *Potential of ASI for a Possible Application in Smart Grid and Renewable Energy Resources Systems*

The transformation of swarm intelligence to Artificial Swarm intelligence is interesting. Translating the collective intelligence phenomena into real-time systems such as Smart Grid Management, Optimum utilization of Renewable Energy Resources requires certain key components or interfacing units. Intelligent algorithms, networking paradigms, and specific interface systems will act as such components. This chapter discusses intelligent algorithms.

These efficient algorithms such as Ant Colony Optimization Algorithm (ACO), Particle Swarm Intelligence Algorithm (PSO), Artificial Bee Colony Algorithm (ABC) and many others govern the transformation process of SI to ASI. The scope of this chapter permits the reader to understand about SI algorithms. These intelligent algorithms facilitate the nature-inspired technology transfer, which is detailed in Sect. 3.

Optimization strategy, dynamic deployment of networks, and fusion of data are exceptional in Swarm based Intelligent Optimization algorithms. For example, Intelligent Optimization algorithms approach a problem with the bionics principle (biologically inspired engineering systems design) as the kernel. This bionics principle is in concurrence with the behavioral habits of nature and biology. There is a striking difference between natural swarm intelligence and artificial swarm intelligence.

Ability to control the overall system without any exclusive external control unit is another salient feature of Artificial Swarm Intelligent Systems. This is being achieved by the coordinated effort of individual units. Swarm as a whole will exhibit intelligent behavior by consistent coordination with the neighbors. In short Artificial Swarm Intelligence exhibits inherent randomness that reflects as the certainty in overall response. This property is the most fascinating yet practically accurate in terms of prediction.

Swarm intelligence algorithms offer advantages over traditional optimization algorithm. There are certain characteristics of ASI inherited from SI, made it a competent choice for Smart Grid Management. Decentralization, robustness, indirect interactions, autonomous agents, Self-Organized Structures, and Division of Labour are making ASI as an opportunity. Here structures appear at the global level i.e. at the system level, possess well-defined local level interactions.

Stabilization, Coordination, and Optimization are important points to be considered while applying ASI for Smart Grid Management. Apart from the possible application in Smart Grid Management, there are various potential applications of ASI. Current and possible areas of applications of ASI are depicted in Fig. 3.

These applications include weather and financial markets forecasting, routing, medical data prediction, and decision-making that is crucial in effective diagnosis, telecommunication networks, military, process optimization, and robotics.

The application of intelligent algorithms in Smart Grid Management and Renewable Energy Systems is the topic of interest. The efficient use of swarm intelligent

Fig. 3 ASI applications

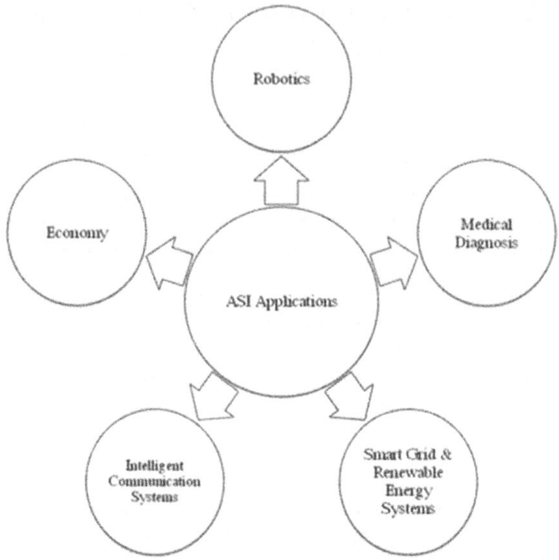

algorithms towards enhancing performance of Smart Grid and Renewable energy Systems is discussed in Sect. 3.

3 Collection to Convergence

The basic fabric of ASI opens up the possibilities of fetching feasible solution for a problem. This journey is well supported from Collection to Convergence stages of problem-solving. For example, in Ant Colony Optimization (ACO) the search for global information, a start with the collection till the convergence point has arrived, a similar approach can also be observed in Particle Swarm Optimization (PSO). In fact, the underlying nature of almost all these intelligent algorithms starts with the collection of information and aims at the global or local convergence.

The necessity for the transition to Smart Grid from the conventional grid has valid reasons such as reducing carbon emissions and automation at every possible stage of energy generation and distribution. ASI adds value to this energy chain with the biologically inspired powerful yet simple intelligent algorithms. In this section, this is addressed with the introductory information about Smart Grid and Renewable Energy Systems, followed by ASI algorithms and their possible application.

3.1 Smart Grid and Renewable Energy Systems

Smart Grid is the modernized version of traditional electric grid that a twenty-first century is experiencing [12]. The electric grid is operated, monitored, and efficiently managed from remotely operated computerized mechanisms and reliable communication systems. "Smart Grid" does intelligent integration of generation and consumption. This integration helps load forecasting and in delivering socio-economic needs of energy with sustainability [13]. The concept of Smart Grid throws light on integration of sustainability, reliability, and economy as the epicenters of energy generation and distribution [14]. Infusion of technology and intelligent systems will ensure that the grid becomes more flexible and effectively manageable.

Research is going on towards the development of smart applications related to grid management; the future of smart grid management and associated techniques is promising [15]. Need for grid modernization and way ahead to future was reported by Michael I. Henderson, Damir Novosel, and Mariesa L. Crow in 2017 [16], smart grid control and management by Mohamed Zahranet al., in 2013 [17], enhancing grid flexibility, robustness using intelligent converters, demand response, energy management are few notable advancements by Mohammad Asaad in 2019 [18].

Role of Renewable Energy in Smart Grid is vital. Optimal use of renewable energy resources has become increasingly important for an efficient electric grid. Effective utilization of renewable energy resources contributes to environmental sustainability. Clean energy is driving to a large extent, almost all recent technological developments in the energy sector. Industry 4.0 insists on clean energy as one of the focal points of development. Interestingly, the generation of clean energy, where solar and wind make major contributions is the best example of decentralization, which is also one of the key components of ASI. This is quite interesting to explore the possible application of ASI in the optimum utilization of clean energy. Understandably the application of ASI shall indirectly contribute to the reduction of carbon footprints.

ASI has the potential to change the way in which the electric grid has been functioning. The transition of the electric grid to the smart grid becomes more vibrant and dynamic in terms of delivery mechanism with the fusion of intelligent algorithms inspired by Swarm Intelligence. The prediction, optimization, decision-making and diagnosing a problem or fault are the wonderful bi-products of ASI; besides scalability, robustness, and ease in management of the smart grid. Smart Grid offers observability, flexibility, controllability, integration, and management when compared to the conventional or traditional grid [19]. Few but not limited, factors contributing to the transition of the traditional grid to smart grid have been listed in Table 1. Smart Grid possesses digital systems in the place of electromechanical as we found in the conventional grid [20].

Smart Grid offers bi-directional communication, distributed generation, Self-monitoring, and pervasive control instead of limited control offered by the conventional grid.

Self-healing is another feature of Smart Grid, which improves the reliability factor. Consumer contribution in effective energy management and in decreasing the load

Table 1 Comparison of smart grid and traditional grid features

Traditional grid	Smart grid	Contribution factor for this transition
One way communication	Bi-directional communication	Developments in communication field
Analog and electrical machinery	Digital systems	Digital revolution
Flow control is limited	Wide and remote control	Wireless communication
Reliability is a challenge	Reliability factor has been predominantly improved	Developments in system modeling
Consumer connect is less	Consumer input is significant in energy management	Customer-centric evolution of industry

effect of grid is crucial. Smart grid provided enough scope for the effective interaction between generation and consumption units.

Real-time analysis is possible with efficient forecasting/prediction of operations of the grid and its resources. ASI exactly steps in addressing this particular issue. Artificial Swarm Intelligence creates bright opportunity to expand horizontal controllability and self-sustainability of the Smart Grid.

A suitable analogy of Smart Grid overall operations represented as a reference model, comprising of five layers is presented in Fig. 4. This summarizes the collection to convergence mechanism imbibed in the operation of a Smart Grid. The Smart Grid

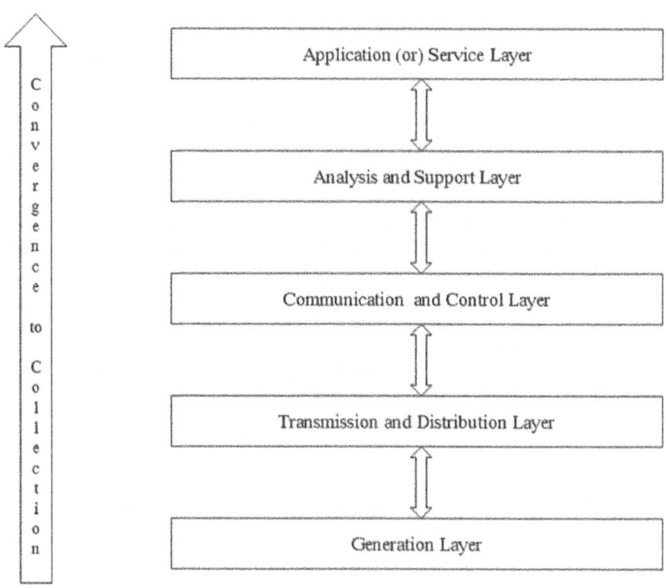

Fig. 4 Five layer model of smart grid operation

operation, balancing of energy resources in serving consumers has been laid down into five layers for better understanding the complex nature of Smart Grid Operation.

3.1.1 Generation Layer

Generation layer represents generation of energy with striking balance between conventional and renewable energy sources. Optimum utilization of renewable energy through prediction is the driving point of efficiency in this first and fundamental layer.

3.1.2 Transmission and Distribution Layer

The energy transmission and distribution mechanism is represented at second layer. The loss incurred in the transmission and distribution should be estimated in advance to take necessary proactive steps. Estimation and reduction of energy loss, which is very important for Distribution Management System (DMS) governs the process in this layer. Effective DMS is the challenge at this juncture. Controllability is achieved only through reliable communication systems.

3.1.3 Communication and Control Layer

With the advent of revolutionary developments in telecommunications arena, this layer is strengthening. Remote controlling and taking decisions at ground level are the key challenges of this layer.

3.1.4 Analysis and Support Layer

Analysis and Support layer comprises of huge set of data analytics, outage, and asset management. This layer is a crucial segment of Smart Grid, arriving at optimized solutions, data generation supporting crucial decision-making process, predictive analysis is some of the outcomes here, which are also potential challenges to be addressed with better computing structure.

3.1.5 Application and Service Layer

The topmost layer comprises the consumer segment, i.e. application and service layer. Smart Grid versatility lies in consumer (residential, commercial, and industry) interaction for better Demand Response (DR), Energy Storage (ES), Efficient Energy Management Systems (EMS), and Demand Side Management (DSM). This is in fact one of the notable paradigm shift of Electric Grid operation. Management is the core principle and customer-centered approach is the challenge to handle at this layer.

Interestingly, not all but most of the challenges of these layers shall be mitigated with the ASI, which is elaborated in Sect. 4. SI algorithms facilitate this process, which is included in Sect. 3.2.

3.2 Swarm Intelligence Algorithms

Natural Scientists and Ethologists interest in the behavioral study leads to the discovery of social intelligence in insects and animals. This ability is exceptional while procuring, sharing knowledge about surroundings and sources of food. Observations from nature such as the movement of flock of birds, ants foraging, bees, and bats lifestyle are the basis for these intelligent algorithms. Few of them are as follows:

- Ant Colony Optimization Algorithm (ACO)
- Particle Swarm Intelligence Algorithm (PSO)
- Artificial Bee Colony Algorithm (ABC)
- Glowworm Swarm Optimization (GSO)
- Cuckoo Search Algorithm (CSA)
- Differential Evolution (DE)
- Bat Algorithm (BA)
- Firefly Algorithm (FA)
- Artificial Fish Swarm Algorithm (AFSA)
- Shuffled Frog Leaping Algorithm (SFLA)
- Fruit Fly Optimization Algorithm (FOA)
- Bacterial Foraging Optimization (BFO)
- Chicken Swarm Optimization Algorithm (CSO)
- Wolf Pack Algorithm (WPA)
- Artificial Plan Optimization Algorithm (APOA)

This sub-section is constrained to the explanation of a few apt algorithms in the context of Smart Grid Management and Renewable Energy Systems. ACO, PSO, and ABC are explained in detail, whereas few algorithms concept and operational aspects are briefed.

3.2.1 Ant Colony Optimization

Interesting questions kindle mind when we observe the social behavior of insects. For example, ants foraging through the shortest paths remained as Nature's surprise to man for many years. Later, it was found by researchers that "Stigmergy" is the reason for exploring paths to the destination by ants.

Every ant while moving releases pheromone at different levels. This leaves a specific trace or mark on the surface, which is known as a pheromone trail. An ant that has identified this trace or pheromone concentration will follow this path and reinforce the trace with its pheromone. Follower ants do this with high probability, whereas isolated ants may be moving at random [5].

This is natural that more number of ants following the same trace or trial exhibit collective behavior. This collectiveness helping them in reaching their target (source of food), even though their contributions are tiny and independent. However, the observation suggests that the probability of an ant choosing a trial depends on the number of predecessor ants chosen the same trial. A fine example of collective behavior leads to "Collective Intelligence".

Ant System (AS) inspired from ant colonies was reported in a research paper by Dorigo et al. in 1996 [5]. They suggested this as a "computational paradigm" [5]. Ant Colony optimization was proposed as an approach by them to solve combinatorial optimization problems. This approach was applied to the traveling salesman problem towards reaching optimal solutions [21]. Other applications of this algorithm are network routing, data mining, etc. The very nature of this algorithm offers the possibility of application to the most challenging problems, where optimization is a goal.

Beginner ant returns to the point of starting quickly and other ants those choose this path also returns fast. This is possible because of "pheromone trials", which means the best routes enjoy more concentration or intensity of the pheromone.

Pheromone will be secreted by ants whenever they travel from one place to another in the search of food, i.e. foraging behavior. Pheromone leaves the information for the following ants. Based on the concentration level of the Pheromone the best path to reach the food is identified and followed. This is useful for other ants to finish the target in less time and using the shortest path. Optimization is inherent in this approach.

The optimization problem shall be framed as a pathfinding problem. This problem consists of weighted graphs, which are updated with time. This is a meta-heuristic approach and shall be brought down to three different segments; Construct Update and Daemon Actions. These actions are used to make an effective decision about the necessity of further pheromone secretion. These Daemon actions help in gathering information that cannot be carried out by single agent or ant. This is a collective intelligence used to reach an optimal solution or convergence.

This collective intelligence is adopted and developed as ACO algorithm, which is used in various applications demanding combinatorial optimization. Calculation of probability, pheromone deposition, and decay rate guides the problem-solving procedure [1].

Reaching out to the best solution to start with choosing the next node; this is done by calculating the probability of finding the next best node.

The probability of the kth ant, which is at node i and heading to node j through the pheromone trail τ_{ij} is given Eq. 1.

$$P^k_{(i,j)}(t) = \frac{\left([\tau_{ij}(t)]^\propto \cdot [\eta_{ij}]^\beta\right)}{\left(\sum_{k \in J_k} [\tau_{ij}(t)]^\propto \cdot [\eta_{ij}]^\beta\right)}. \quad (1)$$

η_{ij} is the visibility of ant between i and j nodes.

$\tau_{ij}(t)$ is the amount of pheromone at time t between i and j nodes.

$\propto$ and β are control parameters. If $\propto$ is higher, the ant depends on pheromone for search, and if β value is higher, and depends on its visibility to move forward. The nodes that ant is permitted to travel are represented by J_k. Pheromone deposition is calculated using Eq. 2.

$$\Delta\tau^k_{ij}(t) = \begin{cases} \frac{Q}{L_k}(t) \\ 0 \end{cases}. \quad (2)$$

Here, Q is a constant, L is the length of the generated trial, k is the ant and t is the iteration number. $\Delta\tau^k_{ij}(t)$, indicate the rate of pheromone deposition between i and j nodes. If no trial is chosen i.e. in the case of an isolated ant, this value becomes zero.

Pheromones will be decayed after sometime or in other words, pheromones evaporate. This evaporation rate reflects the exploring capacity of the ant or agent. Higher values represent that ant lost the pheromone trial or path; low values indicate its inability to explore the trial. In this pheromone evaporation (decay) equation i.e. Equation 3, m indicates the number of ants in the search space, p is the pheromone evaporation rate.

$$\tau_{(i,j)}(t+1) = (1-p) \cdot \tau_{(i,j)(t)} + \sum_{k=0}^{m} \left[\Delta\tau^k_{(i,j)}(t)\right]. \quad (3)$$

The evaporation rate provides the necessary feedback towards reaching the target or objective function. ACO Flow Chart is self-explanatory as depicted in Fig. 5.

Summary: To apply ACO in an optimization problem, the following mapping should be done.

- Represent the problem as a network; the solution will be a specific trial or path in the network
- Model the problem in such a way that it has a start node and a finish node
- The best path connecting start and finish nodes provides the solution
- The quality of the solution depends on the pheromone trial intensity or concentration or most chosen path

Pheromone tables are used for optimal solutions in communication systems, decision making in routing problems.

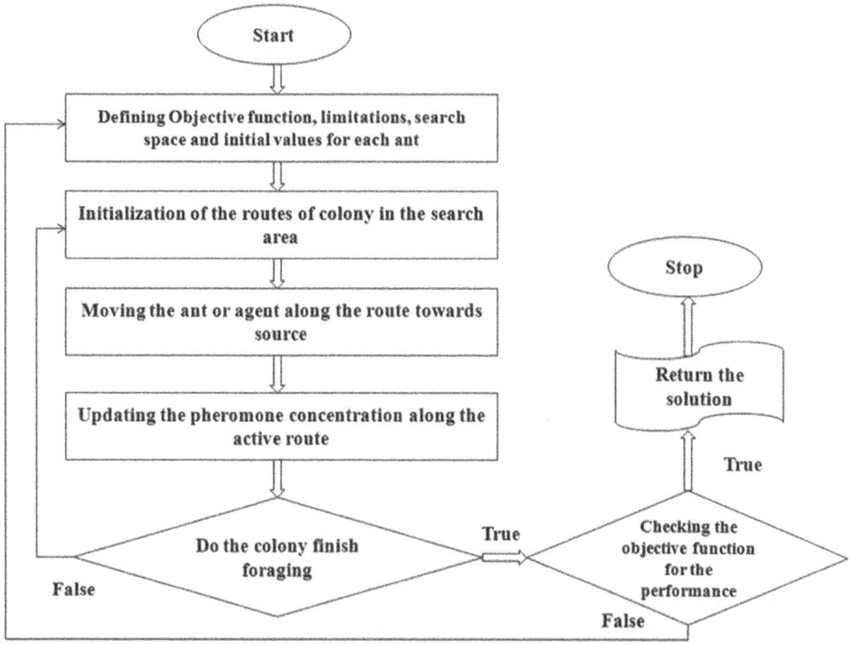

Fig. 5 Flow chart of Ant Colony Optimization Algorithm *t. Source* ACO by Dorigo [5]

3.2.2 Particle Swarm Optimization (PSO)

PSO algorithm was proposed in 1995 by Kennedy and Eberhart [6]. The inspiration was derived from the movement of birds, insects, and fishes. This is simple to implement, easy for concurrent processing, and one of the efficient global search optimization algorithms.

When the social behavior of a flock of birds is observed, one can watch for three striking behavioral patterns. Separation is the first pattern, where a typical behavior of deviating from the local flock is observable. This portrays the courage to lead or create a separate identity. Creativity is possible when we think differently from others is the best match for this situation. Alignment follows the Separation. Movement in the average direction of the local flock is alignment. This resembles collective behavior. The third and crucial pattern is Cohesion, moving towards the average position of the local flock. This is a step towards convergence. PSO also portrays collectiveness to convergence behavior.

Some of the applications include energy distribution network optimization, structural optimization, neural networks, and biochemistry. The velocity of the particle is calculated, each particle moves to better space in the area of the problem. Therefore, overall population will be moving towards better positions of the problem space. This is a fine example of Social Intelligence. The success of the neighbor contributes to the movement of other neighbors. Interdependency, coordinated movement, and solving a complex problem by wise and timely wise are the core principles of this Social Intelligence. The global best is obtained by exploring and updating consistently the local best.

This is a simple three-step procedure. After initialization, the velocity of the particle is calculated. Next is to find out the local best position (result of alignment). Updating the local best and achieving global best position is the final step. This process is repeated till entire population move towards global best (convergence).

Mathematical expression [1] of this above process shall be understood using Eqs. 4 and 5. Here, the velocity of the particle is calculated using Eq. 4.

$$V_{id}^{t+1} = V_{id}^{t} + c_1 \cdot \text{rand}(0, 1) \cdot \left(p_{id}^{t} - x_{id}^{t}\right) + c_2 \cdot \text{rand}(0, 1) \cdot \left(p_{gd}^{t} - x_{id}^{t}\right). \quad (4)$$

The position of the particle is calculated using Eq. 5.

$$x_{id}^{t+1} = x_{id}^{t} + v_{id}^{t+1}. \quad (5)$$

Here, V_{id}^{t} is the velocity of the particle, x_{id}^{t} is the position, t represents the iteration, i is the particle in search space of dimension d. p_{id}^{t} and x_{id}^{t} are the particles best position and previous particles best position respectively. p_{gd}^{t} is the entire population global best, c_1, c_2 are the speed of the particle when it is aligning towards the most optimal particle of the swarm and the most optimal particle independently. x_{id}^{t+1} , v_{id}^{t+1} updated values of position and velocity of the particle, whereas rand(0, 1) represents random values between 0 and 1. If $\left(p_{id}^{t} - x_{id}^{t}\right)$ and $\left(p_{gd}^{t} - x_{id}^{t}\right)$ resultant is high or any of them is higher value, exploration occurs. If both these differences are yielding small values then exploitation occurs. Figure 6 represents PSO Algorithm flow diagram.

The PSO algorithm's first step is the initialization of population. The second step is calculating the fitness values of each particle. Next to find the local best of the each particle's P_{best}, followed by updating particles own and global best values (G_{best}). Finally, the velocity and the position of the particles get updated. The second to fourth steps get repeated until the termination condition is reached.

Summary: Simple procedure to follow, for the application of PSO in optimization problems.

- Find the current velocity of the particle or agent
- Find the Personal best (P_{best})
- Find the Global best (G_{best})
- Particle always travel from the current best to personal best
- Global best is the best of the personal best positions of the particles
- Final update of velocity and position shared by all the particles in the search space indicates that the global best position of the swarm or the global optimization is achieved

One of the merits of PSO is the insensitivity to scale the design variables; this is a quite an interesting feature of concurrent process applications. Decentralization of

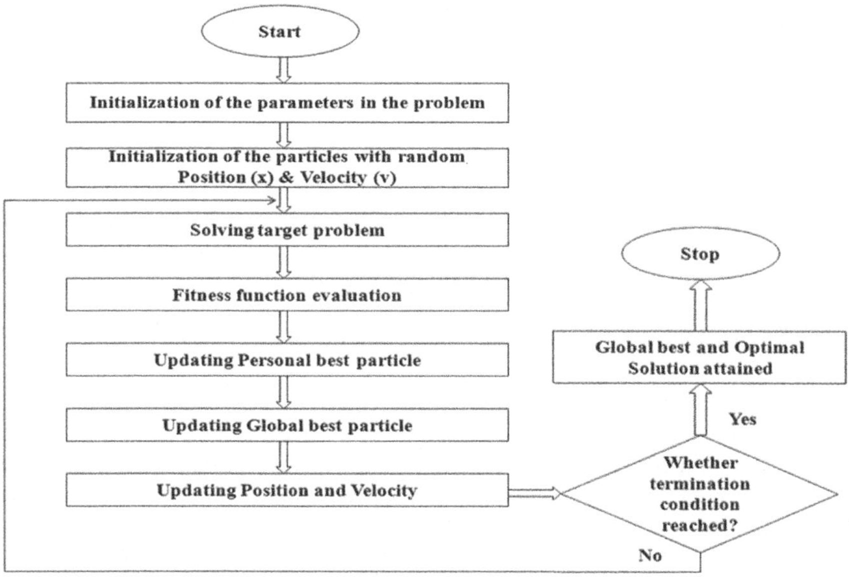

Fig. 6 Flow chart of PSO algorithm. *Source* PSO by Kennedy [6])

objective functions to sub-warms shall be used in multi-objective problem-solving. This in turn enhances the pace of the convergence.

Speed in convergence and finding better solutions is obtained with the inclusion of constriction factor and inertia weight invariant models of PSO. Premature convergence is possible with collapse of swarm, which has been dealt with in further modifications of the algorithm (additional reading of PSO variants is suggested).

It was also reported by Kennedy and Eberhart that PSO shall be used as an alternative to back-propagation used in training artificial neural networks [6].

3.2.3 Artificial Bee Colony Algorithm (ABC)

This is one of the frequently tested Swarm Intelligence algorithms, which was proposed in 2005 by Dervis Karaboga [1]. As the name implies, this is inspired by honeybee's lifestyle. This is one of the best suited intelligent algorithms for Smart Grid applications. The robustness of the method, ease at implementation level, and highly flexible nature that makes this algorithm as best for engineering designs, scheduling operations, and networking. Different layers of Smart Grid shall employ this technique to mitigate the challenges encountered in routing and DMS.

In ABC the striking factor is the coordination of honeybees. This is a similar situation for microgrids in energy management. The search for food is a coordinated act achieved by the honeybees but with an independent effort. The information about the food i.e. the nectar is shared with other bees in the hive, just like microgrids share the information about the energy. Here, the phenomena of nectar search take place with the three categories of honeybees or artificial agents (in terms of ASI), namely, employee bees, onlooker bees, and scout bees.

Initializing x_i, is the population food source. l_i and u_i are lower and upper boundary values of x_i. Optimization of x_i value leads to minimizing the objective function. The size of x_i is 1 to S_n (S_n is the population size). x_i is calculated using Eq. (6). Equations 6–9 are used for parameter calculation in the ABC algorithm [1].

$$x_i = l_i + \mathrm{rand}(0, 1) * (u_i - l_i). \tag{6}$$

Neighborhood food source is calculated using Eq. (7).

$$v_i = x_i + \emptyset_i\left(x_i - x_j\right). \tag{7}$$

x_i, is a random food source, $\emptyset_i$ is a random value of [−*a*, *a*]. Exploration occurs if $\left(x_i - x_j\right)$ is high, otherwise, exploitation happens. Fitness function calculation is carried out using Eq. (8).

$$\mathrm{fit}_i\left(\overrightarrow{x_i}\right) = \begin{cases} \frac{1}{1+f_i\left(\overrightarrow{x_i}\right)} & \text{if } f_i\left(\overrightarrow{x_i}\right) \geq 0 \\ 1 + abs\left(f_i\left(\overrightarrow{x_i}\right)\right) & \text{if } f_i\left(\overrightarrow{x_i}\right) < 0 \end{cases} \tag{8}$$

$\left(f_i\left(\overrightarrow{x_i}\right)\right.$ is the objective function of $\overrightarrow{x_i}$. The probability value is obtained using Eq. (9).

$$p_i = \frac{\mathrm{fit}_i\left(\overrightarrow{x_i}\right)}{\sum_{i=1}^{\mathrm{SN}} \mathrm{fit}_i\left(\overrightarrow{x_i}\right)}. \tag{9}$$

In ABC, three different types of bees are assigned with different tasks. Employee bee carries out the search for the food, memorizes the location, and transmits the information to onlooker bee. The receptor of the information about nectar is the onlooker bee, from which one food source is selected. The scout bee task is to search for new food location. This is a repetitive process, which is useful for scheduling, coordination, and exploration. As per the stages of honey bees operation, the algorithmic steps are designated as initialization phase, employed bees phase, onlooker bee phase, scout bee phase, best fit, and termination checking phase. Figure 7 explains this process in a flowchart.

Phase 1: Initialization phase, where the scout bee initializes search for new food.

Phase 2: Employed bees start searching for the location of the food, once found will be shared with the neighbors and the neighbors' data will be updated.

Phase 3: Onlooker bees make the choice of the food location based on the information shared by the employed bees. Fitness function is used to calculate a suitable source.

Phase 4: Scout bees chose their sources of food randomly. Every new search for new food location is random in nature. These are also called unemployed bees.

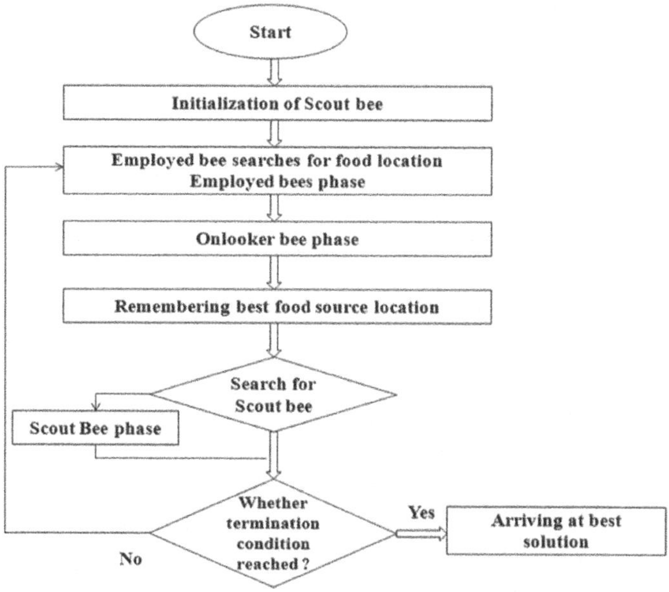

Fig. 7 Flow chart of ABC algorithm. *Source* ABC by Karaboga [1]

Phase 5: Best fitness function values are memorized, if the termination is achieved the process stops, otherwise the process gets repeated from phase 2.

Summary: This algorithm is a good choice to implement in decision-making situations.

- Assign the task of search for resources or solutions to an agent or team.
- Instruct the second team or agents to store the information about available resources.
- Designate the third team or local supervisors to choose the best fit.

There are few demerits such as relatively slow process and performance enhancement with entirely new fitness tests. Inspite of these demerits, research is in progress for the usage of ABC and its variant models extensively in various engineering systems.

3.2.4 Other Intelligent Algorithms

A short description about a few algorithms [1] is provided here.

Glowworm Swarm Optimization (GSO) was proposed by Krishnanad and Ghose proposed GSO in 2005. The agents here are glowworms. Three important parameters in GSO optimization are movement of the glowworm, update of luciferin level and update about neighborhood position or range. After Glowworms initial random placement in workspace, the fitness function of the glowworm or agent is calculated to find the luciferin level.

The update in search space takes place; later the glowworm tries to explore its neighbor. If there exists a scenario of multiple neighbors, in such case glowworm makes use of the probability function to arrive at the decision or to select the neighbor. After this process, glowworms moves one step towards the neighbor, later the neighborhood range is calculated.

Higher efficacy in exploitation shall be achieved using GSO. In this manner, GSO is good enough to explore multiple sources or neighbors based on the luciferin level. The demerits include slow convergence and less accuracy. The application of GSO is best for optimization problems apart from usage in Swarm robots, image processing, and routing.

Cuckoo Search Algorithm (CSA) was proposed in 2009 by Yang andDeb. As the name suggests, the inspiration is from Cuckoo birds. In this method, three operations exist.

Step 1: Nest is chosen randomly by cuckoo to lay a single egg (limit per iteration).

Step 2: Quality is the parameter to carry forward nests and eggs to the next generation.

Step 3: Host's nests are fixed. An egg layed by the bird is chosen using probability function.

At step 3, the host has a choice of throwing the egg and/or shall also build entirely a new nest. This algorithm finds application in problems with multimode objective functions such as neural networks and electromagnetism related fields.

Differential Evolution (DE) was proposed in 1997 by Storn and Price. This is almost same as Genetic Algorithm. The basic difference is DE uses mutation, whereas Genetic Algorithm uses crossover. The search is carried out using mutation in this algorithm. Three vectors, namely, target, mutation, and trail vectors are used in completing the search and in exploring new population. The specialty of DE lies in not considering the fitness function for parent selection in the population. This is an unstable process, slow in convergence but is good for local search problems. DE is used in machine learning and image processing applications.

Firefly algorithm was proposed in 2008 by Yang, operates like GSO. Less brighteous fireflies' move towards brighteous fireflies. In the absence of brighteous fireflies, these less bright ones possess random movement. The fitness function in this case is the brightness of the firefly.

PSO and Batalgorithms have similarities. Both are optimization techniques, Bat algorithm was proposed in 2012 by Yang and Gandomi. The search for food is governed by position and velocity calculations.

Artificial Fish Swarm Algorithm (AFSA) is an optimization algorithm proposed by Li in 2002, inspired by the fish social behavior in search of food and in anticipating dangers. Fast convergence, accuracy, and flexibility are the advantages of this method.

The applications include intrusion detection, energy distribution networks, and in financial markets forecasting [22].

Other optimization algorithms are Shuffled Frog Leaping Algorithm (SFLA), Fruit Fly Optimization Algorithm (FOA), Bacterial Foraging Optimization (BFO), Chicken Swarm Optimization Algorithm (CSO), and Wolf Pack Algorithm (WPA). All these algorithms are used in solving combinatorial optimization problems.

Systems developed using ASI models are scalable. Scalability enables us to use the control architecture for multiple agents. Systems offer flexibility and adaptability. These algorithms are helpful in building up robust systems as any single node or agent failure will not affect the final outcome.

4 Optimization—Prediction, Decision-Making and Diagnosis

Researchers and other domain experts are working on the challenges of Smart Grid at every stage of operation and providing innovative solutions. These significant contributions are driven by a thirst to achieve among innovative thinkers besides scope for improvement in engineering systems. ASI algorithms are suggested by various works to strengthen and boost towards providing feasible solutions to these challenges.

The four critical dimensions contributing to enhancing the efficiency of the Smart Grid are Optimization, Prediction, Decision-Making, and Diagnosis. The concept of Smart Grid has given scope to these dimensions. There exists scope for imparting better procedures, to ensure these four dimensions remain dynamic. This section provides such insights, where ASI has a significant role to play.

4.1 Possibility and Probability

These two P's always put forward fascinating questions when a new concept or technology is introduced. Probability of success and Possibility of implementation of ASI algorithms in Smart Grid and Renewable Energy systems shall be viewed from the dimensions of Optimization, Prediction, Decision-Making, and Diagnosis.

4.1.1 Optimization

Optimization in this context can be understood as knowing the status of every module in the system. Demand-side management (application layer)is one of the important challenges in the operation of Smart Grid [12]. Glowworm optimization (GSO) along with Support Vector Machine (SVM) was used in decision-making process of battery storage. This led to a reduced electricity tariff [23]. Performance optimization is possible through intelligent algorithms such as PSO, ACO, and AFSA in wireless networks [8].

DSM is related to plan, implement, and monitor the customer's consumption of energy. In other words, DSM brings needed control in customer side energy consumption. GSO is used for the optimization of battery energy storage. This also helps in understanding the capacity of the battery. This application leads to knowing the total amount of energy transfer in and out of the storage battery [23].

Optimum utilization of renewable energy resources is an increasing demand at the energy generation layer.

4.1.2 Prediction and Decision Making

Initially, the proportion of fossil fuels and other conventional resources was predominant in generation of electricity. Renewable energy resources inclusion in the grid gave scope to dynamic tariff facility for the consumers. But the challenge is the prediction, decision-making, and utilization of RER that shall be addressed by ASI algorithms. For example, Ant Colony Optimization explores the best possible solution in exploring alternatives, which are best in terms of time and energy-saving, thereby improving resource utilization. Forecasting of renewable energy resources and load demand is possible with the prediction capacity of ASI.

Smart Grid decision-making capacity to utilize wind energy on a cloudy day, solar energy on sunny day, and hydro energy on deluge or excess water at reservoirs is crucial for the effective operation of the grid. ASI adds value at this juncture. Smart grids will be useful in making power management, where ASI shall be used. Intelligent algorithms shall be used at DSM.

- Market Prediction analysis
- Providing accurate data to consumers on peak and less energy consumption
- Strategies to adopt in effective EMS

Researchers emphasize on the necessity of highly scalable, decentralized, and computing systems for future Smart Grids [17]. ASI offers and has the capability to meet these requirements.

DMS software consists of mathematical models for distribution of energy. This software predicts the impact of outages, generation discontinuities, and consistency in energy transmission. Intelligent algorithms will provide new computing models that empower such software modules. Insufficient intelligence has been the cause of concern for Smart Grid, particularly in prediction and decision making, which shall be addressed by ASI.

4.1.3 Diagnosis

Operation, Control, and Protection of crucial software and hardware installations of Smart Grid is possible only with a proactive approach in the diagnosis of problems. Starting from renewable energy resources to EMS, this diagnosis is an inseparable component of Smart Grid and Renewable Energy Systems.

Large, Medium, and Small scale renewable energy resources at mass production to micro grid-level require diagnosis. This is more evident in the Communication and Control layer of the five-layer models. Advancements in communication systems naturally demand integration and interdependency, which are hard challenges to achieve and sustain. This is not possible without proper diagnostic systems in place. ASI brings reliability in the operation and resolving the conflicts in balancing the integration and interdependency of various communication systems used in the IT wing of the Smart Grid.

This intelligent diagnosis not only predicts the possible failures in communication systems but also provides other routing solutions with shortest paths and minimum loss solutions. For example, ACO is a suitable candidate for such situations. Supervisory control is provided by these intelligent algorithms. Network analyses, simulation of large energy grid enable us to forecast, diagnose, and provide pragmatic solutions.

Reliable wireless communication systems, cybersecurity, designing fault-tolerant systems and their implementation, IoT enabled installations for automation requires proper diagnosis mechanism. Fault identification and reduction are possible only with suitable procedures, which shall be empowered by the application of intelligent algorithms. ASI assists in the self-healing process and self-optimization of Smart Grid systems.

Load Profile Study: Thorough profile studies help in successful generation on par with raise in demand. Unnecessary load on gird shall be avoided with intense study pertinent to load profiling. Intelligent algorithms shall be used to generate load profiles, anticipating reduced load peaking, and subsequent cost involved. Optimization process shall be applied to RER profiles and consumption profiles (both RER and Consumption profiles are mostly random in nature). For example, battery energy storage is calculated by GSO algorithm.

Customer Utilization Statistics: Analyzing Customer Statistics related to energy consumption and finding the probabilities in understanding this random nature of consumption is significant. This is in specific case when energy consumption is dependent on the sporadic weather variations. Load forecasting in DSM is an interesting area of study. Probability studies involved in estimating success or failures in information transfer using wireless communication devices, mathematical modeling of large scale grid structures are essential.

These studies are more relevant to Analysis and Support layer of Smart Grid systems where outage management, asset management took place. Estimation calculations of DR, ES capacity, EMS, and DSM have performance parameters that are subjected to randomness. These parameters are associated with psychological, social,

Table 2 Possible areas of application in smart grid and algorithms suggested

Dimension	Possible area of application	Suggested intelligent algorithms
Optimization	Renewable energy utilization, DSM	ACO, PSO, GSO
Prediction	Weather forecasting, load forecasting, DR	ACO, ABC, AFSA
Decision making	Scheduling, routing, DMS	ABC
Diagnosis	Fault-tolerant systems development, failure analysis, Telecommunication networks error modeling	DE, GSO

and environmental factors towards demand side of the spectrum. In Smart Grid, sufficient knowledge on demand-side will have a tremendous effect on supply-side i.e. generation, which curtails unnecessary burden on the grid. This is very clear and important at the Application or Service layer, which is the topmost layer of Smart Grid Management. Table 2 provides the possible application of intelligent algorithms in the appropriate systems of Smart Grid Management. In a Smart Grid, the significant areas of importance as mentioned in the Five Layer model are Generation, Transmission and Distribution, Communication, Computation, and Consumption Management.

At Generation, availability of clean energy resources and their strategic usage as per the increase in demand shall be effectively managed using ACO and ABC. Similarly, DMS shall be managed by employing PSO algorithm. GSO and DE algorithms shall be used at Communication layer. ABC algorithm shall be used for predictive analysis, decision making at Analysis and Support Layer. Efficient EMS and DSM (Application Layer) is possible using the ACO, GSO, and PSO algorithms.

4.2 Research Perspective of ASI

Smart Grid and Renewable Energy Systems open up scope for many new and novel technologies. The opportunities mentioned here shall be addressed using ASI, besides these are equally open to other emerging technologies and computing methodologies, which can contribute to a paradigm shift in the energy industry.

The favorable environment offered by Smart Grid for research has factors such as:

- Clean energy utilization and related environment issues
- Complexity of the grid
- Safety and efficiency of grid operation
- Open market
- Quality of Service to Consumers

The possible components for productive research shall be self-healing, corporation and consumer information, reliability and controllability, optimization, integration, and hardware.

- Self-healing (Automatic Recovery Capacity)—Good amount of research is going on in this area and the possibility for novel avenues is vivid using ASI. The iteration mechanism of modifying objective function in intelligent algorithms contributing to self-healing is an interesting area of study. Self-healing is a possibility with the development of efficient control systems, where closed-loop feedback prevails.
- Corporation—Consumer information flow has chances for more interactive systems. Novel techniques of data mining, analytics provide fruitful interpretations that may lead to the right decisions at right time. ASI algorithms are helpful to arrive at such interpretations.
- Reliability and Controllability—However, efficient systems are in place, there is a demand for improvisation in these areas. Any advancement here leads to the safety and security of our energy sector. ASI inherently possesses this internal control rather than external control guiding the coordination process. The application of these algorithms will provide wonderful results. Modularity in design also improves reliability. Better testing procedures shall be designed to improve reliability and controllability. Automatic Test Equipment development is one of the promising areas for innovation.
- Optimization—presupposes that the most investigated research area, is widely applied to many fields. In the particular case of Smart Grid Management, it can be used for reducing power loss and optimizing renewable energy utilization. The utility factor always needs correction, in other words, better percentages of utilization and intelligent utilization of RER shall be obtained. This most awaiting and anticipating factor to industry from mathematicians, data mining professionals, and engineering brains. ACO and PSO shall be applied in the appropriate scenarios.
- Standardization of integration procedures, refined management systems, and study of compatibility issues pertinent to intelligent integration has scope for more studies. Distributed power generation for centralized targets to achieve demands that level of innovation at every stage of integration. Mathematical (such as intelligent algorithms provided in this book chapter) and engineering solutions are always invited in this case.
- Electronics–Data Acquisition from Sensors is vital for decision making. Design of high precision sensors and transducers is in demand. Advanced power electronics technology, especially semiconductor power devices are most sought in electronic instrumentation. Gallium Nitride (GaN) power devices, amplifiers, and detectors research is active using high electron mobility transistors (HEMTs). Application of ASI in developing mathematical models is a good prospect.
- NanoScience—Nano-robotics with AI, nano-sensors design and development, device modeling, and fabrication techniques are scaling up for new generation applications. Quantum wires may be used in the rewiring of transmission grid for providing higher currents; supercapacitors modeling are emerging areas.

5 Conclusions

To summarize, the application of Artificial Swarm Intelligence in Smart Grid and Renewable Energy Systems is very promising. This computing intelligence can be successfully used in the areas of power generation and consumption.

The strength of intelligent algorithms is in optimization, prediction, decision-making, and diagnosis. It was also clear that these algorithms offer scalability, robustness, and flexibility. This will ensure that the RER, DMS, DSM, DR, and ESM stages are strengthened and more equipped towards optimum utilization of renewable energy resources; in better decision making, and in predicting or forecasting that helps in minimization of loading effect, loss incurred in energy transmission and distribution.

This can be achieved by using ACO, PSO, ABC, and GSO especially and other algorithms in appropriate hardware and software modules to yield encouraging results. Employing proper predictive and diagnostic systems in this energy supply chain shall indicate as well as support in reducing the intermittent errors of instrumentation.

The social, environmental, economic factors influencing the Smart Grid management will be effectively balanced. The efficient use of renewable energy will have an impact towards providing progressive outcomes. Considerable amount of quantitative as well as qualitative results shall be achieved at all five functioning layers of Smart Grid System using intelligent computing algorithms.

This chapter also provided insights into the research perspective of ASI concept and intelligent algorithms. It was also proposed and discussed the possible areas, and potential usage of these algorithms. Research gap and future prospects of ASI in energy sector are highlighted. This will help enthusiastic researchers and innovative thinkers to step into this fusion of fields or ideas that will lead to possible interdisciplinary innovation.

Acknowledgements We express our deep sense of gratitude to Prof. N. S. R. Ayengar, English, Professor (Rtd.) for assisting in pruning and scrutinizing the chapter of this book.

References

1. Ab Wahab, M.N., Nefti-Meziani, S., Atyabi, A.: A comprehensive review of swarm optimization algorithms. PLoSONE **10**(5), e0122827. https://doi.org/10.1371/journal.pone.012282 (2015)
2. Bonabeau, E., Dorigo, M., Theraulaz, G.: Swarm intelligence: from natural to artificial systems. J. Artif. Societies Soc. Simul. **4**, 320 (1999)
3. Boden, M.A.: Creativity and artificial intelligence. Artificial Intelligence, vol. 103, pp. 347–356. Elsevier (1998)
4. Spector, L.: Evolution of artificial intelligence. Artificial Intelligence, pp. 1251–1253. Elsevier. https://doi.org/10.1016/j.artint.2006.10.009 (2006)

5. Dorigo, M., Maniezzo, V., Colorni, A.: Ant system: optimization by a colony of co-operating agents. IEEE Trans. Syst. Man Cybern. Part B Cybern. **26**(1), 29–41. https://doi.org/10.1109/3477.484436 (1996)
6. Kennedy, J., Eberhart, R.: Particle swarm optimization. In: IEEE International Conference on Neural Networks, pp. 1942–1948 (1995)
7. Hüttenrauc, M., Šošić, A., Neumann, G.: Deep reinforcement learning for swarm systems. J. Mach. Learn. Res. **20**, 1–31 (2019)
8. Cao, L., Cai, Y., Yue, Y.: Swarm intelligence-based performance optimization for mobile wireless sensor networks: survey, challenges, and future directions. IEEE Access **7**, 161524–161553. https://doi.org/10.1109/ACCESS.2019.2951370 (2019)
9. Rosenberg, L., Pescetelli, N., Willcox, G.: Artificial swarm intelligence amplifies accuracy when predicting financial markets. In: IEEE 8th Annual Ubiquitous Computing, Electronics and Mobile Communication Conference (UEMCON), pp. 58–62. https://doi.org/10.1109/UEMCON.2017.8248984 (2017)
10. Oliveira, M., Pinheiro, D., Macedo, M., Bastos-Filho, C., Menezes, R.: Uncovering the social interaction network in swarm intelligence algorithms. Appl. Netw. Sci. **5**(24). https://doi.org/10.1007/s41109-020-00260-8 (2020)
11. Rosenberg, L., Willcox, G.: Artificial swarm intelligence. In: Bi, Y., Bhatia, R., Kapoor, S. (eds) Intelligent Systems and Applications. In: IntelliSys 2019. Advances in Intelligent Systems and Computing. Springer, Cham, vol. 1037. 10.1007%2F978-3-030-29516-5_79 (2020)
12. Lakshminarayana, S., Anjul: Smart grid technology & applications. In: Power and Energy Systems: Towards Sustainable Energy, pp.1–6. https://doi.org/10.1109/PESTSE.2014.6805268 (2014)
13. Nigam, A., Kaur, I., Sharma, K.K.: Smart grid technology: a review. Int. J. Recent Technol. Eng. (IJRTE) **7**(6S4), 243–247 (2019)
14. Indrajeet Prasad, M.I.: Smart grid technology: application and control. Int. J. Adv. Res. Electr. Electron. Instrum. Eng. **3**(5) (2014)
15. Anderson, R.N., Ghafurian, R., Gharavi, H.: Smart grid: the future of the electric energy system. IEEE Special Issue on the Smart Grid. https://www.researchgate.net/publication/325742525 (2010)
16. Henderson, M.I., Novosel, D., Crow, M.L.: Electric power grid modernization trends, challenges, and opportunities. IEEE Advanced Technology for Humanity, Nov 2017
17. Zahran, M.: Smart grid technology, vision management and control. WSEAS Trans. Syst. **12**(1) (2013)
18. Asaad, M., Ahmada, F., Alama, M.S., Sarfraza, M.: Smart grid and Indian experience: a review. Resour. Policy. https://doi.org/10.1016/j.resourpol.2019.101499 (2019)
19. European Commission European Smart Grids Technology Platform: Vision and strategy for Europe's electricity. https://ec.europa.eu/research/energy/pdf/smartgrids_en.pdf (2006)
20. Xue-song, Z., Li-qiang, C., Ma, Y.: Research on smart grid technology. International Conference on Computer Application and System Modeling, vol 3, pp. 599–603 (2010)
21. Chowdhury, S., Marufuzzamann, M., Tunc, H., Bian, L., Bullington, W.: A modified ant colony optimization algorithm to solve a dynamic traveling salesman problem: a cases study with drones for wildlife surveillance. J. Comput. Des. Eng. **6**, 368–386 (2019)
22. Neshat, M., Sepidnam, G., Sargolzaei, M., Toosi, A.N.: Artificial fish swarm algorithm: a survey of the state of-the-art, hybridization, combinatorial and indicative applications. Artif. Intell. Rev. **42**, 965–997 (2014)
23. Puttamadappa, C., Parameshachari, B.D.: Demand side management of small scale loads in a smart grid using glow-worm swarm optimization technique. https://doi.org/10.1016/j.micpro.2019.102886 (2019)

Robust Optimal Tuning of UPFC Using Single Objective Optimization Algorithm

G. Kannayeram, P. S. Manoharan, N. B. Prakash, T. Sivakumar, and R. Saravanan

1 Introduction

During the last decade, the Flexible AC transmission system (FACTS) is a novel expertise for reliable operation of power systems. A UPFC is a FACTS device that controls the power system parameters such as impedance, voltage, and current in unified manner [1]. It includes two voltage source converters (VSC). The UPFC aims to upgrade the transient steadiness of the framework and also provide voltage support thereby damp the power system oscillations under system uncertainties [2]. In the power system stability studies, conventional controllers such as lead-lag controllers are used [3, 4]. The tuning of control parameters of lead-lag controllers is a difficult task when conventional controllers are used, hence suitable evolutionary algorithms and intelligent control methods are used for online tuning of control parameters. Nowadays, for little sign soundness and investigation, the direct Phillips-Heffron model has can be used [5, 6]. In this work, a lead-slack damping controller incorporated with UPFC is utilized for the steadiness examination of intensity framework

G. Kannayeram (✉) · N. B. Prakash · T. Sivakumar
National Engineering College, Kovilpatti, Tamilnadu, India
e-mail: g.kannayeram@gmail.com

N. B. Prakash
e-mail: nbprakas@gmail.com

T. Sivakumar
e-mail: sivakumareee@nec.edu.in

P. S. Manoharan
Thiagarajar College of Engineering, Madurai, Tamilnadu, India
e-mail: psmeee@tce.edu

R. Saravanan
Balaji Institute of Technology and Science, Warangal, Telangana, India
e-mail: rsharankiruthiga@gmail.com

B. Vinoth Kumar et al. (eds.), *Intelligent Paradigms for Smart Grid and Renewable Energy Systems*, Algorithms for Intelligent Systems,
https://doi.org/10.1007/978-981-15-9968-2_2

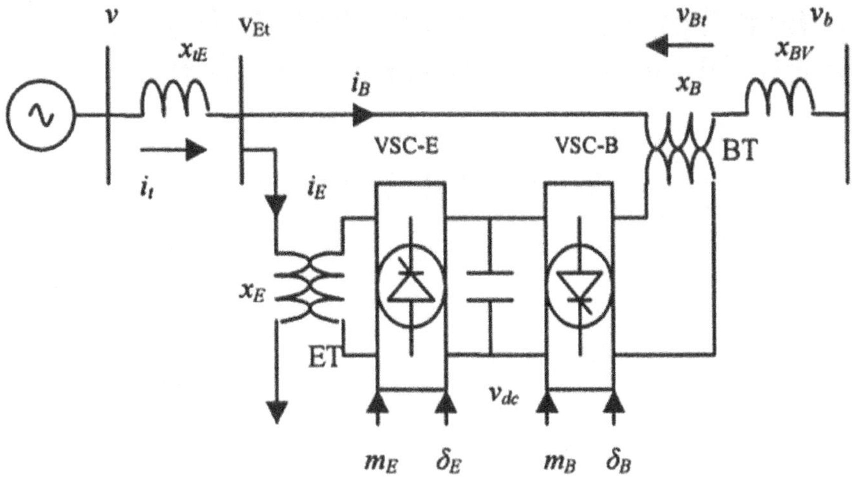

Fig. 1 UPFC damping controller installed in SMIB

[6, 7]. The controller parameters for damping oscillations in UPFC is highly difficult task due to nonlinearities associated with UPFC [8]. Many literatures reported that conventional techniques like the phase compensation technique increase the computational complexity, slow convergence, and time-consuming [3, 6, 7, 9, 10]. Hence for tuning of damping controller parameters, heuristic algorithms such as GA and PSO are most widely used in single and multimachine power systems [11–13]. Sidhartha Panda et al. has been compared PSO and GA for the design of FACTS based controllers with speed deviation as objective [11]. Many literatures reported PSO based UPFC for damping power system oscillations [3, 11, 12, 14]. Gravitational search algorithm based GA is also used for tuning controller parameters [15]. In literature, single objective optimization algorithm is used to regulate controllers of UPFC with objective function as change in speed deviation under uncertainties. In this work, PSO is used for the optimal design of UPFC damping controller by minimizing Integral Squared Error of objective functions to obtain optimum performance under nominal, light, and heavy operating conditions. For the purpose of understanding the benefits of PSO, the UPFC damping controller is also tuned with lead-lag compensator. For comparison, conventional phase compensation method is also considered. Stability is also ensured by eigenvalue analysis.

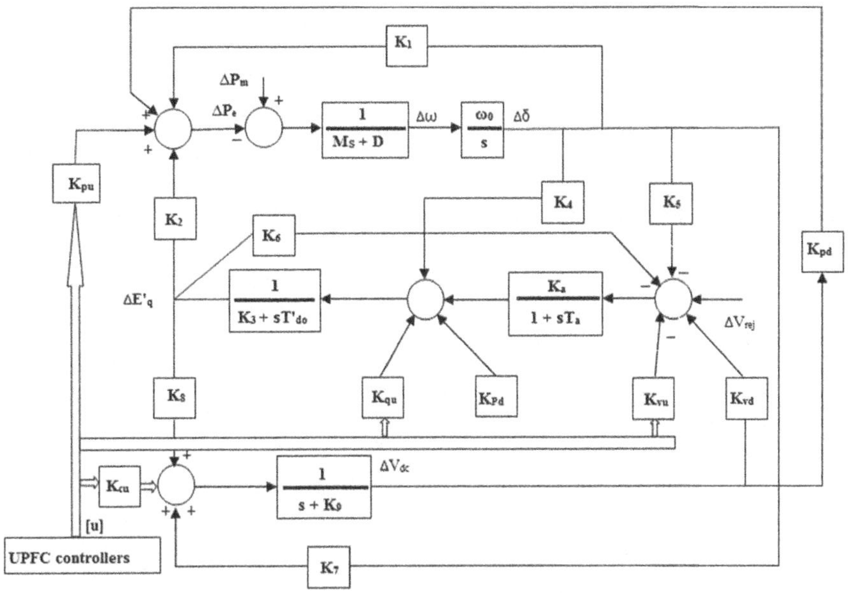

Fig. 2 Linearized Heffron–Phillips model installed with UPFC

2 UPFC

The UPFC with SMIB damping controller is shown in Fig. 1. IEEE-ST1A type excitation system is considered.

The Linearized modified Heffron–Phillips model taken is depicted in Fig. 2. The state-space expression of the proposed work is as follows.

$$\dot{x} = Ax + Bu \tag{1}$$

where

$$x = \left[\Delta\delta \ \Delta\omega \ \Delta E'_q \ \Delta E_{fd} \ \Delta V_{dc} \right]^{\mathrm{T}}$$

$$u = [\Delta m_E \ \ \Delta\delta_E \ \ \Delta m_B \ \ \Delta\delta_B]^{\mathrm{T}}$$

3 Proposed UPFC Damping Controller

The speed deviation is compensated by producing the required electrical torque and it has been done by tuning the UPFC control parameters. The controller parameters

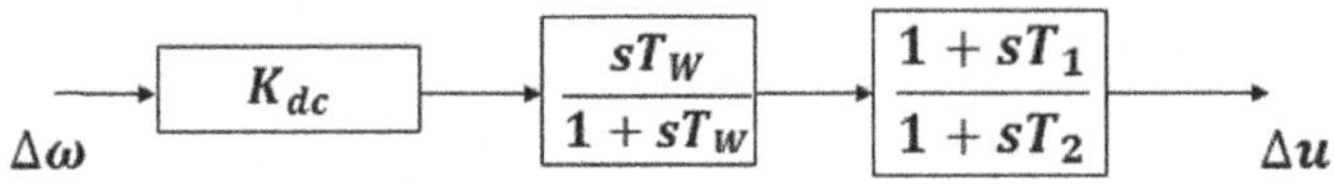

Fig. 3 UPFC controller with washout filter

are tuned by using a solo purpose optimization algorithm such as PSO and lead-lag compensation method respectively. The objective considered for minimization using single objective optimization is ISE of speed deviation. For multi-objective optimization, The two differing objectives are speed deviation and control signal considered. Figure 3 shows the UPFC damping controller structure with washout filter.

For single objective PSO, the Fitness function is given by Eq. (2)

$$F = \int_{0}^{t1} e_1^2(t)\mathrm{d}t \tag{2}$$

The constraints considered for PSO are gain and time constants which are expressed as

$$K_{\mathrm{dc}}^{\min} \le K_{\mathrm{dc}} \le K_{\mathrm{dc}}^{\max} \tag{3}$$

$$T_1^{\min} \le T_1 \le T_1^{\max} \tag{4}$$

$$T_2^{\min} \le T_2 \le T_2^{\max} \tag{5}$$

e_1 is the small change in speed deviation ($\Delta\omega$) and e_2 is the small change in control input signal.

4 Particle Swarm Optimization

In conventional optimization techniques, there is restriction on mathematical representation of the problem in computer and also in real-world nonlinear complex problems, there are a number of controller parameters and optimization of these parameters using conventional algorithms may have a problem of solutions stuck into local optima rather than global optima. Hence the heuristic-based algorithms are used for overcoming the above limitations. These heuristic algorithms such as evolutionary algorithms, GA, PSO, GSA, and simulated annealing may give a global optimal solution, easy implementation, and robust. In this work, PSO is used as an optimization algorithm where it is a solid stochastic improvement calculation dependent on the development and collaboration of swarms [12]. Social cooperation to critical thinking rule is included. It utilizes various particles that include a swarm moving around in the quest space watching for the best arrangement. Every molecule is taken as a point in an$N$-dimensional space which controls its mounting as indicated by its own raised understanding just as the taking off an understanding of different particles. Every particle monitors its co-ordinates in the arrangement space which are related to the best arrangement that was accomplished for by that molecule. First best worth is called individual best, 'pbest'. Second best worth acquired so far by any molecule in the area of that molecule. This subsequent worth is called worldwide best, 'gbest'. The position of molecules is updated using gbest and pbest.

Tuning of UPFC controller is carried out by diminishing Integral Squared Error of Fitness functions to obtain best performance under different loading conditions. For the purpose of appreciative the welfares of PSO, Eqs. (6) and (7) used in PSO for updating control variables in every iteration,

$$v_i^{k+1} = wv_i^k + C_1\text{rand}(1)\{pbest - x_i^k\} + C_2\text{rand}(1)\{gbest - x_i^k\} \tag{6}$$

$$x_i^{k+1} = x_i^k + v_i^{k+1} \tag{7}$$

Where, w is the inertia weight, C_1, C_2 are socio and cognitive component coefficients respectively, rand is random number, i is iteration, *pbest* is the personal best position of particle, gbest is the best position of the particles, x_i is particle position.

5 Results and Discussion

In these works, optimal parameters of UPFC damping controller are determined by subjecting the system to disturbance. The optimal parameters obtained using PSO tuned UPFC. The results obtained in the above-described method are compared with the conventional phase compensation method. The population size is taken as 200 and the functional evaluation is taken as 20,000. The performance of the proposed controller is evaluated by applying disturbance at different operating conditions as shown in Table 1.

The Parameter Settings of PSO is also given in Table 2.

Table 1 Operating conditions

Operating condition	P (p.u)	Q (p.u)
Nominal load	0.8	0.167
Light load	0.2	0.01
Heavy load	1.2	0.4

Table 2 Parameter setting of PSO

Parameters	PSO
Population size	200
Number of iteration	20
Functional evaluation	20,000
C_1	1
C_2	1
C_3	0
w_1 and w_2	0.7–0.9

5.1 Nominal Load Operating Condition

In this normal load operating condition, the damping controller performance is valued by giving a mechanical disturbance of Pm = 0.01 p.u. The optimal values of the lead-lag compensator gain and time constants obtained for this operating condition is reported in Table 3. In PSO, the objective function considered (Speed deviation) is having minimum value with higher control input provides better damping performance with minimum control energy. The AC voltage regulator (m_E) type damping controller provides outstanding performance with minimum objective function values thereby damp out oscillations quickly. In single objective tuning using PSO, δ_{E} controller furnishes better execution as contrasted and other controllers. Figure 4 shows the speed deviation, control input, and rotor angle curves obtained for different controllers at nominal operating conditions.

5.2 Heavy Load Operating Condition

The damping controller performance is investigated by applying mechanical disturbance under heavy operating conditions. The evolutionary algorithm tuned damping controller parameter; error and control input values are shown in Table 4. It is inferred that at heavy operating conditions DC voltage regulator (δ_{E}) based controller gives better performance with single objective optimization. Figure 5 delineates the dynamic reaction of the framework under substantial burden working conditions.

5.3 Light Load Operating Condition

The lead-lag compensator parameters and objective function values obtained by using PSO and phase compensation method for light operating condition are discussed in Table 5. It reveals that DC voltage regulator type damping controller provides better performance in light operating condition with minimum error.

Table 3 Nominal load operating condition (Pm = 0.01p.u)

Controller/Algorithm		δ_E		m_B		m_E	
		PSO	Lead-lag	PSO	Lead-lag	PSO	Lead-lag
Parameter	K	25.8959	18.0960	38.9826	31.1257	19.8908	12.4617
	T_1	0.4990	0.2296	0.4927	0.4144	0.4987	0.5247
	T_2	0.2012	0.2516	0.2109	0.2176	0.2100	0.2146
	F_1	2.0458e−8	3.4983e−5	6.779e−8	1.191e−6	2.144e−8	8.0777e−7
	F_2	7.646e−4	0.0102	0.0023	2.323e−3	6.001e−4	3.5897e−4

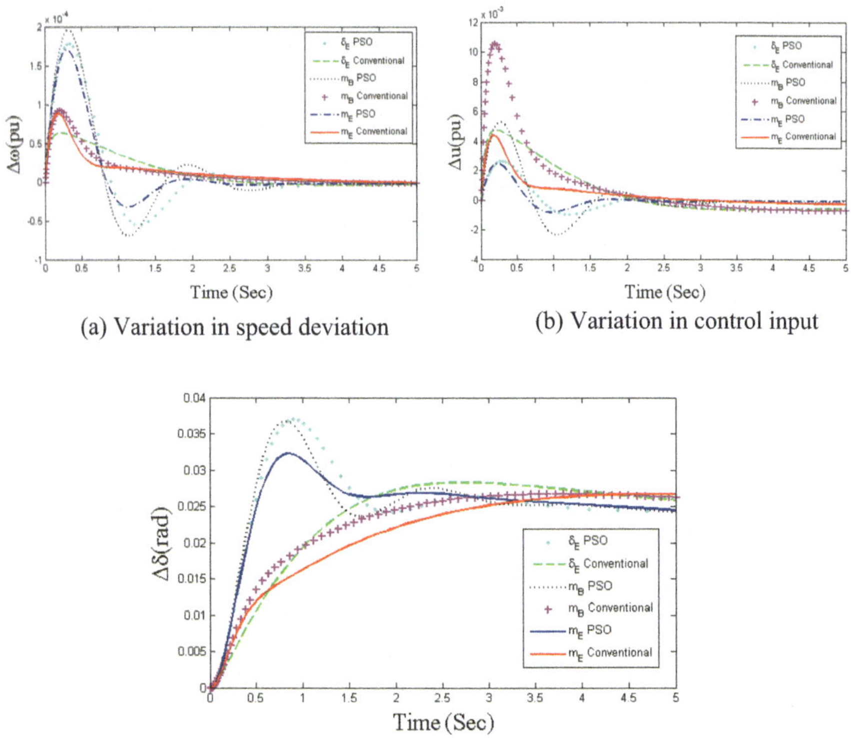

(a) Variation in speed deviation (b) Variation in control input

(c) Variation in rotor angle

Fig. 4 Response of different UPFC controllers at nominal operating condition

5.4 *Stability Analysis*

The parameter called Eigen esteems assists with understanding the steadiness of the framework. The Eigen esteems got for ostensible burden working condition is given in Table 6. The 50% of the eigenvalues are in left half of s-plane in PSO based controller and hence stability increases.

6 Conclusion

In this chapter, single objective algorithm based UPFC damping controller tuning is carried out by narrowing the ISE of the change in speed deviation and input control signal. So as to evaluate vigorous execution of proposed controller, the system is subjected to different operating conditions and scrutinized. Simulation results reveal that the single objective optimization gives minimum speed deviation with higher

Table 4 Heavy load operating condition

Controller/algorithm		δ_E		m_B		m_E	
		PSO	Lead-lag	PSO	Lead-lag	PSO	Lead-lag
Parameter	K	59.9749	27.9776	61.8428	31.1257	19.5562	12.4617
	T_1	0.5999	0.3371	0.4986	0.4144	0.4998	0.5247
	T_2	0.1126	0.3735	0.2001	0.2176	0.2004	0.2146
	F_1	1.286e−7	9.502e−7	4.039e−7	1.191e−6	1.437e−7	8.0777e−7
	F_2	0.0043	6.478e-4	0.0171	2.323e−3	0.0055	3.5897e−4

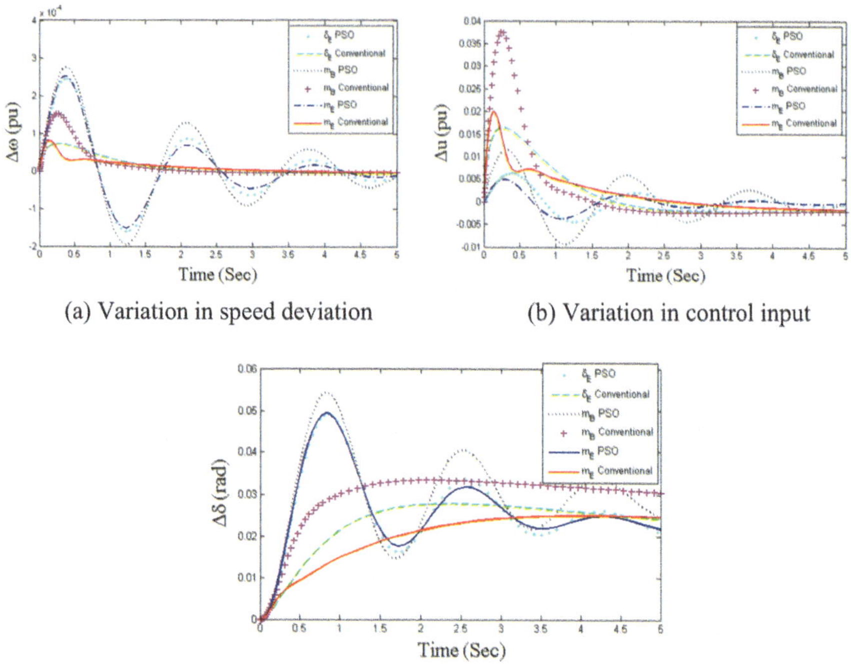

(a) Variation in speed deviation

(b) Variation in control input

(c) Variation in rotor angle

Fig. 5 Responses of different UPFC controllers at heavy load operating condition

input control signal. The dynamic responses obtained shows that DC voltage regulator type damping controller δ_E damp out oscillations quickly compared with other controllers in single objective optimization compared to conventional phase compensation based controller. The eigenvalue analysis proves that the UPFC controller performed well in terms of stability when PSO controller is used.

Table 5 Light load operating condition

Controller/algorithm		δ_E		m_B		m_E	
		PSO	Lead-lag	PSO	Lead-lag	PSO	Lead-lag
Parameter	K	29.9655	15.9172	40.9251	39.4335	28.8055	14.0452
	T_1	0.4990	0.2322	0.4965	0.4202	0.4723	0.5348
	T_2	0.2005	0.2156	0.2	0.2749	0.2172	0.2041
	F_1	2.937e−8	8.931e−7	4.158e−7	9.659e−7	2.086e−7	9.3551e−7
	F_2	4.447e−4	7.7856e−5	0.0182	2.608e−1	0.0091	5.7848e−4

Table 6 Eigenvalues of nominal load operating condition for UPFC

Eigenvalues in phase compensation method	Eigenvalues in PSO
−15.31, −5.76, −0.24, 0.408 ± 3.541i	−88.3526 −54.287 −2.0185 −0.646 + 0.8891i −0.646 − 0.8891i −0.218 −0.037

References

1. Hingorani, N.G., Gyugyi, L.: Understanding FACTS: flexible AC transmission systems. A book (Chap. 1). IEEE press, USA (2000)
2. Mohan Mathur, R., Rajiv, K.V.: Thyristor-based FACTS controllers for electrical transmission systems (Chap. 1). IEEE Press. Wiley (2002)
3. Abido, A.T., Awami, A., Abdel Magid, Y.L.: A particle-swarm-based approach of power system stability enhancement with unified power flow controller. Electr. Power Energy Syst. **29**(2), 251–259 (2007)
4. Wang, H.F.: Damping function of unified power flow controller. IEE Proc. C **146**(1), 81–87 (1999)
5. Smith, K.S., Ran, L., Penma, J.: Dynamic modeling of a UPFC. IEE Proc-Gener. Trans. Distrib. **144**(1), 7–12 (1999)
6. Tambey, N, Kothari, M.L.: Damping of power system oscillations with unified power flow controller (UPFC). IEE Proc. Gener. Trans. Distribut. 129–140 (2003)
7. Tambey, N., Kothari, M.L..: Unified power flow controller (UPFC) based damping controllers for damping low frequency oscillations in a power system. IE(I)EL **84**, 35–41 (2003)
8. Dash, P.K., Mishra, S., Panda, G.: A radial basis function neural network controller for UPFC. IEEE Trans. Power Syst. **15**, 1293–1299 (2000)
9. Albatsh, F.M., Mekhilef, S., Ahmad, S., Mokhlis, H.: Fuzzy logic based UPFC and laboratory prototype validation for dynamic power flow control in transmission lines. IEEE Trans. Ind. Electr. (2017)
10. Kannayeram, G., Manoharan, P.S., Willjuice Iruthyarajan, M., Sivakumar, T.: UPFC damping controller design using multi-objective evolutionary algorithms. Int. J. Bus. Intell. Data Mining **13**(1/2/3), 52–74 (2018)
11. Panda, S., Padhy, N.P.: Comparison of particle swarm optimization and Genetic algorithm for FACTS based controller design. Appl. Soft Comput. **8**(4), 1418–1427 (2008)
12. Kannayeram, G., Manoharan, P.S., Willjuice Iruthayarajan, M., Sivakumar, T.: Tuning of unified power flow controller (UPFC) using PSO and NSGA-II Investigations. J. Electr. Eng. **18**(2), 488–495 (2018)
13. Kannayeram, G., Manoharan, P.S., Prakash, N.B.: PI-tuned UPFC damping controllers design for multi-machine power system. J. Measur. Eng. **6**(2), 81–92 (2018)
14. Shayeghi, H., Shayanfar, H.A., Jalilzadeh, S., Safarim A.: A PSO based unified power flow controller for damping of power system oscillations. Energy Convers. Manag. **50**(10), 2583–2592 (2009)
15. Khadanga, R.K., Satapathy, J.K.: A new hybrid GA-GSA algorithm for tuning damping controller parameters for a unifed power flow controller. Electr. Power Energy Syst. **73**, 1060–1069 (2015)

Hybrid Sensorless MPPT Algorithm for Detecting Fast Irradiation Changes and Partial Shading Conditions on PV Systems

Balaji Veerasamy, Takaharu Takeshita, Viswanath Anbazhagan, Angamuthu Ananth, and Selvagopinath Mayandi

1 Introduction

Solar-powered PV generators rank top in the list of renewable power sources and total PV power generation capacity around the globe reaches 700 GW. Photovoltaic (PV) power plants operating at Mega-Watt (MW) levels consist of a huge number of individual solar modules connected together. Small PV generators are becoming more familiar in the distribution network. These PV generators are installed on the rooftops of houses, buildings, industries, and also on water bodies close to the load centers. The power output of the PV panels is directly proportional to the intensity of sunlight incidence on the modules. During uniform irradiation on all the modules in the array causes a single Maximum Power Point (MPP) to occur on the Power versus Voltage (*P–V*) curve. However, the sunlight incidents on the PV modules are not uniform in all the instants. Shading due to clouds and nearby objects, dust deposits, aging of particular cells, and bird dropping reduces the power output of the PV system. In this case there exist multiple numbers of MPPs on the *P–V* curve depending upon the position, length, and level of shading. There exist a single Global Maximum Power Point (GMPP) and the position of the GMPP depends on the shading patterns. The mismatch loss in the PV arrays depends on the shading patterns, causing the system to reduce the power extracted.

The PV system is always composed of an MPPT controller for detecting the MPP of the PV array. Perturb and Observe (P&O), Incremental Conductance (INC) and Hill Climbing (HC) algorithms are well-known MPPT methods to find the maximum point on the *P–V* curve. These methods have proved the highest efficiency for their

B. Veerasamy (✉) · V. Anbazhagan · A. Ananth · S. Mayandi
PSG College of Technology, Coimbatore, India
e-mail: vbi.eee@psgtech.ac.in

T. Takeshita
Nagoya Institute of Technology, Nagoya, Japan

B. Vinoth Kumar et al. (eds.), *Intelligent Paradigms for Smart Grid and Renewable Energy Systems*, Algorithms for Intelligent Systems,
https://doi.org/10.1007/978-981-15-9968-2_3

accuracy in tracking MPP, fast-tracking during changes in solar irradiation, and low cost [1–3]. The above methods lack training in detecting GMPP when partial shading occurs. Additionally, light-detecting sensors are employed along the PV array to compare the light incident on the panels and power extracted from the PV system. The controller identifies the presence of partial shading by calculating the imbalance between the solar incident on the panels and the PV power output. This increased the cost of the overall PV system.

Modified INC and P&O MPPT algorithms are developed to scan the *P–V* curve at regular intervals for detecting partial shading. This reduces the overall efficiency of the system. And implementing these algorithms on the PV converter system has limitations on the operating voltage. To solve this, problem Artificial Intelligence (AI) methods are introduced to track and find partial shading and GMPP, respectively. Innovated AI's embedded with Field Programmable Gate Arrays (FPGAs) created new design applications and problem-solving techniques in PV systems. The historical data of the PV system is fed into the AIso to train itself and find the solution during abnormal conditions. AI algorithms like fuzzy logic, neural networks, fuzzy-neural algorithms, particle swarm optimization, and genetic algorithm are the most common methods used for detecting the presence of partial shading on the PV modules [4–8]. Conventional MPPT algorithms P&O and INC are hybridized in combination with AIs for overcoming the challenges and problems faced in PV system. Although these methods have high accuracy, the overall control system gets complex and much knowledge is required for further changes in the PV configurations [9, 10].

The PV system consists of single or multiple power electronics converter circuits connecting the PV modules with the utility grid through different configurations. Figure 1 show the three commonly used PV converter configurations used for converting the *DC* power into *AC* power. Figure 1a shows the centralized *DC/AC*

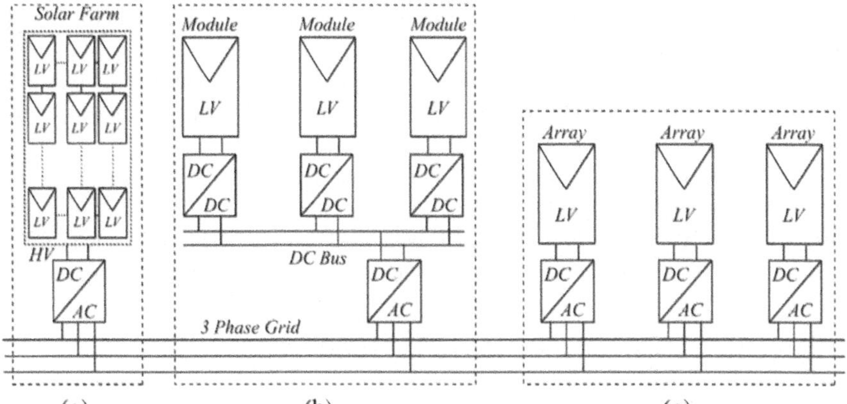

Fig. 1 PV inverter topologies. **a** Centralized DC/AC PV inverter, **b** DC/DC module integrated converter system and **c** string PV inverter

inverter system where all the modules are connected together and the PV bulk power is injected into the grid through a single-stage PV inverter. Centralized inverter configuration has very low efficiency during partial shading conditions or during the failure of certain PV modules in the array. To improve the efficiency, PV modules integrated with *DC/DC* boost converter, as shown in Fig. 2b, can extract maximum power from the PV system during partial shading disturbances. This improved the array efficiency but reduced the overall efficiency due to an increase in power electronic switches. Rooftop PV array with few modules in a string is directly connected to the grid through a single-stage *DC/AC* string inverter. All the converters are integrated with the MPPT controller for detecting the current/voltage at which the maximum power to be extracted from the PV array.

All the MPPT algorithms are categorized under two control modes, voltage-based control, and current-based control [11]. The PV panels are connected to a load or to the utility grid through a converter stage which converts the *DC* form of electricity

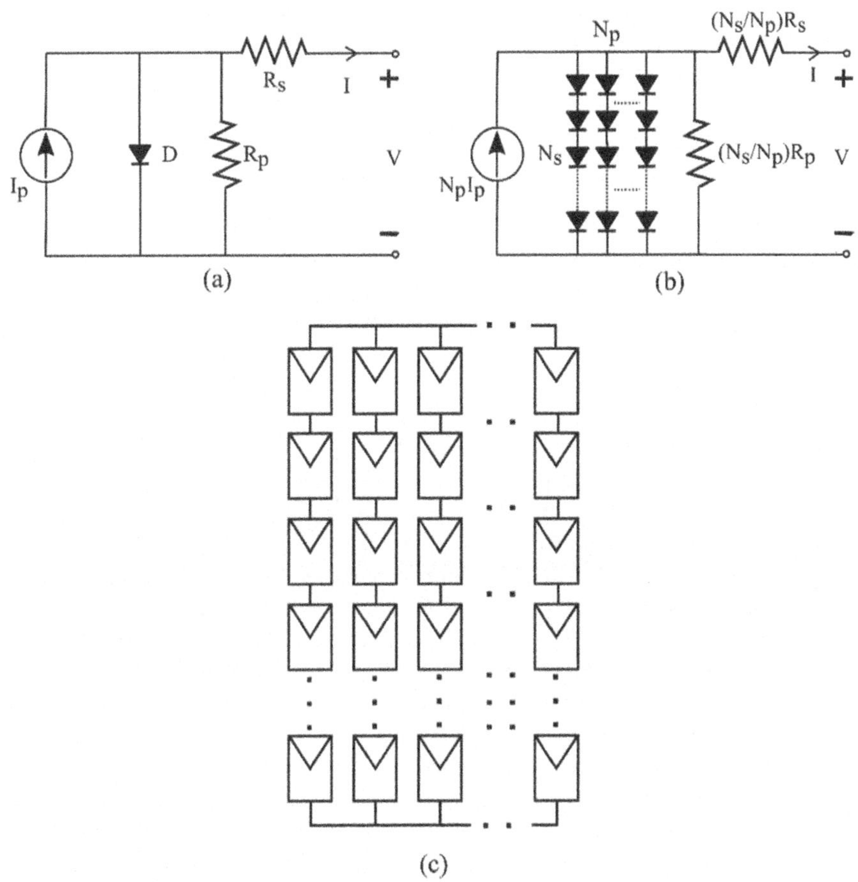

Fig. 2 Equivalent circuit. **a** PV cell, **b** PV module and **c** PV array

from PV panels to the form required for the load. The MPPT controller generates a reference signal to the converter based on which the operating point (voltage or current) of the PV system is determined [12]. Both control methods have its advantages and disadvantages in detecting the maximum operating point of the PV array. In this chapter, a hybrid MPPT controller is proposed which operates in both voltage and current control modes to detect the MPP under uniform radiation and tracking the GMPP during partial shading condition. The proposed algorithm is validated in the simulation environment with a single-stage PV inverter capable of operating on both voltage and current control modes. The proposed algorithm is testing under various levels of solar radiation and various patterns of shading on PV array.

2 PV Characteristics

The PV cell is a current source with a parallel diode, series, and shunt resistance [13]. Figure 2a shows the equivalent representation of a PV cell. The current from a PV cell is expressed in the following Eq. (1).

$$I_{\text{cell}} = I_{\text{p}} - I_{\text{o}}\left(e^{\left(\frac{q(V_{\text{cell}}+\text{IR}_{\text{s}})}{V_{\text{cell}}\text{TA}}\right)} - 1\right) - \frac{V_{\text{cell}} + I_{\text{cell}}R_{\text{s}}}{R_{\text{p}}} \tag{1}$$

The cell parameters I_{p}, I_{o}, R_{s}, R_{p}, T, and A in (1) are given in the datasheet provided by the manufacturer. The PV cells combine together to form a module. Figure 2b shows the circuit of the PV module. N_{s} is the number of cells in series and N_{p} is the number of the cell connected in parallel. The mathematical model of PV module is given in Eq. (2).

$$I_{\text{m}} = \left(I_{\text{p}} \times N_{\text{p}}\right) - N_{\text{p}} \times I_{\text{o}}\left(e^{\left(\frac{q\left(\frac{V_{\text{cell}}}{N_{\text{s}}} + \frac{\text{IR}_{\text{s}}}{N_{\text{p}}}\right)}{(V_{\text{cell}} \times N_{\text{s}})\text{TA}}\right)} - 1\right) - \frac{V_{\text{cell}} \times \left(\frac{N_{\text{p}}}{N_{\text{s}}}\right) + I_{\text{cell}}R_{\text{s}}}{R_{\text{p}}} \tag{2}$$

The array has N number of PV modules connected in different configurations. The total number of modules $N = m \times n$, where m is number of modules in a string, and n is number of strings. The PV voltage is determined by the number of series-connected modules m in the string and the PV current is dependent on number of strings n in the array. The modules are connected in the series–parallel configurations to increase the power capacity of the PV array.

Figure 2c shows the PV array connected in series–parallel configuration. The SP configuration has m number of modules in a string and n number of parallel strings. The current from the module is directly proportional to the solar irradiation. The short circuit current of a module, in terms of the solar irradiation is given in Eq. (3).

$$I_{\mathrm{m(sc)}} = I_{\mathrm{mSTC}}\left(\frac{G}{G_{\mathrm{STC}}}\right), \quad G_{\mathrm{STC}} = 1000\ \mathrm{W/m^2}\ \mathrm{at}\ 25\,^{\circ}\mathrm{C} \tag{3}$$

where STC denotes the standard test condition. The open circuit is less dependence on solar irradiation and expressed as

$$V_{\mathrm{m(oc)}} = V_{\mathrm{mSTC}} + V_{\mathrm{m}}\left(\frac{I_{\mathrm{m(sc)}}}{I_{\mathrm{mSTC}}}\right) \tag{4}$$

The photovoltaic array consisting of $m \times n$ number modules has short circuit current $I_{\mathrm{sc(array)}} = n \times I_{\mathrm{m(sc)}}$ and PV array open circuit voltage $V_{\mathrm{oc(array)}} = m \times V_{\mathrm{m(oc)}}$.

3 Partial Shading

The capacity of the PV system is increased by integrating more number of modules into the PV array. Irrespective of any number of modules in the system the *P–V* characteristic curve has a single peak power called Maximum Power Point (MPP) for uniform solar radiation on all the modules. The PV array voltage is determined by the number of modules connected in series and PV array current is determined by the number of modules connected in parallel. The current output from the PV array is directly proposal to solar irradiation. Figure 3 shows the *I-V* and *P–V* of a PV array varying from 100 to 1000 W/m^2. In all the conditions the output power is directly proportional to the solar radiation. The maximum power $P_{\mathrm{MPP}} = V_{\mathrm{MPP}} \times I_{\mathrm{MPP}}$, and from Fig. 3 the maximum voltage V_{MPP} occurs at a narrow region around $0.85 \times V_{\mathrm{oc}}$. But the maximum current I_{MPP} is varying with respect to solar radiance (*G*). The Maximum Power Point Tracking (MPPT) algorithms work on voltage control reference mode rather than current control reference mode because the tracking MPP through voltage reference is quick and easy than current reference.

When a group of modules in a PV array receives a different level of solar radiation due to shading, multiple peaks on the *P–V* curve occurs [14]. Figure 4 shows the *I-V* and *P–V* characteristic curves under uniform solar irradiation and when influenced during partial shading conditions. There exist multiple power peaks based on the shading level and its position on the array [15, 16]. Out of the multiple peaks there exists one power peak greater than other power peaks. The maximum power peak is the Global Maximum Power Point (GMPP) and remaining power peaks are Local Maximum Power Points (LMPPs). Rather than connecting all modules in series, the modules are connected in Series–Parallel (SP) configuration to improve the power out-put during partial shading conditions [17]. Total Cross Tied (TCT), Bridge Link (BL), Honey Comb (HC), and Sudoku are different PV array topologies brought in existence to increase the power output during partial shading [18–21].

Figure 5 shows the *I–V* and *P–V* curves of 4 different shading patterns on a PV array. The GMPP occurs at different voltage levels. This concludes that the GMPP

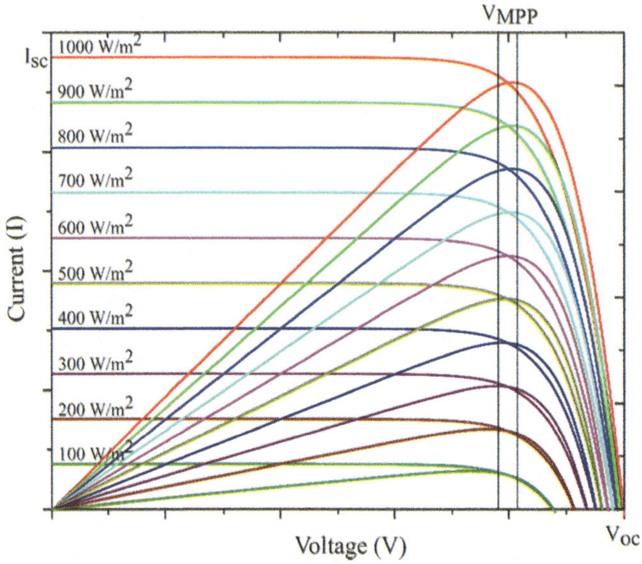

Fig. 3 *I-V* and *P–V* characteristic curves of PV array for solar irradiance varying from 100 to 1000 W/m²

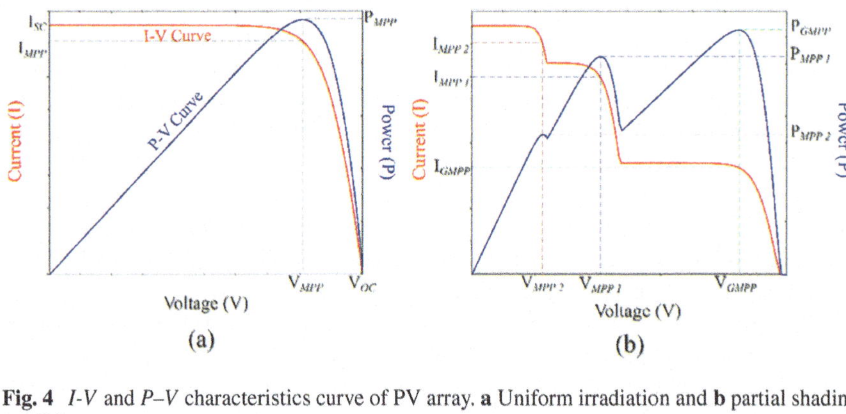

Fig. 4 *I-V* and *P–V* characteristics curve of PV array. **a** Uniform irradiation and **b** partial shading conditions

can occur out of the narrow voltage range marked in Fig. 3. The classical MPPT algorithms have been modified for tracking for GMPP. The voltage-based MPPT controllers are suitable for uniform solar radiation. Current based MPPT algorithms are better than voltage based MPPT algorithm for detecting the partial shading. But current based MPPT tracker loses stability for sudden solar radiation levels.

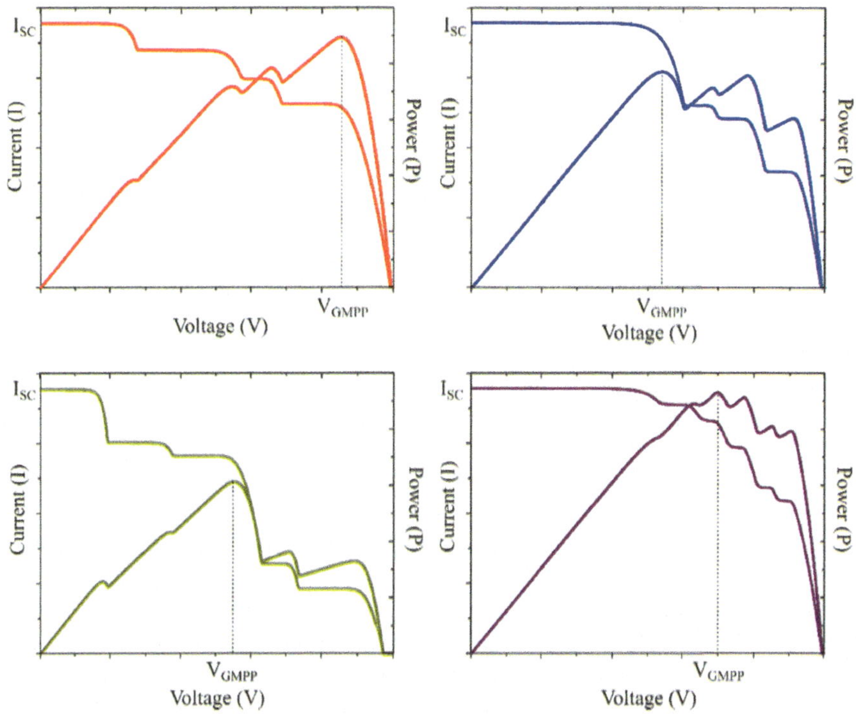

Fig. 5 *I-V* and *P–V* characteristic curves of partially shaded PV arrays

4 MPPT Algorithms

The MPPT controller is the main unit in the PV system for tracking the maximum power point of a module or an array. Due to non-linearity of the PV source a MPPT controller is mandatory for all PV systems for finding the maximum electrical power for the corresponding solar irradiation incident on the array. Various MPPT algorithms have been proposed to scan the *P–V* curve of the characteristic curve of PV array output. In a two-stage power converters, the MPPT controller is integrated into the *DC/DC* converter and in single-stage power converters the MPPT controller is integrated with the PV inverter. The MPPT controller generates the duty cycle ratio *D* for the *DC/DC* boost converter and generates a PWM reference signal for a *DC/AC* inverter. For all the MPPT methods, the voltage and current of the PV module/array is sensed and sampled for calculating the change in voltage and power. Based on the changes in power, voltage and current the MPPT controller generates the reference signal for the converter. The reference value is incremented or decremented till the MPP is reached. The most reliable conventional MPPT algorithms are Incremental Conductance (INC) and Perturb and Observe (P&O).

4.1 Perturb and Observe (P&O) MPPT Algorithm

The two parameters voltage *(V)* and current *(I)* are measured using the sensors. The third parameter Power *(P)* is calculated using the values of V and I. The maximum power P_{MPP} occurs at the particular value of voltage V_{MPP} which produces the corresponding maximum current I_{MPP} from the PV array. The concept behind tracking the maximum power is the algorithm increments or decrements the values of voltage (ΔV) to find the sign of change in power ΔP. The value of ΔV is perturbed to observe the change in the value of ΔP. The P&O algorithm detects the MPP based on the sign of the ratio $\Delta P/\Delta V$. The conditions of polarity of the ratio are $\Delta P/\Delta V > 0$, $\Delta P/\Delta V = 0$ and $\Delta P/\Delta V < 0$.

Figure 6 shows the variations of $\Delta P/\Delta V$ on the characteristic P–V curve. The sign of change in power to change in the voltage $(\Delta P/\Delta V)$ is positive on the left side of the MPP and the sign of the ratio $\Delta P/\Delta V$ is negative on the right side of MPP. At MPP the ratio $\Delta P/\Delta V$ is zero. The slope of the $\Delta P/\Delta V$ is very low on the left side of the curve and slope of $\Delta P/\Delta V$ is high on the right side of MPP.

Figure 7 shows the flowchart of P&O MPPT algorithm. The sign of change in power ΔP for the change in voltage ΔV determines the direction of perturbation. If $\Delta P/\Delta V > 0$ the controller perturbs in positive direction and if $\Delta P/\Delta V < 0$ in negative direction. The controller stops to perturb when $\Delta P = 0$ for the change in voltage ΔV. The main advantage of P&O algorithm is it can be easily implemented in the FPGA controller. The disadvantage is oscillations around the MPP. In order to overcome this disadvantage, ΔV is set to a small value. This in turn increases the number of iterations to reach MPP. Modified P&O algorithm came into existence by changing the step size of ΔV based on the magnitude of the ratio $\Delta P/\Delta V$. This increases the speed of scanning the P–V curve of the array and finding the MPP.

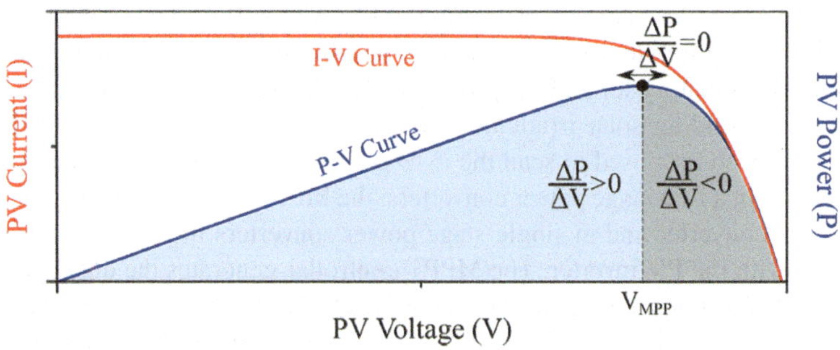

Fig. 6 Concept of perturb and observe algorithm

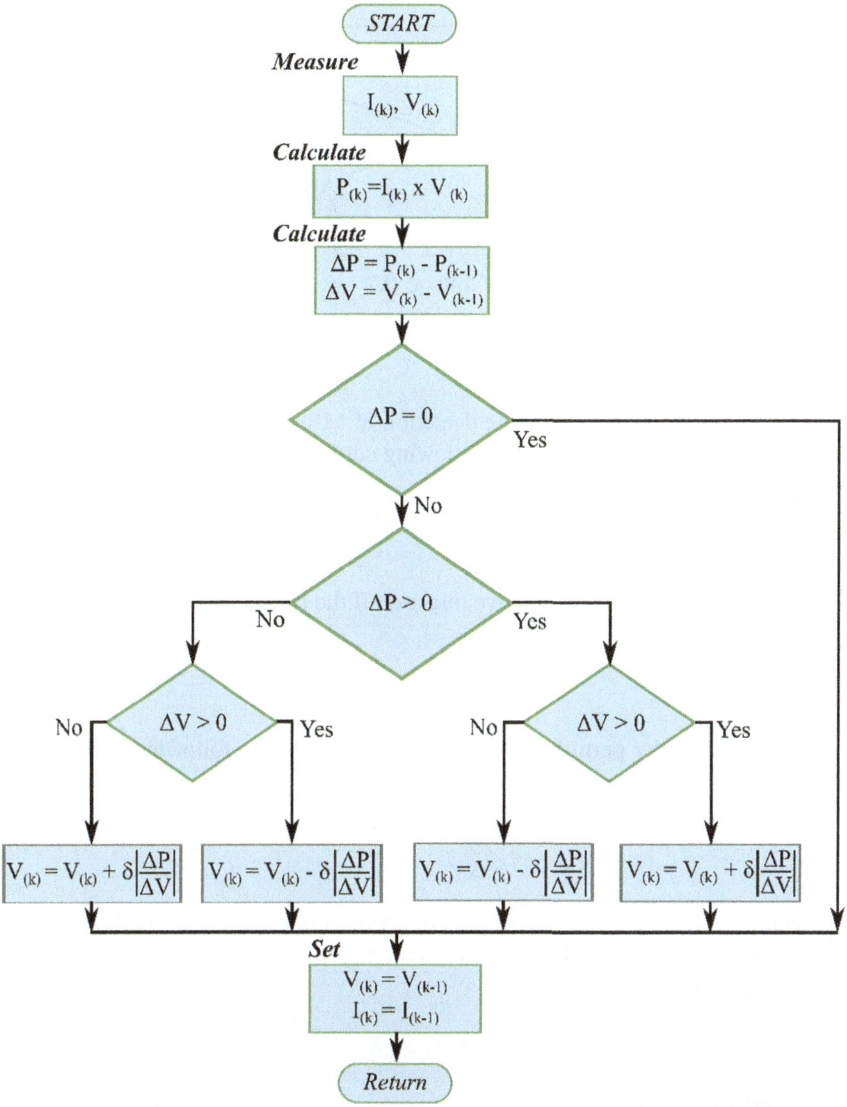

Fig. 7 Perturb and observe MPPT algorithm

4.2 Incremental Conductance (INC) MPPT Algorithm

The P&O algorithm utilizes only the *P–V* characteristic curve to track the MPP. But the INC algorithm utilizes both *P–V* and *I–V* curve of the arrayto track the MPP. This makes it faster to reach the MPP when compared to P&O algorithm. In INC method, the sign of $\Delta P/\Delta V$ is determined using incremental conductance of $\Delta I/\Delta V$.

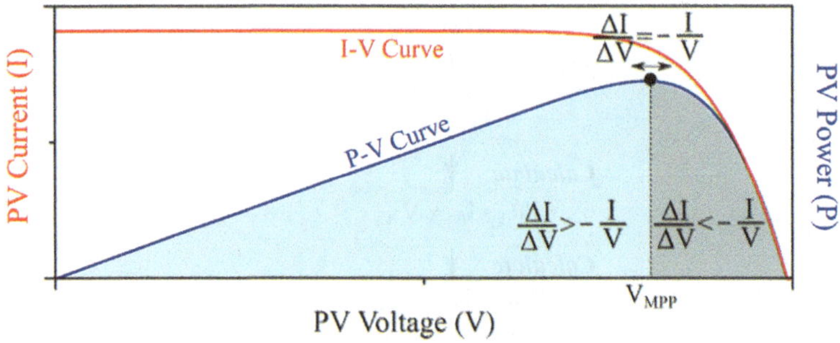

Fig. 8 Concept of incremental conductance MPPT algorithm

Figure 8 shows the logic behind the detection of MPP on *I–V* and *P–V* curve. The algorithm stops to perturb when the following equation is satisfied.

$$\frac{\Delta I}{\Delta V} = -\frac{I}{V} \tag{5}$$

The controller perturbs in positive direction if the following condition occurs,

$$\frac{\Delta I}{\Delta V} > -\frac{I}{V} \tag{6}$$

And the controller perturbs in negative direction for the following condition of *I/V*.

$$\frac{\Delta I}{\Delta V} < -\frac{I}{V} \tag{7}$$

Figure 9 shows the flowchart of INC MPPT algorithm. The INC method is quick in detecting the MPP and the main disadvantage is complexity in control when compared to P&O.

4.3 *Proposed Ideology*

The solar dependent PV modules are current source generators operating on a particular voltage reference for maximum power extraction. The MPPT controller involves either in controlling the voltage or the current of the PV array. As mentioned earlier both voltage and current based approach has its own advantages and disadvantages. Voltage based control has better performance for uniform irradiation on the PV array. The MPP occurs at a voltage range of between 0.75 V_{oc} and 0.85 V_{oc} for all levels of irradiation, as shown in Fig. 3. So the voltage based controllers are designed for operating at the particular voltage range for easily detecting the MPP. As shown in

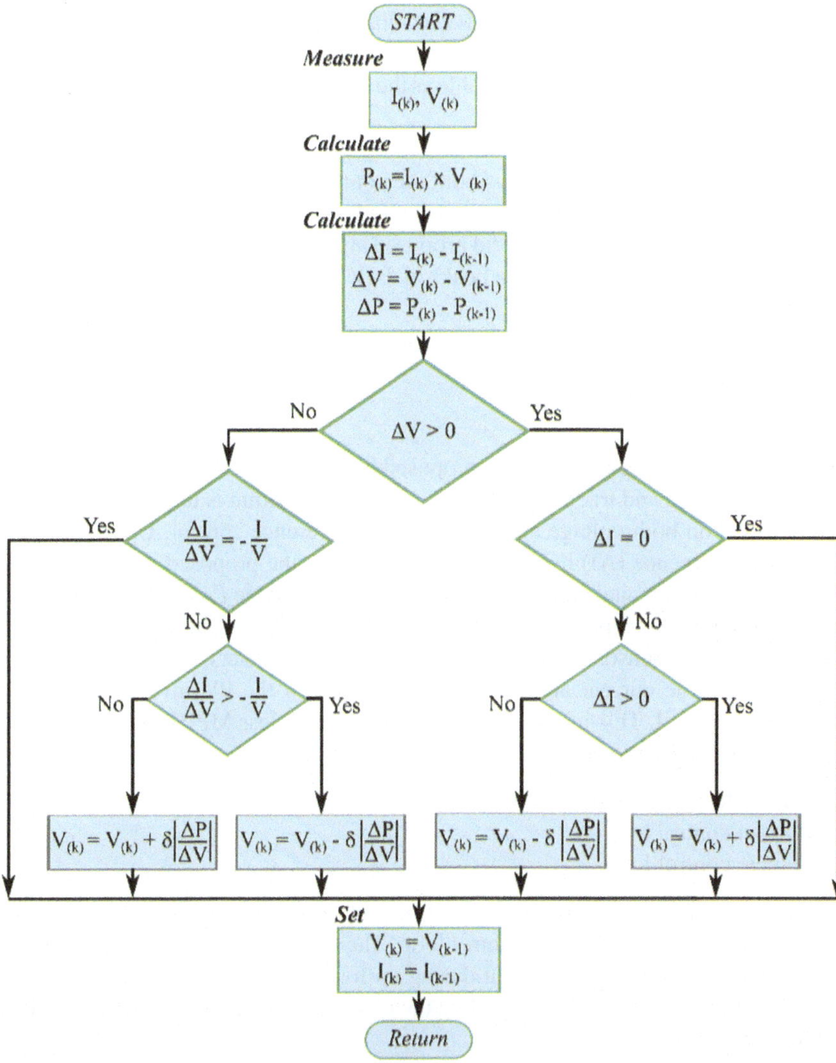

Fig. 9 Incremental conductance MPPT algorithm

Fig. 5, during the presence of shading on the PV modules, multiple peaks exist on the *P–V* curve. The voltage V_{GMPP} for the global maximum power point occurs at any voltage range from a low level to $0.9 \times V_{oc}$. So the controller has to scan the whole *P–V* curve for detecting the multiple peaks. The voltage-based approach lags behind in differentiating whether the drop in voltage is caused because of uniform irradiation drop or due to partial shading.

When compared to voltage based approach, very little literature is available based on control MPPT algorithms. Current based control has a better ability to predict the power variations caused during intensive irradiation changes. The main disadvantage of the current based approach is, if the reference current is greater than the PV array capacity the PV system becomes unstable as the controller moves towards the positive side of *P–V* curve.

By considering the advantages of both voltage and current based controllers, the proposed MPPT algorithm employs both the control strategies for finding the MPP during uniform irradiation. The proposed method instantly detects the partial shading occurring on the PV array. The hybrid proposed algorithm not only detects the partial shading but also scans the multiple peaks of *P–V* array and track the GMPP.

5 Proposed Algorithm

Figure 10 shows the flowchart of the proposed hybrid MPPT algorithm for detecting the partial shading and tracing the *P–V* curve. The algorithm is termed to be hybrid as it is based on both voltage and current based reference control. Unlike in other Artificial Intelligence (AI) hybrid MPPT algorithms, the proposed method doesn't need other environmental data or irradiation level to trace the *P–V* curve. These data require additional sensors to get information. The proposed hybrid algorithm doesn't employ additional sensors to get the environmental parameters. The controller only employs the basic current and voltage sensor used in the PV inverter circuit for closed-loop control. This makes the name hybrid sensorless MPPT algorithm.

The proposed hybrid controller has three control loops:

1. GMPP detection loop
2. Current control loop
3. Voltage control loop

The decision block is the main control unit of the system which makes the controller to slide towards each control loop. The algorithm performs the following steps to find maximum power point during uniform irradiation, detecting the presence of shading on the panels and finding the global maximum power during shading conditions.

Step 1: The controller initializes the iteration by sensing the PV current I_k and PV voltage V_k. The sampling period of the iteration $s = 0.005$ s. The present iteration is denoted as k and $k - 1$ is the value of the previous sampling period.

Step 2: The PV power P_k is calculated for the current iteration.

$$P_k = V_k \times I_k \tag{8}$$

Step 3: The change in current, voltage, and power is calculated between the iterations.

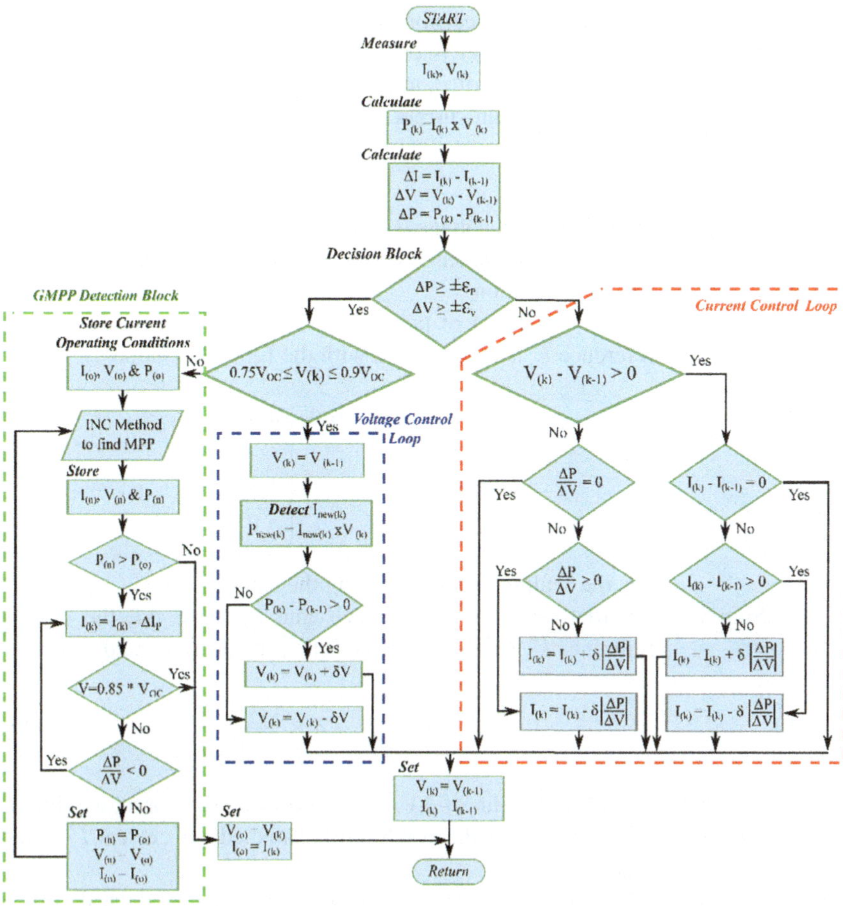

Fig. 10 Proposed MPPT algorithm

$$\Delta I = I_k - I_{k-1} \tag{9}$$

$$\Delta V = V_k - V_{k-1} \tag{10}$$

$$\Delta P = (I_k \times V_k) - (I_{k-1} \times V_{k-1}) \tag{11}$$

Step 4: The controller checks for the following condition of change in power ΔP and ΔV.

$$\left\{ \begin{array}{l} \Delta P > \pm\varepsilon_{\mathrm{p}} \\ \Delta V > \pm\varepsilon_{\mathrm{v}} \end{array} \right\} \tag{12}$$

The above condition in the decision block is the key point in differentiating whether the change in power is caused by uniform irradiation or due to partial shading. The values of ε_p and ε_v is fixed based on the open-circuit voltage V_{oc} and short circuit current I_{sc} of the PV array.

Step 5: If the above condition is not satisfied the controller enters the current control loop.

Step 5.a: If the change in voltage is positive the algorithm checks whether the current reference I^* is incremented or decremented in the previous iteration.

Step 5.b: For the condition of positive ΔV and positive ΔI the current reference I^* is incremented with the following value.

$$I_k = I_k + \text{Step} \tag{13}$$

$$\text{Step} = \delta \times \left|\frac{\Delta P}{\Delta V}\right| \tag{14}$$

δ is the scaling factor for tuning the step size.

Step 5.c: Alternatively, for positive ΔV and negative ΔI, the current reference I^* is reduced by the following decremented value.

$$I_k = I_k - \delta \times \left|\frac{\Delta P}{\Delta V}\right| \tag{15}$$

Step 5.d: For negative value of ΔV, the algorithm checks for the conditions $\Delta P/\Delta V = 0$. If the condition is satisfied the current I^* is left unaltered.

Step 5.e: Alternatively, if the above condition is not satisfied the algorithm checks for $\Delta P/\Delta V > 0$ or $\Delta P/\Delta V < 0$. If $\Delta P/\Delta V > 0$ the current I^* is incremented as given in Eq. (14) and if $\Delta P/\Delta V < 0$ current I^* is decremented as in (15).

Step 6: If the condition in step 4 is satisfied, the controller checks for the following condition $0.75V_{oc} \leq V_k \leq 0.9V_{oc}$. This condition is key block in sliding between the voltage control loop and partial shading detection loop. If V_k is between the limits $0.75V_{oc}$ and $0.90V_{oc}$ there is sudden drop in irradiation level.

Step 7: The PV inverter works on voltage reference V^* instead of current reference I^*. The voltage reference is set to the previous sampling value $V_k = V_{k-1}$.

Step 7.a: The new value of current is detected from the sensor and the new value of power P is calculated.

Step 7.b: The power difference $\Delta P = P_k - P_{k-1}$ is calculated.

Step 7.c: If ΔV is positive the voltage reference is incremented by a value of $V_k = V_k + \delta V$, where δ is the scaling factor.

Step 7.d: If ΔV is negative the voltage reference is decremented by a value of $V_k = V_k - \delta V$.

Step 7.e: The tracker returns to step 1 and senses the PV array power P_k and current I_k for V_k.

Step 8: In step 6 if the condition $0.75V_{oc} \leq V_k \leq 0.9V_{oc}$ is satisfied the partial shading occurrence is detected and the controller moves in GMPP detection block to trace the *P–V* curve with multiple peaks.

Step 8.a: The instant power, current, and voltage is stored as P_o, I_o, and V_o, respectively.

Step 8.b: The incremental conductance is performed to find the MPP.

Step 8.c: The new maximum power P_n, current I_n, and V_n is stored.

Step 8.d: If $P_n > P_o$, the tracker stops searching for further maximum points as the GMPP is P_o.

Step 8.e: If $P_n < P_o$, the tracker decrements in $I^* = I^* - \Delta I_p$.

Step 8.f: The current is decremented in constant values of ΔI_p till the condition $\Delta P/\Delta V < 0$.

Step 8.g: When the condition $\Delta P/\Delta V < 0$ is reached the tracker moves to step 8.b to find the new MPP using INC algorithm.

Step 8.h: Alternatively the current ΔI_p is decremented till the condition $V = 0.85\ V_{oc}$, as this voltage value no MPPs occur.

Step 8.i: Finally the controller settles down with the maximum power GMPP and controller starts from step 1.

6 Dual-Mode Single-Stage PV Inverter

Figure 11 shows the three-phase single-stage PV inverter used in converting the DC power into three-phase AC power. The inverter composes six IGBT switches

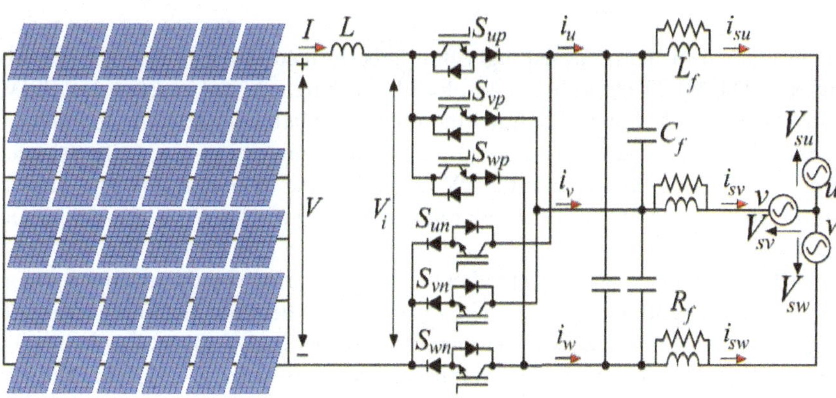

Fig. 11 Transfromerless single-stage PV inverter with 36 modules

and Pulse Width Modulation (PWM) technique is employed to control the switching sequence of six IGBTs [22]. The inverter topology has a coupled inductor L connected in the input side, thus making it a current source inverter. The output of the inverter is connected to the grid through the filters R_f, L_f, and C_f. The inverter doesn't have any galvanic isolation as the authors claim it be a transformerless single-stage high boost PV inverter [23].

The hybrid proposed MPPT algorithm operates in dual control mode (current/voltage). In current control mode, the reference current I^* from the MPPT controller is given to the control unit of the PV inverter. And during the voltage reference mode, the inverter should be capable of operating at the specific DC reference voltage V^*. The duality of the inverter makes it suitable for operating in both modes of operation and extracts continuous power from the PV panels [24].

The PV array voltage V and array current I are the output DC values from the PV array and fed into the inverter. The inverter input voltage V_i is controlled by the PWM technique for realizing the PV voltage. The currents i_u, i_v, and i_w are the three-phase inverter output current. The outputs currents i_{su}, i_{sv}, and i_{sw} are the three-phase sinusoidal current injected into the corresponding grid phases of voltages V_{su}, V_{sv}, and V_{sw}, respectively. The main feature of the topology is the operating DC voltage range of the PV inverter is stated as;

$$0 \leq V_i \leq \frac{\sqrt{3}}{\sqrt{2}} E \cos \varphi^* \tag{16}$$

As the GMPP occurs between wide ranges of voltages from low-level to $0.9 \times V_{oc}$, the above inverter is suitable for operating at voltages given in (16). The inverter maintains unity power factor between the phase current with the corresponding grid voltage by varying the power factor reference φ^* [25].

7 Performance Validation

The evaluation of the proposed algorithm is performed in the PowerSIM (PSIM) platform. The MPPT algorithm is developed in C language and interfaced in PSIM using C code block. Figure 12 shows the PSIM simulation model with 36 modules connected is series–parallel configuration with an array size of 6×6. The single-stage PV inverter is connecting the PV array with the three-phase voltage source considered as the grid. The PV inverter has six IGBT switches $S_{up} - S_{un}$ connected as shown in Fig. 12.

A DC link inductor is connected between the PV array and the inverter system. The three-phase grid filter R_f, L_f, and C_f is connected between the three-phase grid and the PV inverter to reduce the harmonics caused by high-frequency switching of IGBT's. Table 1 summarizes the overall specifications of the PV system. The PV

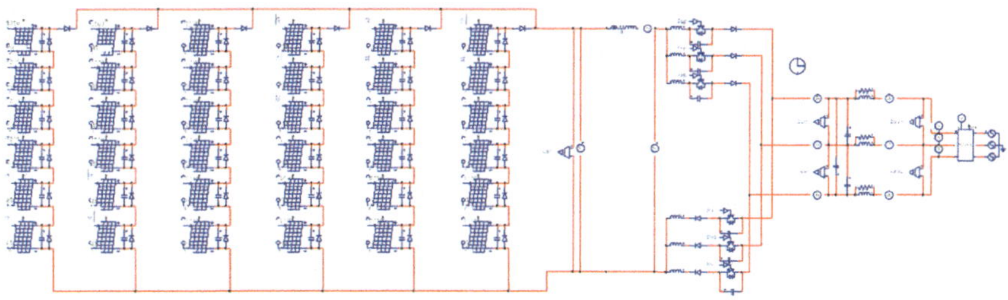

Fig. 12 PSIM simulation model

Table 1 Specifications of PV array and inverter

Short-circuit current I_{sc}	3.616 A
Open-circuit voltage V_{oc}	28.13 V
Current at MPP I_{MPP}	3.537 A
Voltage at MPP V_{MPP}	22.81 V
Solar radiation, temperature T	1000 W/m^2, 25 °C
Maximum power P_{MPP}	80.70 W
Number of modules	36 (6 × 6)
Maximum power output P_{out}	3 kW
Grid voltage E, ω	200 V, 2π × 60 rad/s
Switching frequency f_s	10 kHz
Grid-connected filter R_f, L_f, C_f	47Ω, 1.0 mH, 10.47 μF
Inductance L	20 mH

array composes of 36 modules connected in the combination series–parallel configuration with string length of 6 modules and a total of such 6 strings are connected in parallel. Simulation is performed under two irradiation cases, (i) Continuously changing uniform solar radiation and (ii) Dynamically changing partial shading condition.

7.1 *Case 1*

Case 1 is study of the proposed MPPT algorithm for uniform solar irradiation changes on all the modules varying from 100 to 1000 W/m^2 at an atmospheric temperature of 25 °C. Figure 13 shows the I–V and P–V characteristic curves for uniform solar irradiation changing from 100 to 1000 W/m^2. The controller begins in voltage control mode with a voltage reference $V^* = 136$ V which is 0.8 × V_{oc} of PV array. The corresponding current $I = 1.67$ A is stored as the initial value. For the next iteration, the controller slides towards the current control loop with a current reference of $I^* = 1.67$ A. The algorithm tracks the MPP at the current reference of $I^* = 2.068$ A, and

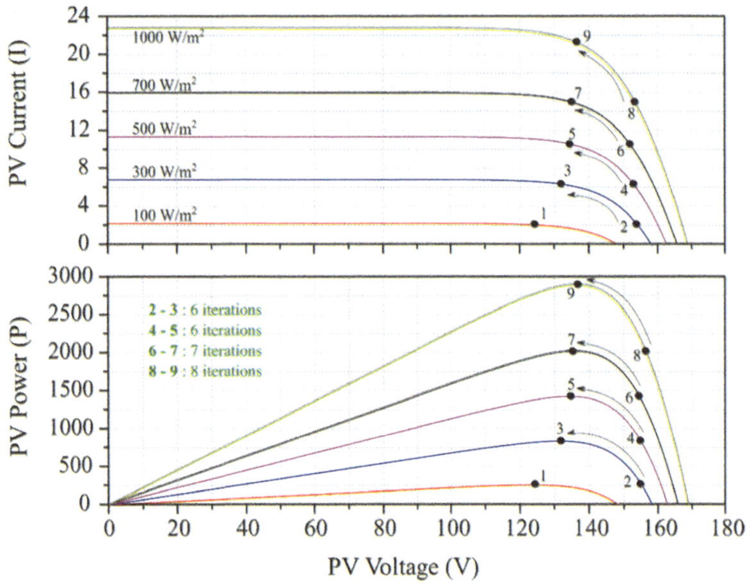

Fig. 13 Case 1: Uniform shading characteristics curve for solar radiation level changing from 100 to 1000 W/m^2

output power of 258 W and settles down at point 1 as shown in Fig. 13. At $t = 0.2$ s the solar radiation changes from 100 to 300 W/m^2. At this instant, the operating point shifts to point 1 to 2, the ratio $\Delta P/\Delta V$ is negative and controller is on the negative side of P–V. The controller increments the current and tracks the new MPP at point 3. The controller takes 6 iterations to find the new MPP. Figure 14 shows the output waveform of PV power *(P)*, PV current *(I)*, and PV voltage *(V)* for case 1. The solar level changes to 500 W/m^2, 700 W/m^2 and 1000 W/m^2 at $t = 0.4$ s, $t = 0.6$ s and $t = 0.8$ s, respectively.

As shown in Fig. 13, the proposed controller tracks the MPP for every solar radiation change and the operating reference values move from point 4 to 9. Corresponding changes in power, current, and voltage are confirmed in the output waveform in Fig. 14. The controller finally settles down at point 9 for a maximum power of $P_{\mathrm{MPP}} = 2.901$ kW and $V_{\mathrm{MPP}} = 137.13$ V at the current reference $I^* = 21.15$ A. The controller requires 6 iterations to track the new MPP when the irradiation changes from 100 to 300 W/m^2. The number of iterations required for finding the new MPP for change in irradiation is given in Fig. 13.

Figure 15 shows the overall waveform of the whole PV system. The step-change in irradiation level is manually created in the simulation environment and the MPPT tracks the maximum power for the change in irradiation. The output waveform of PV power P, PV current I, and PV voltage V is varying with respect to the reference value of the MPPT controller. The voltage $V_i$is the voltage on the input side of the inverter. The current i_u is the pulsed current of phase u caused by the PWM

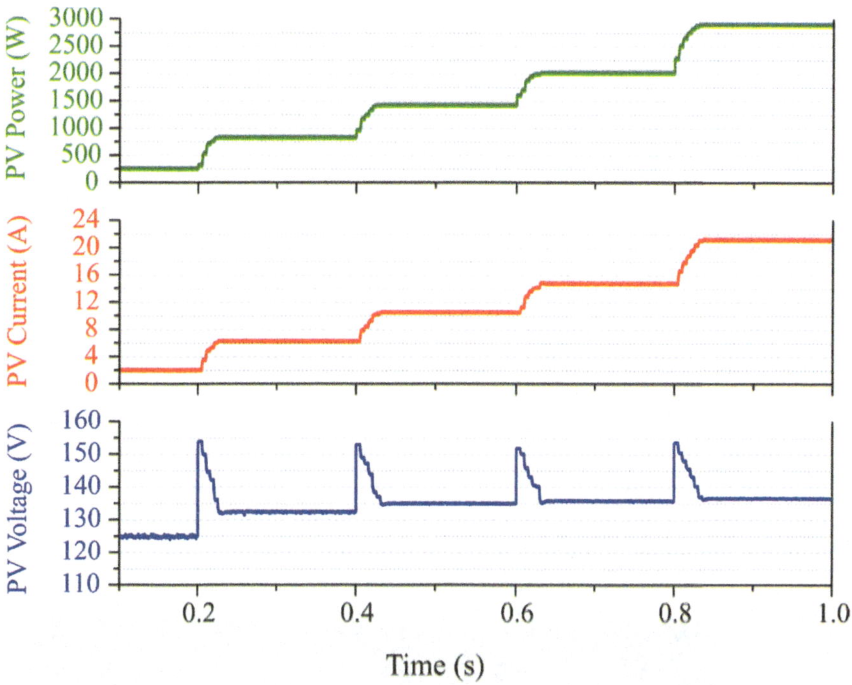

Fig. 14 PV output waveforms of the proposed method for step change in solar irradiation varying from 100 to 1000 W/m^2

switching. The current i_{su} is injected into the phase u of corresponding voltage V_{su}. The inverter continuously inject three-phase current into the grid when there is change in irradiation level and while tracking the MPP.

7.2 *Case 2*

The second case is validating the proposed method for detecting partial shading from uniform solar radiation conditions. Figure 16 shows the shading patterns considered for testing the controller. Figure 16a shows the initial condition of case 2, where the maximum solar radiation of 1000 W/m^2 incidence on all the 36 modules. Figure 16b is the shading pattern that occurs at the ends of the PV array with 4 PV modules receiving 300 W/m^2 and the other four modules receiving 500 W/m^2.

Gradually the intensity of shading level changes as shown in Fig. 16c with four modules with 300 W/m^2 radiation, four modules with 400 W/m^2 radiation, and three modules with 500 W/m^2 solar radiation.

Figure 17 show the I–V and P–V characteristic curves for uniform irradiation of 1000 W/m^2, shading pattern 1, and shading patterns 2. In the first condition, the

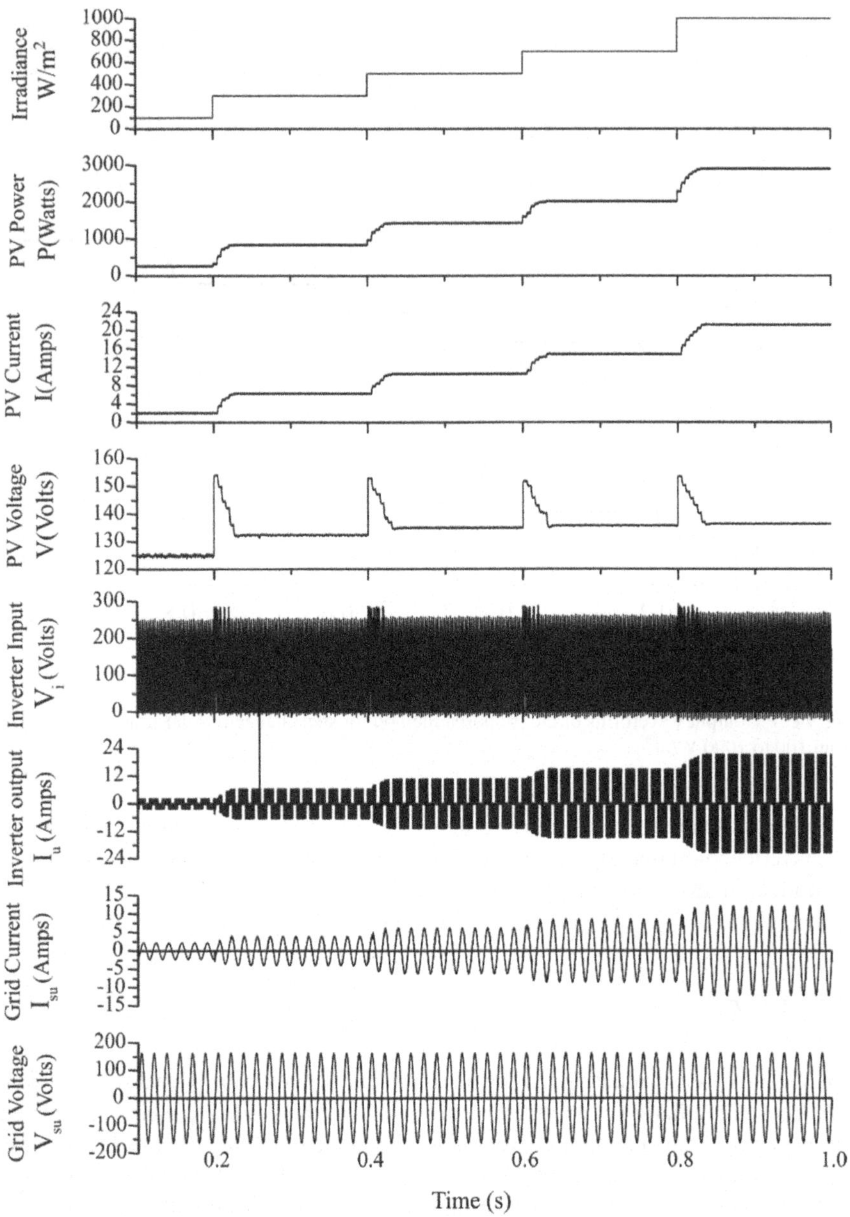

Fig. 15 Overall waveforms of the proposed algorithm for step change in solar irradiation varying from 100 to 1000 W/m^2.

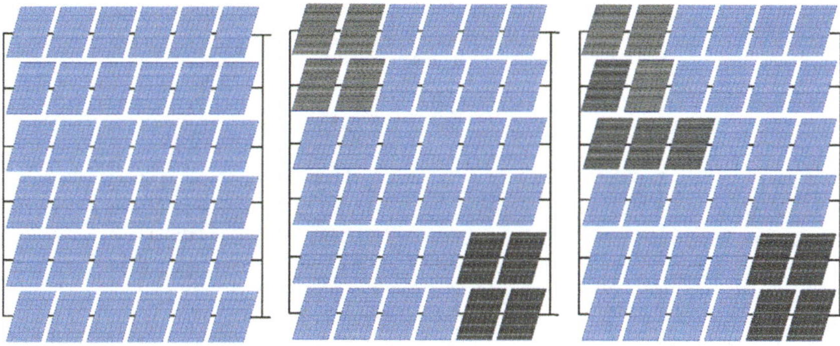

(a) Uniform Radiation (b) Shading Pattern 1 (c) Shading Pattern 2

Fig. 16 Shading patterns

maximum power point is 2901 W at 137.13 V for the current reference $I^* = 21.15$ A. It is marked as point 1 on the characteristic curve in Fig. 17. In the simulation environment, the shading pattern 1 is applied to the 8 modules as discussed above at a time t = 0.2 s. The *I–V* and *P–V* curves shift to long dotted lines as in Fig. 17. As the controller is working at a reference value of $I^* = 21.15$ A, the operating point moves to 2. At point 2, the power $P = 1.987$ kW and voltage V = 93.95 V, The value of ΔP is greater than ε_p and ΔV is greater than ε_v. Thus the decision blocks slide towards the condition of $0.75V_{\mathrm{oc}} \le V_k \le 0.9V_{\mathrm{oc}}$. As the value of voltage is outside the limits the controller detects the occurrence of partial shading on the modules. The controller slides into the GMPP detection loop to scan the *P–V* curve for other power peaks. The controller performs INC method to reach point 3. The Power at this instant is stored as P_0. Now the controller decrements the current by constant values of ΔI_p till it reaches the positive condition of $\Delta P/\Delta V$. At point 6 the controller satisfies the condition of $\Delta P/\Delta V > 0$. The controller again performs INC to detect the new MPP at point 7. The power at this instant is 1.802 kW. This power is less than the previous maximum power P_0. So the controller moves back to point 8 which is the GMPP for the shading pattern 1. The point 8 is nothing but the value of point 3 with the GMPP of 1.987 kW.

At time $t = 0.6$ s, the shading moves from pattern 1 to pattern 2 as shown in Fig. 16. The *I–V* and *P–V* curve for the shading pattern 2 is shown in short dotted lines in Fig. 17. At the instant when shading pattern 2 occurs the current reference is $I^* = 21.21$ A. For this reference current the controller shifts to point 9 on the *P–V* curve of shading pattern 2. Here again, the power is 1.640 kW and voltage V = 77.25 V. Thus both the power and voltage change are greater than ε_p and ε_v, respectively. The decision block is triggered and looks for the condition $0.75V_{\mathrm{oc}} \le V_k \le 0.9V_{\mathrm{oc}}$. As the condition is not satisfied the controller slides into GMPP detection loop. The controller performs INC method to find the maximum power of 1.640 kW at point 10 and this value is stored as P_0. Now the controller decrements a value by ΔI_p and moves to point 11. At point 11 the condition $\Delta P/\Delta V > 0$ is satisfied

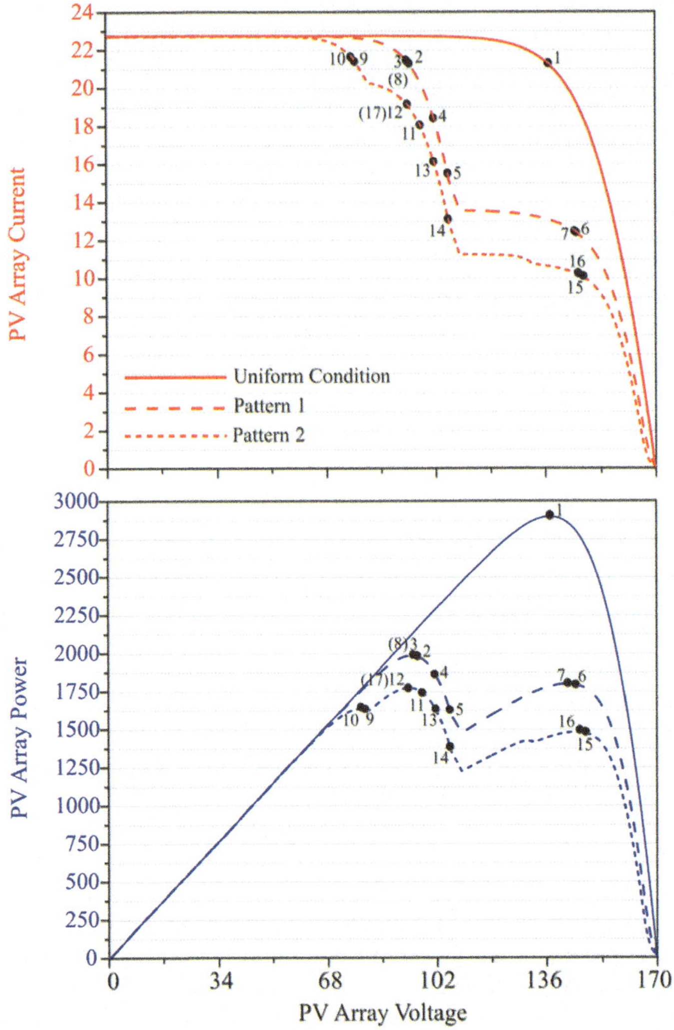

Fig. 17 *I–V* and *P–V* curves for uniform radiation, shading pattern 1 and shading pattern 2

and the algorithm tracks the new MPP. At point 12 the new MPP is detected and the power $P_n = 1.775$ kW. Here $P_n > P_0$, so now the new MPP power is set to P_0 and controller starts to decrease the current in the value of ΔI_p. After three instants the controller stops at point 15 when $\Delta P/\Delta V > 0$ is satisfied. At point 16 the new MPP of $P_n = 1.483$ kW is tracked using the INC algorithm. As $P_0 > P_n$ the controller stops searching for further MPPs and the controller moves to point 17 and continues to operate at $P = 1.775$ kW.

Figure 18 shows the output waveforms of array power, current, and voltage for the proposed method. The shading pattern 1 occurs at time $t = 0.2$ s and the shading

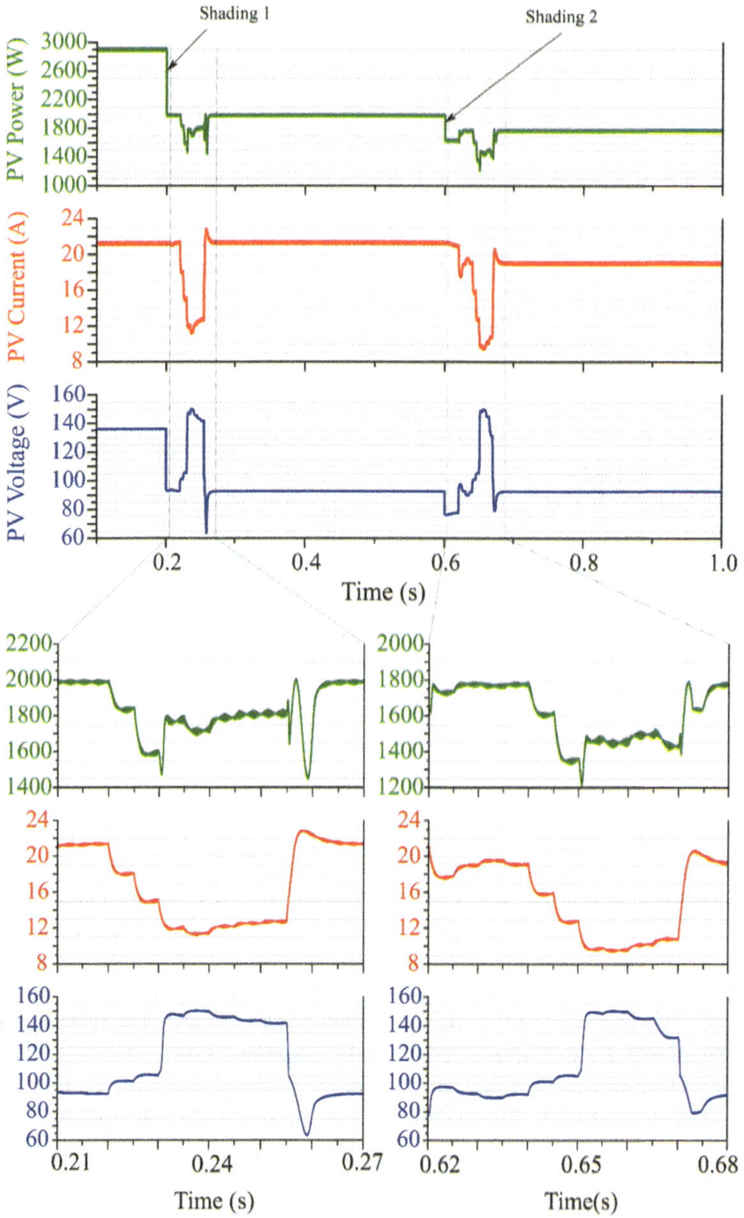

Fig. 18 PV output waveforms of proposed algorithm for shading patterns 1 and 2

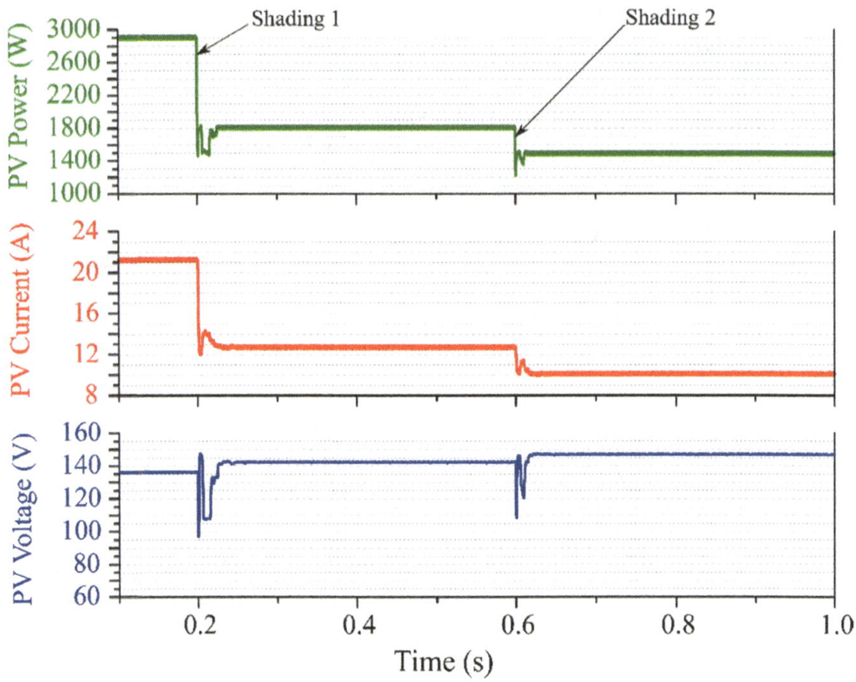

Fig. 19 PV output waveforms of P&O algorithm for shading patterns 1 and 2

pattern 1 is shifted to pattern 2 at time $t = 0.6$ s. The changes in the 3 parameters are clearly seen when the controller responds to their conditions in the algorithm. The results are compared with the P&O and INC MPPT methods as discussed in Sect. 4. Figure 19 show the output waveforms of P&O MPPT control algorithm. The controller detects the maximum power of 1.802 kW during shading pattern 1 and settles down for the maximum power of 1.482 kW during shading pattern 2. Similarly, Fig. 20 show the output waveforms of INC MPPT algorithm. During shading pattern 1 the INC algorithm operates at the maximum power $P_{\text{MPP}} = 1.810$ W and during shading pattern 2 the algorithm operates at the maximum power of $P_{\text{MPP}} = 1.485$ W.

Figure 21 shows the overall waveforms of the proposed MPPT algorithm for change in shading patterns. The maximum power point for the corresponding shading condition is shown in the waveform. The output waveform of PV Power P, PV Current I, and PV Voltage V is varying with respect to the reference value of the MPPT controller. The voltage V_i is the voltage on the input side of the inverter. The current i_u is the pulsed current of phase u caused by the PWM switching. The current i_{su} is injected into the phase u of corresponding voltage V_{su}. Figure 22 shows the partial waveform of Fig. 21 when the inverter at a maximum power of 1775 W. Figures 23 and 24 show the overall waveforms of the PV system for P&O and INC MPPT methods, respectively. The waveforms conclude that the continuous power is injected into the grid even when the controller is tracking the GMPP.

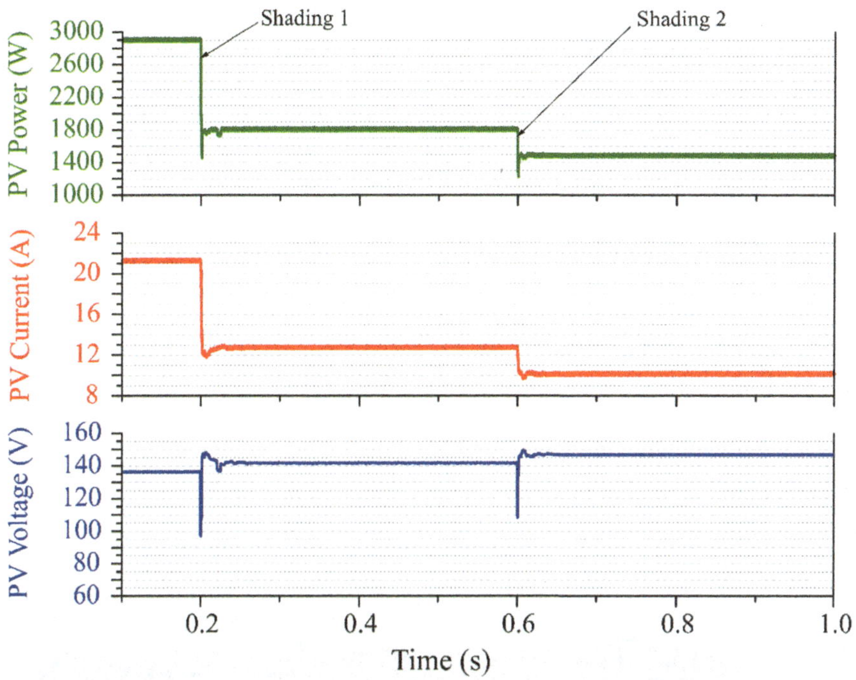

Fig. 20 PV output waveforms of INC algorithm for shading patterns 1 and 2

8 Conclusion

In this chapter a hybrid MPPT algorithm employing both voltage and current based controller for detecting the presence of partial shading on the PV array. The proposed methods track the GMPP without using any AI techniques or additional light sensors. The evaluation of the proposed method is validated using the PSIM software. Dynamic change in shading patterns has been created in the simulation environment and the proposed algorithm is applied to detect the shading occurrence and quickly tracing the *P–V* curve and finding the GMPP. The proposed method is compared with the conventional P&O and INC MPPT algorithms and the results are published.

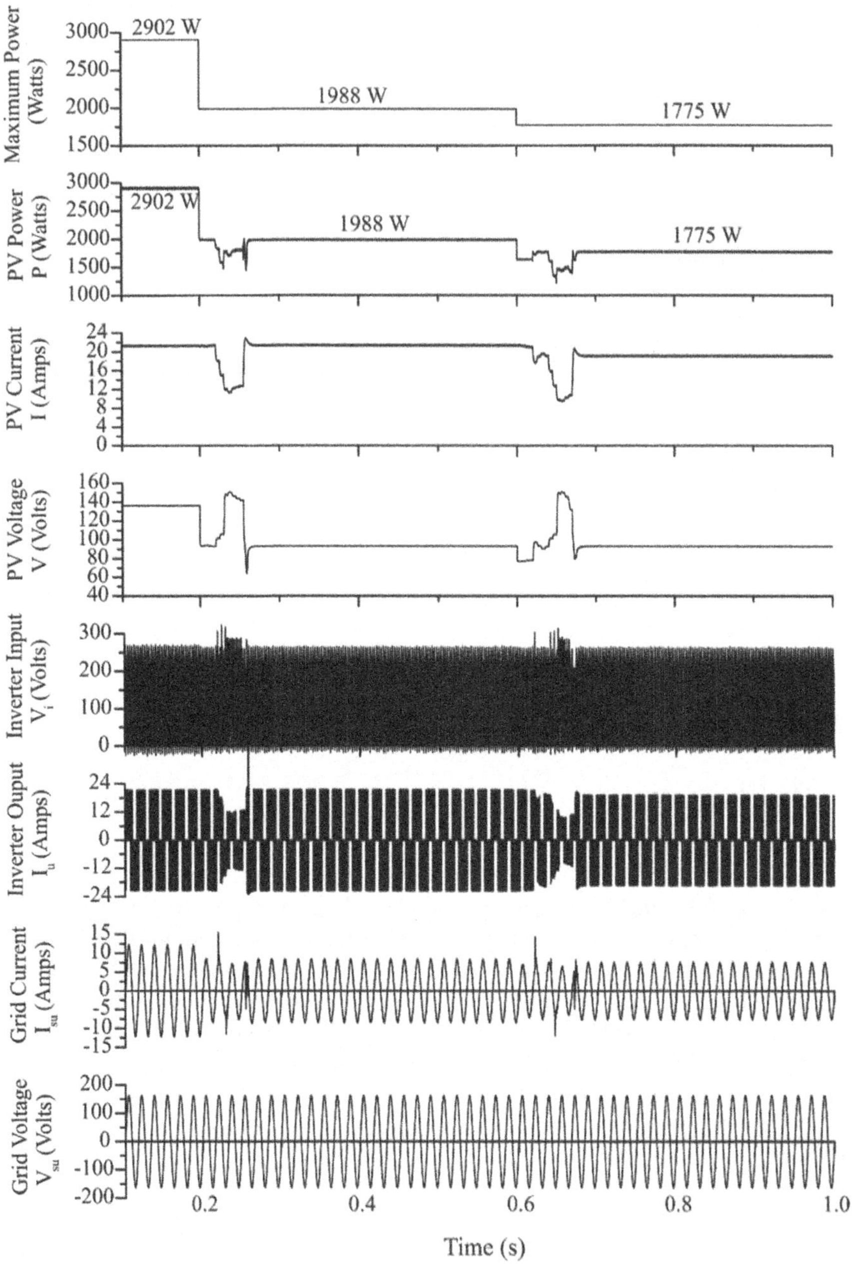

Fig. 21 Overall output waveforms of the proposed MPPT algorithm during change in shading pattern from 1 to 2

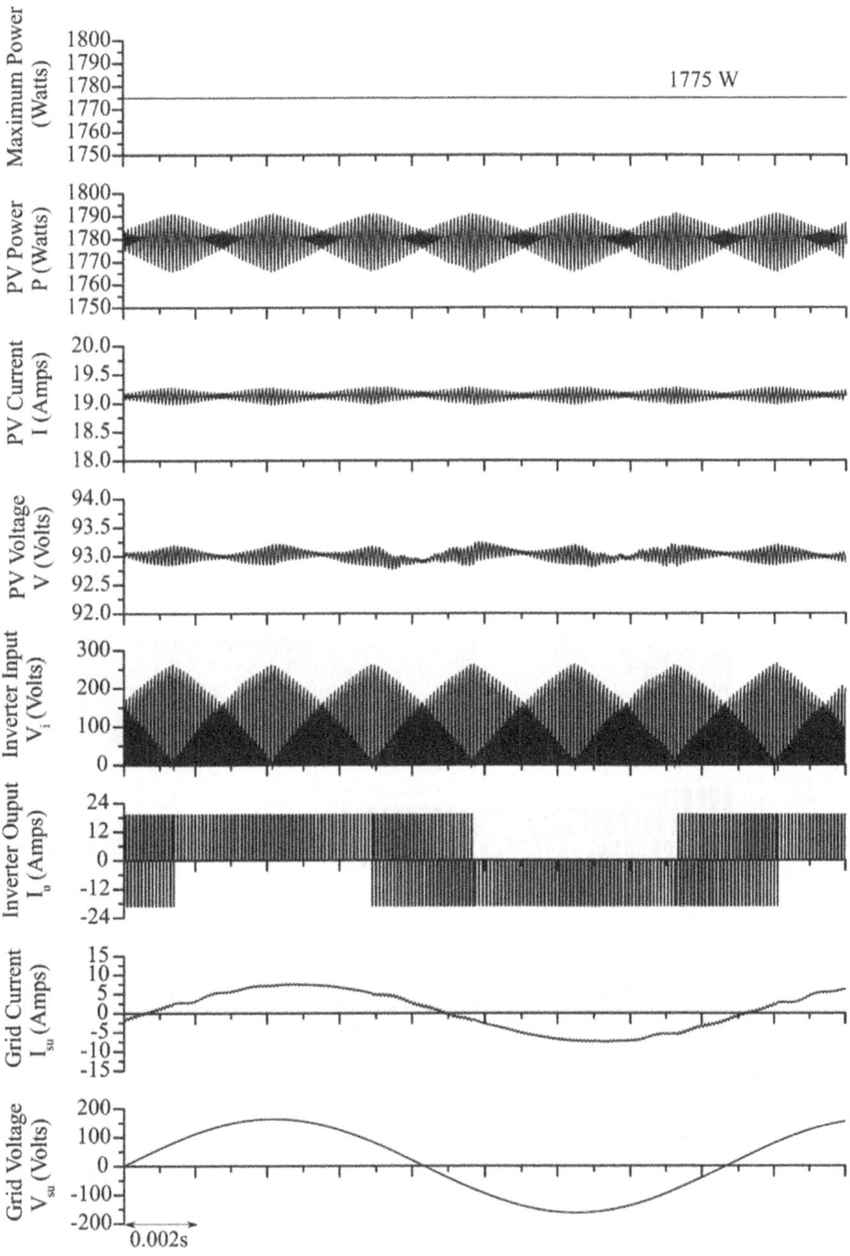

Fig. 22 Partial waveforms of Fig. 21 from $t = 0.80$ s to $t = 0.82$ s

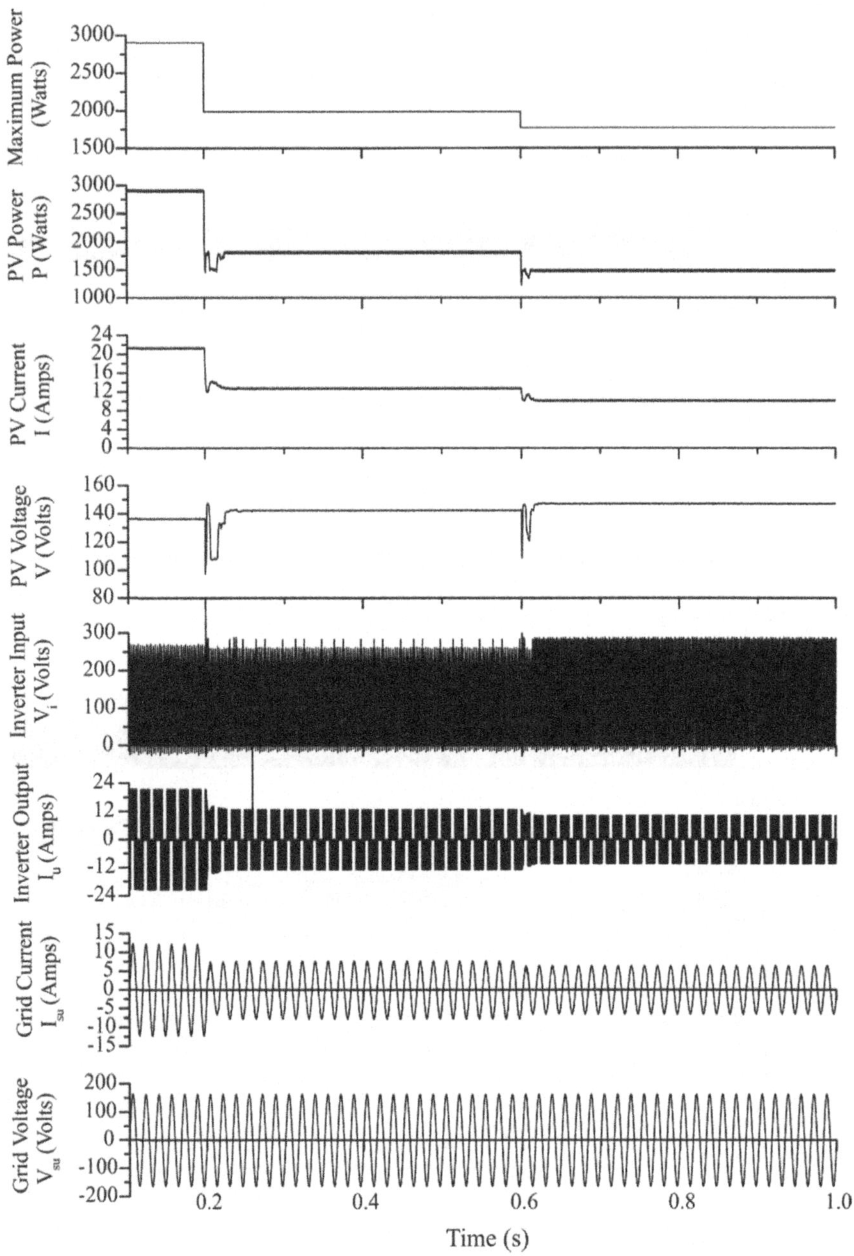

Fig. 23 Overall waveforms of the P&O algorithm for change in shading pattern from 1 to shading pattern 2

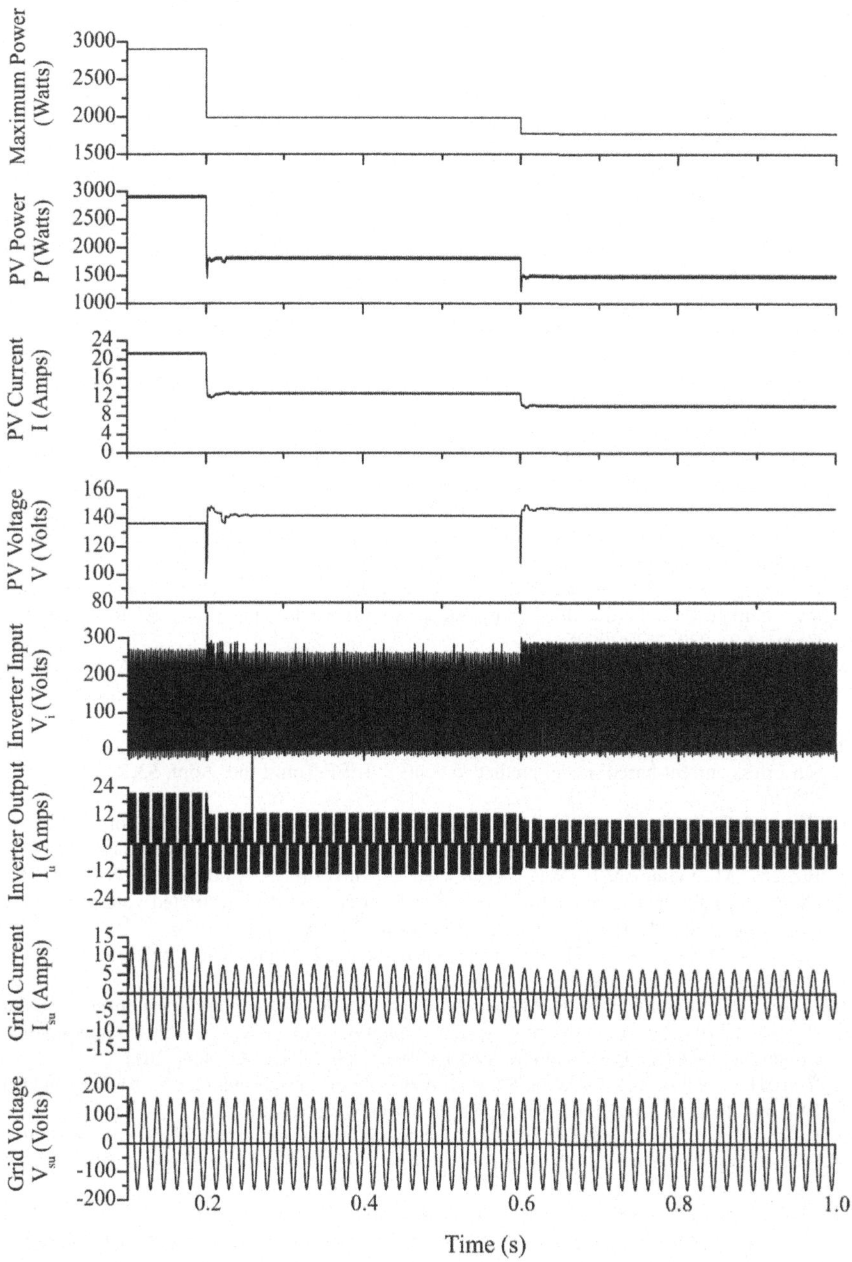

Fig. 24 Overall waveforms of the INC algorithm for change in shading pattern from 1 to shading pattern 2

References

1. Kollimalla, S.K., Mishra, M.K.: Variable perturbation size adaptive P&OMPPT algorithm for sudden changes in irradiance. IEEE Trans. Sustain. Energy **5**(3), 718–728 (2014)
2. Sher, H.A., Murtaza, A.F., Noman, A., Addoweesh, K.E., Al-Haddad, K., Chiaberge, M.: A new sensorless hybrid MPPT algorithm based on fractional short-circuit current measurement and P&O MPPT. IEEE Trans. Sustain. Energy **6**(4), 1426–1434 (2015)
3. Motahhir, S., El Hammoumi, A., El Ghzizal, A.: Photovoltaic system with quantitative comparative between an improved MPPT and existing INC and P&O methods under fast varying of solar irradiation. Energy Rep. **4**, 341–350 (2018)
4. Li, S.: A maximum power point tracking method with variable weather parameters based on input resistance for photovoltaic system. Energy Convers. Manage. **106**, 290–299 (2015)
5. Elobaid, L.M., Abdelsalam, A.K., Zakzouk, E.E.: Artificial neural network-based photovoltaic maximum power point tracking techniques: a survey. IET Renew. Power Gener. **9**(8), 1043–1063 (2015)
6. Kota, V.R., Bhukya, M.N.: A novel global MPP tracking scheme based on shading pattern identification using artificial neural networks for photovoltaic power generation during partial shaded condition. IET Renew. Power Gener. **13**(10), 1647–1659 (2019)
7. Liu, Y.H., Huang, S.C., Huang, J.W., Liang, W.C.: A particle swarm optimization-based maximum power point tracking algorithm for PV systems operating under partially shaded conditions. IEEE Trans. Energy Convers. **27**(4), 1027–1035 (2012)
8. Karatepe, E., Hiyama, T.: Artificial neural network-polar coordinated fuzzy controller based maximum power point tracking control under partially shaded conditions. IET Renew. Power Gener. **3**(2), 239–253 (2009)
9. Bianconi, E., Calvente, J., Giral, R., Mamarelis, E., Petrone, G., Ramos-Paja, C.A., Spagnuolo, G., Vitelli, M.: A fast current-based MPPT technique employing sliding mode control. IEEE Trans. Industr. Electron. **60**(3), 1168–1178 (2012)
10. Metry, M., Shadmand, M.B., Balog, R.S., Abu-Rub, H.: MPPT of photo-voltaic systems using sensorless current-based model predictive control. IEEE Trans. Ind. Appl. **53**(2), 1157–1167 (2016)
11. Hosseini, S., Taheri, S., Farzaneh, M., Taheri, H.: A high-performance shade-tolerant MPPT based on current-mode control. IEEE Trans. Power Electron. **34**(10), 10327–10340 (2019)
12. Boztepe, M., Guinjoan, F., Velasco-Quesada, G., Silvestre, S., Chouder, A., Karatepe, E.: Global MPPT scheme for photovoltaic string inverters based on restricted voltage window search algorithm. IEEE Trans. Industr. Electron. **61**(7), 3302–3312 (2013)
13. Hansen, C.W., King, B.H.: Determining series resistance for equivalent circuit models of a PV module. IEEE J. Photovolt. **9**(2), 538–543 (2018)
14. Hurayb, K., Moumouni, Y., da Silva, F.A., Baghzouz, Y.: Evaluationof the impact of partial shading on the performance of a grid-tied photovoltaic system. In: 2015 International Conference on Clean Electrical Power (ICCEP), pp. 430–434. IEEE, June 2015
15. Sahu, H.S., Nayak, S.K., Mishra, S.: Maximizing the power generation of a partially shaded PV array. IEEE J. Emerg. Sel. Top. Power Electron. **4**(2), 626–637 (2015)
16. Lappalainen, K., Valkealahti, S.: Effects of irradiance transition characteristics on the mismatch losses of different electrical PV array configurations. IET Renew. Power Gener. **11**(2), 248–254 (2016)
17. Vijayalekshmy, S., Iyer, S.R., Beevi, B.: Comparative analysis on the performance of a short string of series-connected and parallel-connected photovoltaic array under Partial Shading. J. Inst. Eng. (India) Ser. B **96**(3), 217–226 (2015)
18. Krishna, G.S., Moger, T.: Improved SuDoKu reconfiguration technique for total-cross-tied PV array to enhance maximum power under partial shading conditions. Renew. Sustain. Energy Rev. **109**, 333–348 (2019)
19. Veerasamy, B., Takeshita, T., Jote, A., Mekonnen, T.: Mismatch loss analysis of PV array configurations under partial shading conditions. In: 20187th International Conference on Renewable Energy Research and Applications (ICRERA), pp. 1162–1183. IEEE, Oct 2018

20. Yadav, P., Kumar, A., Gupta, A., Pachauri, R.K., Chauhan, Y.K., & Yadav, V.K.: Investigations on the effects of partial shading and dust accumulation on PV module performance. In: Proceeding of International Conference on Intelligent Communication, Control and Devices, pp. 1005–1012. Springer, Singapore (2017)
21. Changmai, P., Metya, S.K.: Determination of the best shading pattern tomaximize the power of TCT connected solar PV array during partial shading condition. J. Opt. **48**(4), 499–504 (2019)
22. Veerasamy, B., Maruthachalam, S., Jagannathan, K., Takeshita, T.: Single stage high gain transformerless three phase PV inverter. In: 2019 2nd International Conference on Smart Grid and Renewable Energy (SGRE), pp. 1–6. IEEE, Nov 2019
23. Veerasamy, B., Jaganathan, K., Takeshita, T.: Transformerless single stage high boost PV inverter with reduced leakage current. In: 2019 IEEE 4th International Future Energy Electronics Conference (IFEEC), pp. 1–6. IEEE, Nov 2019
24. Isozaki, J., Veerasamy, B., Kitagawa, W., Takeshita, T.: Duality of PWM strategies between current and voltage source AC/DC converters for suppressing DC ripple. In: 2015 9th International Conference on Power Electronics and ECCE Asia (ICPE-ECCE Asia), pp 1809–1814. IEEE, June 2015
25. Veerasamy, B., Kitagawa, W., Takeshita, T.: Input power factor control of bi-directional ac/dc converter. In: 2013 IEEE 10th International Conference onPower Electronics and Drive Systems (PEDS), pp. 1103–1108. IEEE, Apr 2013

A New Model of Demand Response in Smart Grid

Somayeh Siahchehre Kholerdi and Ali Ghasemi-Marzbali

1 Introduction

With the changes in the energy sector and the emergence of competitive markets, the traditional power systems in most countries of the world have been shattered and their components have become independent and sometimes conflicting targets for independent players. Electricity companies produce and sell energy competitively in wholesale markets, and electricity distribution companies purchase the energy from the above-mentioned market and resell it to retail markets. The retailers then sell to customers after purchasing electricity from the market. This organization is an economic enterprise that trades within the framework of market laws and its goal is to provide cheap and quality electricity to customers [1].

Changes in wholesale prices, rising operating costs, and the development of power grids are some of the factors that are pushing the dense grid of global electricity markets to implement consumption management programs. Factors managing electricity markets quickly realized that it was not possible to solve the above problems without the active participation of the participants in the market. So they looked for ways to re-use existing consumer management programs in line with market operations to encourage the customer to actively participate in the market, not to underestimate the nature of the market (competitive economic approach).

The International Energy Agency (IEA) called the methods "Demand Response (DR)". It should be emphasized that customer participation in the market is not

S. S. Kholerdi · A. Ghasemi-Marzbali (✉)
Department of Electrical and Biomedical Engineering, Mazandaran University of Science and Technology, Babol, Iran
e-mail: ali.ghasemi@ustmb.ac.ir

S. S. Kholerdi
e-mail: s_siahchehre@yahoo.com

B. Vinoth Kumar et al. (eds.), *Intelligent Paradigms for Smart Grid and Renewable Energy Systems*, Algorithms for Intelligent Systems,
https://doi.org/10.1007/978-981-15-9968-2_4

an easy task, because electricity has been provided to household and industrial applications for years with ease and cheapness.

Also, the cost of electricity is a small part of the cost of industrial production or a small portion of a household's expenditure portfolio. In other words, customers are very sensitive to electricity prices. The following activities are used to encourage them to participate more and more in demand response programs [2].

- Possibility of short-term estimation of electricity price by the customer
- Ability to use energy storage equipment
- Ability to use advanced measuring devices
- Ability to select different contracts by customers.

2 Restructuring and Deregulation in the Electricity Industry

Traditional power systems were state-owned, in this way, the generation, transmission, and distribution of this energy in an area were done by one or more units and small and large consumers were buying the energy they needed from the government. In fact, governments used small and large power plants to deliver energy to consumers through transmission and distribution lines [3].

Traditional electricity markets in the United States include Idaho, Kentucky, Florida, Colorado, and Tennessee [4]. Figure 1 shows the economic structure of the electricity industry traditionally.

In recent decades, with the expansion of knowledge and the introduction of subjects such as privatization and improved efficiency, governments have considered reducing their ownership and involvement in macroeconomic associations and increasing the share of the private sector in their implementation.

Thus, the modern electricity industry, like other communications industries, included modifies in economic approaches and technological considerations, so as to allow manufacturers to compete and create market conditions, try to reduce costs of production and distribution of electricity, eliminate inefficiencies. separation of duties and increased customer choice. This shift to a competitive electricity market is called deregulation or restructuring and its most important benefits include [3]:

- Provide the right to choose for customers.
- Provide a suitable platform to provide better services.

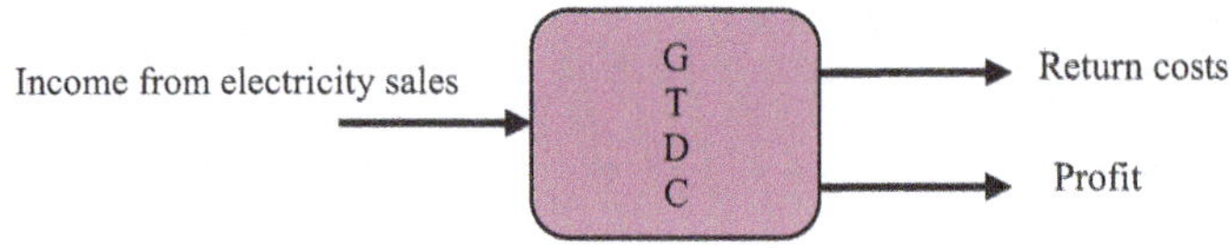

Fig. 1 The economic structure of the electricity industry traditionally

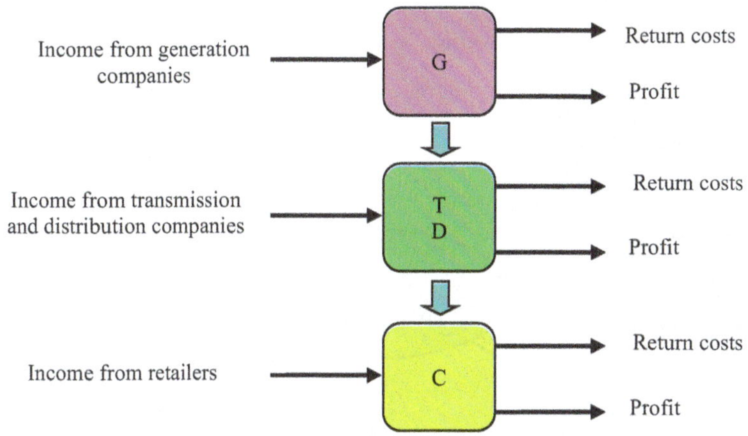

Fig. 2 Economic structure of the electricity industry after restructuring

- Competitive supply of electricity at different levels and consequently determine a reasonable price for the consumer.
- Attracting existing capital in the private sector and directing it in the direction of collective profit and not needing large public investment.
- Increasing the quality of the offered goods according to the existing competition.

Reconstruction in the 1970s in the United States began with the Public Service Act. Truly competitive market after the overall energy policies that constrain prices for the wholesale market was opened in 1992. Figure 2 shows the economic structure of the electricity industry after restructuring or deregulation.

When the traditional environment of the power system is changed to a restructured environment, customers usually do not have an active presence in the market and the main decision-makers were Independent Power Producers (IPPs), Regional Transmission Organizations (RTOs), Independent System Operators (ISOs) and electricity industry legislators. The reason for this is somewhat clear, as consumers not only did not benefit from the benefits of the market but also did not have the information and skills need to participate in complex electricity markets.

For this reason, markets viewed consumers as inactive elements and merely simple burdens to be served, and therefore the customers were interested in receiving fixed-price electricity regardless of market fluctuations [5]. This attitude has led to problems such as the occurrence of peak hours price jumps without the presence of customers in the market and their insensitivity to electricity prices during peak times in many markets, leading to "widespread blackouts" [6].

Electricity policymakers and electrical engineers, in order to establish sustainable development in this industry and solve the problems related to climate change, increase the continuous demand for electricity, preserve the environment, limit fossil resources, and increase economic efficiency, making fundamental changes in the

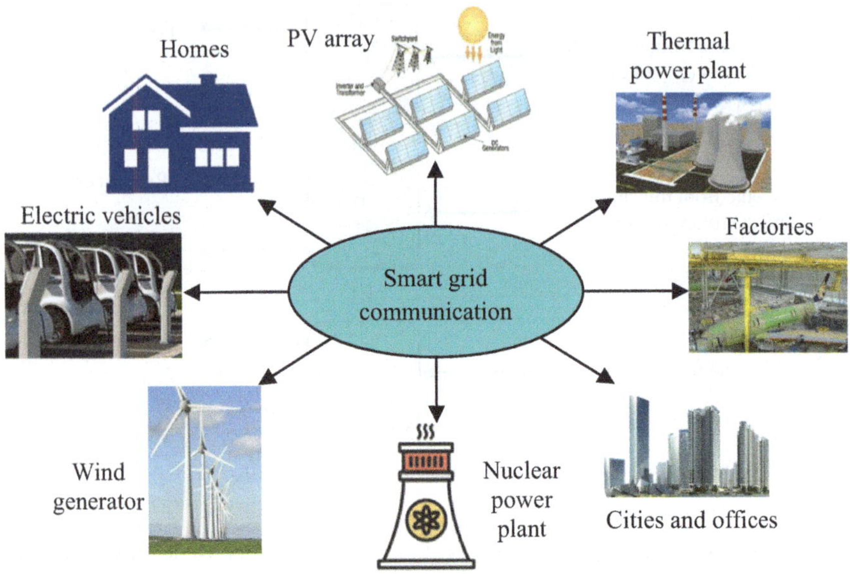

Fig. 3 Schematic of a smart grid

generation and consumption processes of this energy. Including the use of renewable energy sources (RES) and local integration of a variety of distributed energy resources (DER): local storage, distributed generation (DG), electric vehicles (EVs), and general active demand.

The traditional and centralized supply and distribution are pushing the ultimate consumption pattern with the problems mentioned above into a much more complex direction. Therefore, in future power systems, suppliers and consumers are expected to work together to optimize the system as much as possible. This concept led to the emergence of "Smart Grid (SG)" has been in the power industry. Figure 3 shows a schematic view of a smart grid.

In fact, a smart grid (SG) as a new capable electric platform is able to transport electricity in a smart way, secure and controlled from points of generation to customers that can change their consumption models based on the expected information, disincentives, and incentives [7].

3 The Effect of Changes in the Electricity Market and the Formation of Demand-Side Management (DSM)

By competing in the electricity market, electricity producers, transmitters, and distributors are thinking of solving the problems created by activating the presence of electricity customers and consumers in the market. Therefore, demand-side

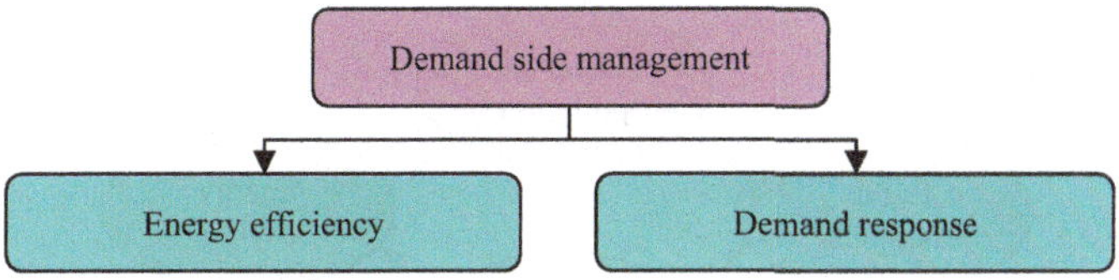

Fig. 4 Classification of demand-side management programs

management (DSM) programs were developed in accordance with the operations and rules of the competitive market. In a competitive market, even a small decrease in demand can lead to significant changes in the overall costs of the power generation, transmission, and distribution system [8].

The main example of such a situation is the peak time. Although peak times are short term, ISOs are forced to use the most expensive generators to balance power supply and demand, causing large changes in the final price of electricity. Demand-side management can reduce demand during peak hours, thus not only reducing the net price in the market but also preventing producers from more interfering in the electricity market [9, 10]. Demand-side management programs are divided into two general forms as shown in Fig. 4.

3.1 Energy Efficiency Programs (EE)

Energy efficiency, which can also be called EE, is the reduction of the amount of initial energy consumed to provide the same amount of products and services [11]. Figure 5 shows the benefits of improving energy efficiency.

3.2 Demand Response Programs (DR)

According to the US Department of Energy (DOE), demand response is the empowerment of commercial, industrial, and residential consumers to progress the pattern of electricity utilization to get sensible prices and improve network reliability [3]. Figure 6 shows the effect of demand response programs on customer load curve. Figure 7 summarizes how the load profile will change under the influence of demand response plans.

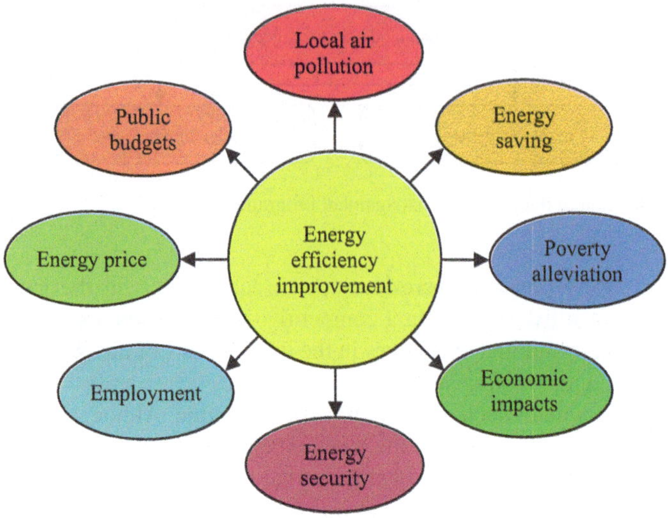

Fig. 5 The benefits of improving energy efficiency

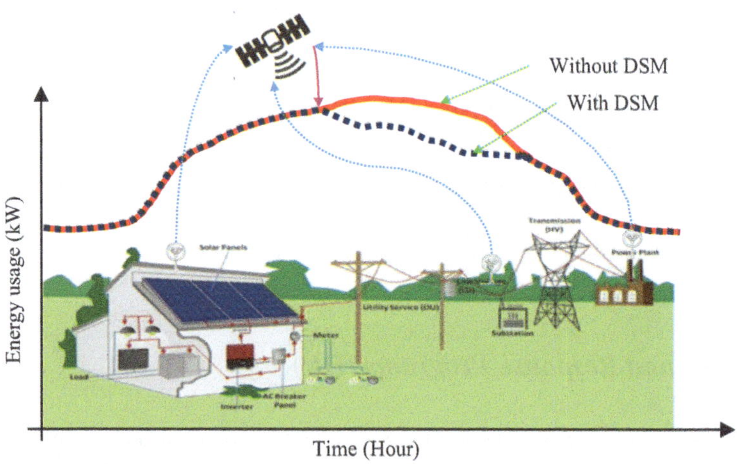

Fig. 6 The effect of demand response programs on customer consumption

4 Types of Demand Response Programs

Figure 8 shows the various categories of demand response programs. As shown in Fig. 8, load response programs are divided into two main branches and several sub-branches [12–15].

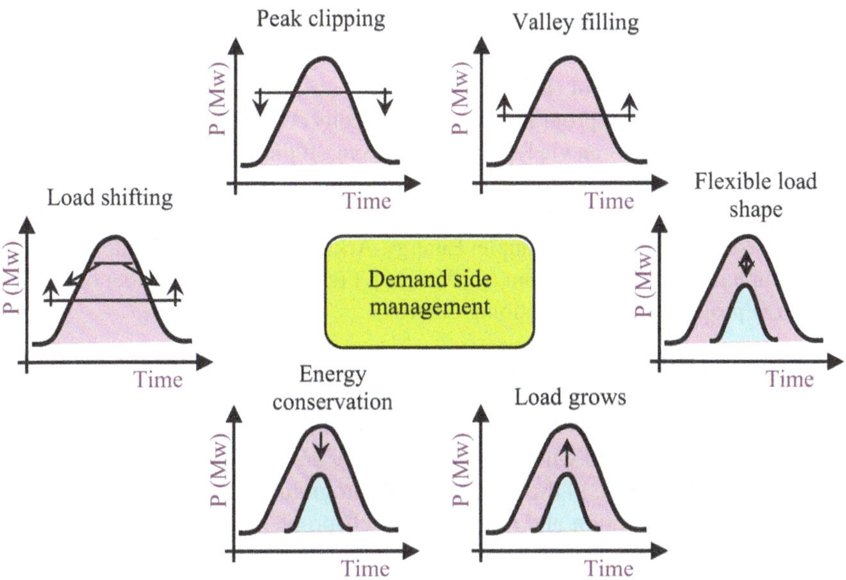

Fig. 7 Demand-side management load shape objectives

Fig. 8 Categories of demand response programs

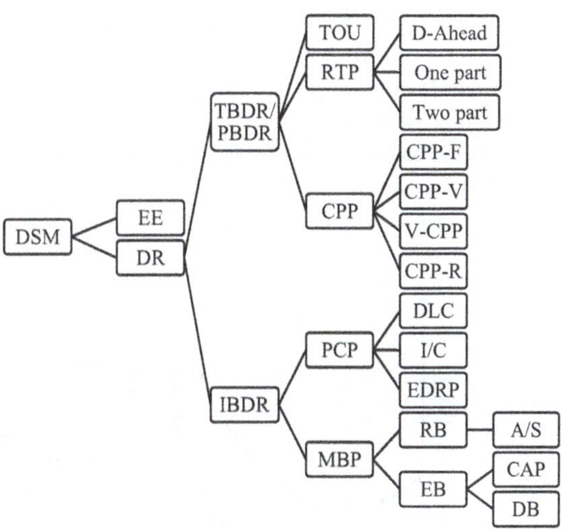

4.1 Incentive-Based Demand Response (IBDR)

IBDR programs actually include those DR programs that are encouraging. These programs also have two subgroups: Power Conservation Program or PCP and Market Based Program or MBP:

4.1.1 PCPs Include the Following Programs:

- **Directed Load Control (DLC)**
  This method includes programs that the electricity company or operator with prior notice and remotely can via Remote Switch, cut off the customer's electricity, and pay for compensation. This is usually done during peak hours and high prices and includes loads such as air conditioning and water heaters in the residential and commercial sectors. For example, Entergy-Arkansas gives $25 rebate each year, for implementing this program in 2018 [16]. Figure 9 shows the effect of running a DLC program on air conditioner control.

- **Interruptible/Curtailable Service (I/C)**
  Customers who participate in this program will receive a discount on their electricity bills or, due to the reduction in their consumption, will have a higher credit rating (Bill Credit) and will be penalized if they do not reduce their consumption an arranged time. One of the features of this program is that it is recommended to reduce the consumption and frequency of self-problem. Major customers (often more than 200 kW) participate in this program. The loading period in the San Diego Electric and Gas Market (SDGE) was between 20 and 60 min in a day and sometimes up to 120 h in 2018 [17]. Figure 10 shows the impact of the I/C program on the consumer consumption curve.

- **Emergency Demand Response Program (EDRP)**
  In this program, customers receive a reward for stopping at the start of an emergency. Of course, power outages are optional and will not be penalized if the customer does not do so. The amount of the bonus or prize is predetermined. Figure 11 shows an example of this program, which was launched in 2002 in the New York market. In this case, the system operator was able to observe the load curve on July 29 and forecast the consumption for July 30 with the simultaneous

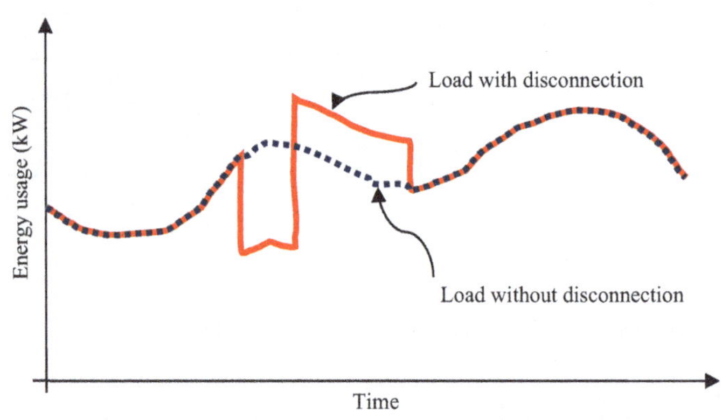

Fig. 9 The effect of DLC program implementation on A/C consumption pattern

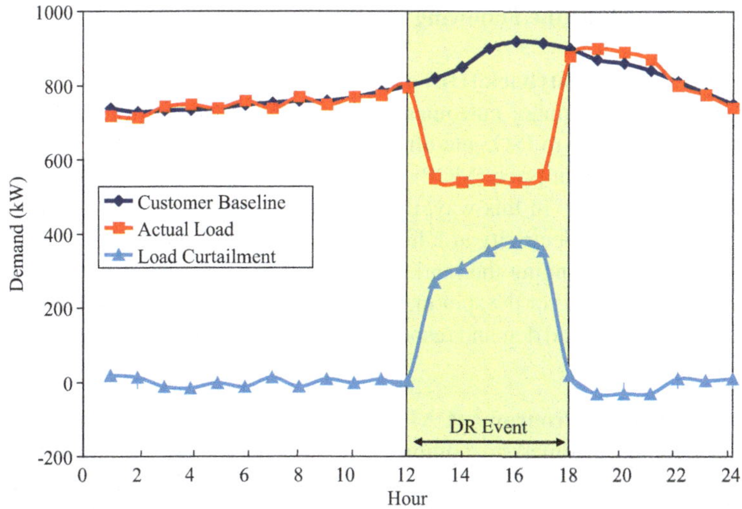

Fig. 10 The effect of I/C program execution on customer load profile

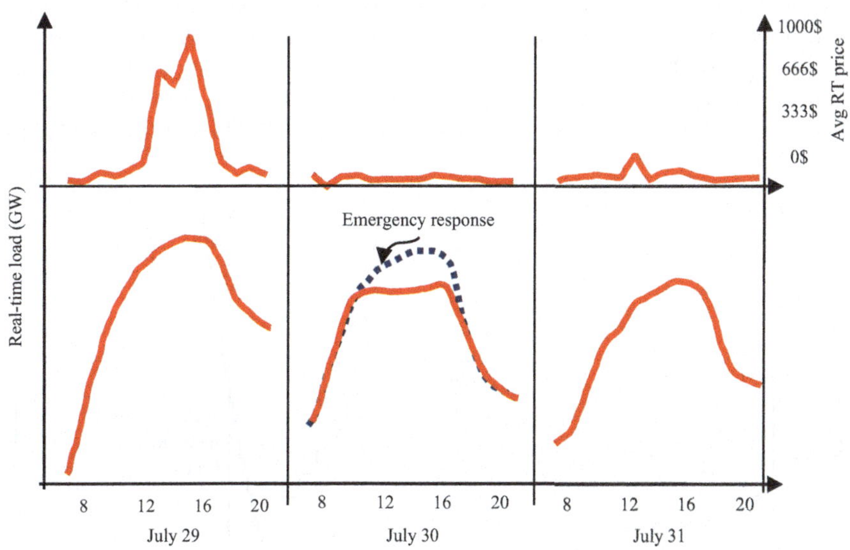

Fig. 11 Implementation of EDRP and CAP programs by NYISO

implementation of the EDRP and CAP program. By cutting the peak consumption, the electricity prices return to normal [18].

4.1.2 MBPs Include the Following Programs:

- **Demand Bidding/Buyback (DB) and Energy Bidding (EB)**
  In these programs, major customers offer a reduced amount of load or energy along with the price to ISO, and after the market operation, if the price is below the market settlement price (MCP), the offer is accepted and the customer must execute the contract. In this way, the customer can actually buy electricity at a low price and sell electricity at a high price. Figure 12 shows the impact of the DB program on changing the market price [18]. This market can be formed the day ahead (D-Ahead) or the spot market. One of the most common types of these programs is day-ahead demand response (DADR), which is used by the New York market.

- **Capacity Market Programs (CAP)**
  In this method, customers are committed to reducing a certain amount of load and will be penalized if they do not. This procedure is usually performed for loads greater than 100 kW, and the reduction time is extended to four hours, and the subscriber is notified 2 h in advance.
  The customer receives a guaranteed amount in return for his commitment. (As if the electricity company has insured itself). The independent system operator identifies these resources and considers them as installed production capacity and regularly checks that the load is ready for cutting. This may not be necessary many times, but incentives such as insurance will be paid. By end of 2015, it is estimated that DR has reduced peak load around 32,875 MW (8703 MW by

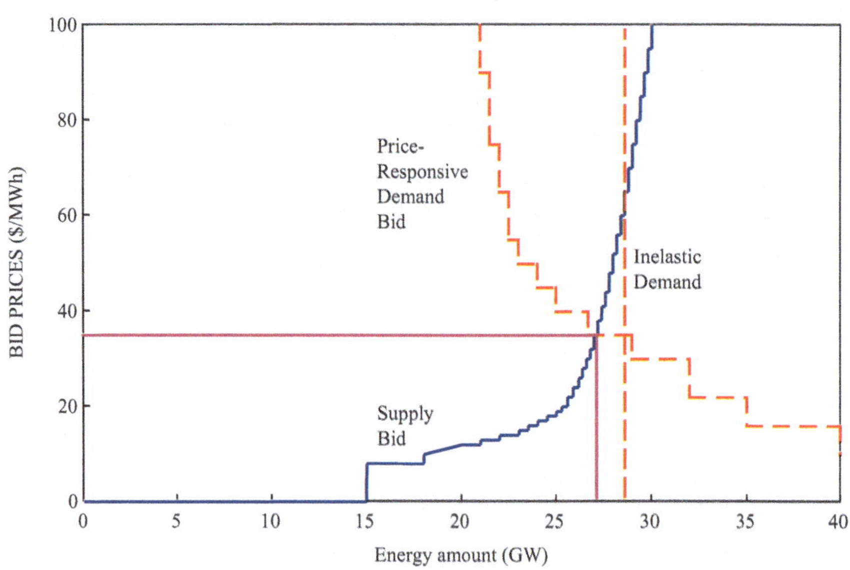

Fig. 12 The effect of DB program on market price

residential sector, 6989 MW by commercial sectors, and 17,169 MW by the industrial sectors [19].

- **Ancillary Service Market (A/S) and Reduction Bidding (RB)**
  In these methods, customers offer ISOs such as the reservation market. If their offer is accepted, they will receive the market settlement price to keep their shipping cut off. Of course, whenever they are called and cut off their load, they may receive the spot market price. These types of loads must be fast to operate in the event of an accident, and must also be large quantities such as large water pumps, electric arc furnaces, and air compressors [20].

4.2 Time-Based Demand Response (TBDR)/Price-Based Demand Response (PBDR)

TBDR or PBDR programs actually include those DR programs which are time-based. These programs are also divided into three categories.

4.2.1 Time of Use (TOU)

In this program, the energy price is calculated and received at least in two or three modes: peak, medium load (mid-peak), and baseload (off-peak) based on different energy prices in each mode.

This tariff can be calculated at different times of the day or on different days of the week or on different days of the year. Figure 13 shows the impact of the TOU program on the consumer load curve [21].

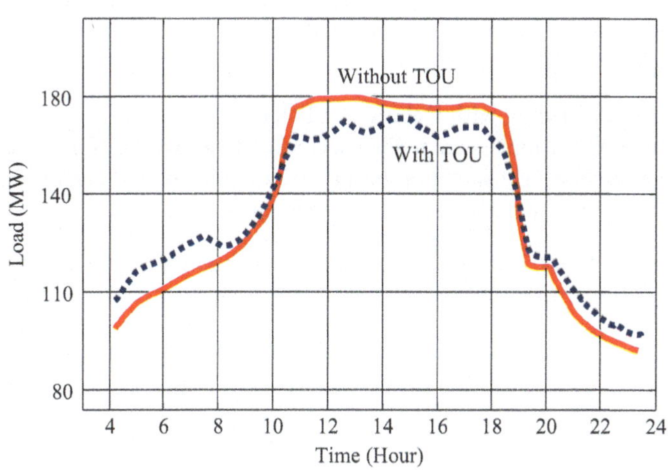

Fig. 13 Impact of TOU program implementation

- **Real-Time Pricing (RTP)**
  In the RTP program, the price is related to the hourly cost of energy. This connection takes place in the real-time market or in the market the day before (Day Ahead or D-Ahead). This program runs in two ways (One Part RTP and Two-Part RTP). In the two-part price method, a consumption ceiling is set for major customers. If the customer consumes below the set ceiling, it is calculated at a lower price, and consumes above a certain ceiling, it is calculated at a higher price. Figure 14 shows the effect of running a one-part and two-part RTP program.

- **Critical Peak Pricing (CPP)**
  The CPP program is obtained by combining the TOU and Flat Rate programs and uses real-time values during peak jumps. However, these mutations may not be more than a few hours a year. Obviously, the price of CPP is higher than the price of regular couriers. But it's not clear when the courier will arrive, so the power industry can't tell the customer in advance.
  The CPP program is implemented in one of the following ways:

- **Critical Peak Price-Fixed (CPP-F)**
  In this method, the time and period of price increase are determined in advance, but the days when the jump in the price of a critical peak occurs is not clear. The maximum number of days that can be called in a year is also predetermined.

- **Critical Peak Price-Variable (CPP-V)**
  In this method, the time of the period and the days when the price increase occurs are unknown, and the event is usually reported the day before. Most telecommunication-controlled thermostats are controlled in this way.

- **Critical Peak Price Variable Period-(V-CPP)**
  In this method, based on the baseload price (non-peak) or intermediate load for a certain period of time, for example, one month or more, the prepayment is taken from the customer and finally settled with the customer based on the final local market price (LMP).

- **Critical Peak Pricing with Rebate (CPP-R)**
  In this method, customers are treated according to a fixed tariff, but because they have reduced their consumption during a critical peak, the electricity industry offers them a discount and returns part of the money [22].

Figure 15 compares the price of electricity under the three TOU, RTP, and CPP programs. Figure 16 shows the different methods of load response programs and their time impact.

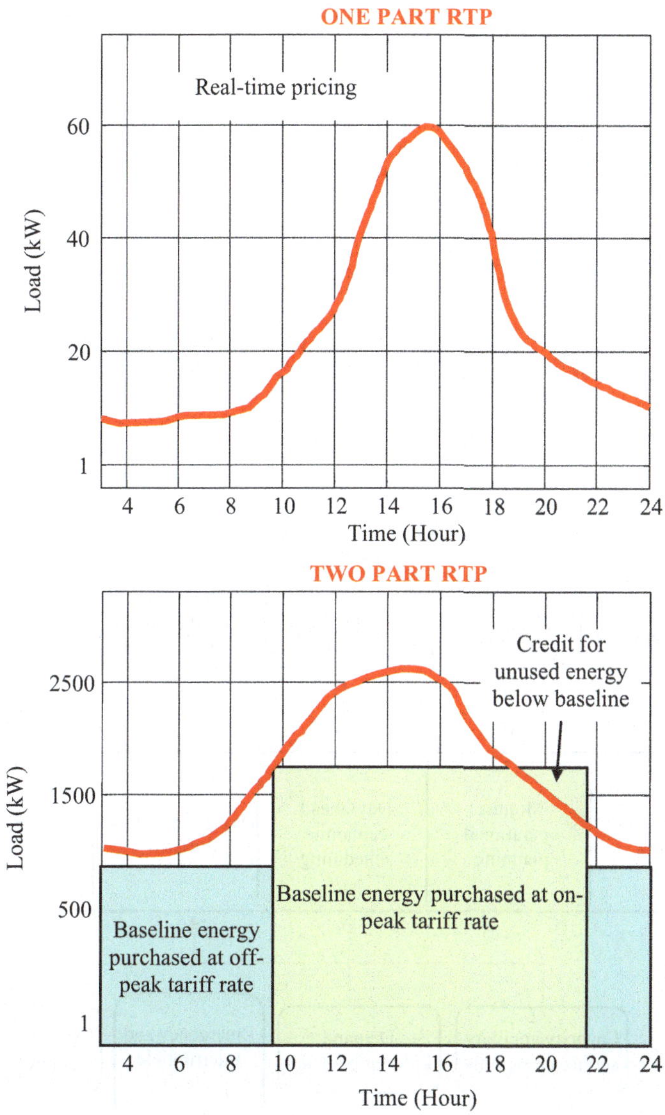

Fig. 14 Price under the influence of RTP program execution

5 The Benefits of Customer Presence in the Electricity Markets of Smart Grid

The customer benefits on the market can be divided into three main groups:

- Customer benefits

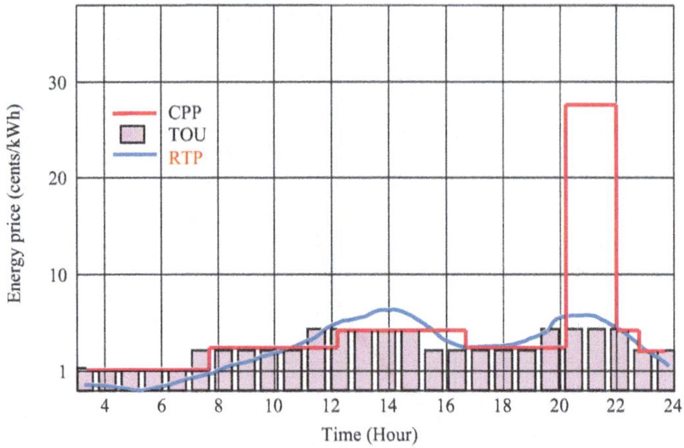

Fig. 15 Comparison of electricity prices in three TOU, RTP, and CPP programs

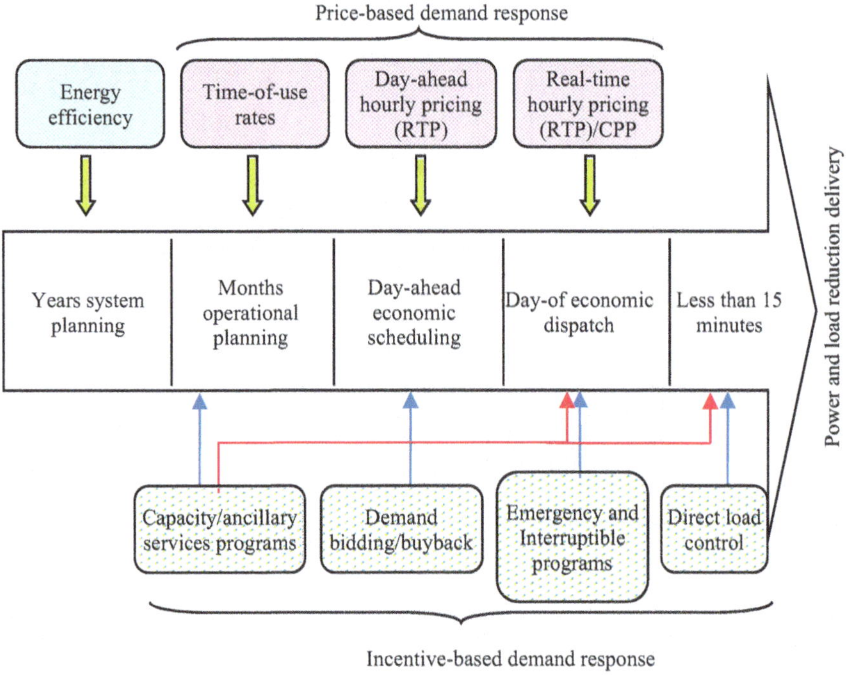

Fig. 16 DR methods and their time impact

- Smart grid benefits
- Other benefits.

5.1 *Customer Benefits*

These include economic benefits such as not buying electricity at expensive prices and the purchase of electricity at low prices and electrical benefits such as increased reliability and continuity of power supply, which is due to the participation in demand response programs is the burden of the participant.

5.2 *Smart Grid Benefits*

The interests of the smart grid are divided into three main groups:

- **Short-term benefits in the smart grid electricity market**
  Short-term effects will include saving on various costs of efficient operation of the power system, helping to stabilize prices in the wholesale market and preventing price spike, reduces prices in the retail market due to falling prices in the wholesale market, and the reduction in the use of expensive generators during peak hours. The effects of responsive loads on market price reduction are shown in Fig. 17. It can be seen that with the participation of the customer in the market and a small reduction in the load, the price of electricity will be greatly reduced. In the California market, for example, a 5% drop in working hours during peak hours reduces prices by 24% and a 10% reduction in load, reducing prices by 50% [23].
- **Long-term benefits in the electricity market of smart grid**

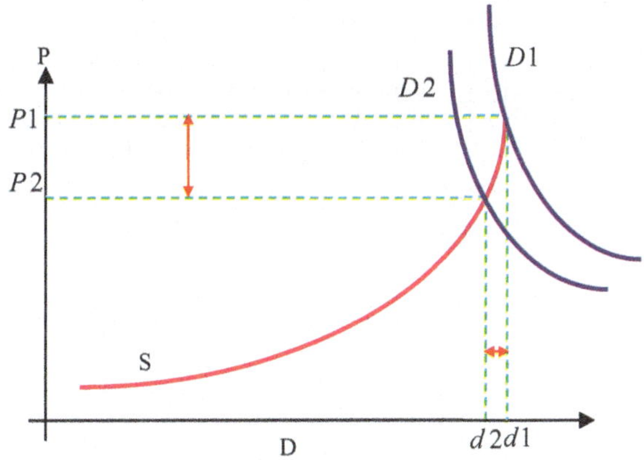

Fig. 17 The effect of responsive loads on market price reduction

Long-term effects include delays in the construction of new power plants, smoothing of the load curve, and high load factor in the network. Saving on investment and generation costs is part of the factors influencing long-term market conditions.

- **Increasing the efficiency and reliability of the power system**
  DR programs are designed to reduce the cost of electrical power systems in intelligent networks. Demand Response programs are also involved in the operation of the power system.

In principle, exploitation is defined in two states: emergency and normal. In an emergency, a certain amount of generation or loss of load (load response methods) must be available that can act quickly, these resources are called spinning reservations. In normal operation for short periods of time, the adjustment service is defined as the task of the system operator, adjusting the network frequency and responding to rapid and small changes in load. In a slightly longer period of time, market equilibrium is introduced, which is responsible for tracking the load (balance of generation and consumption in the slower cycle).

Figure 18 shows the role of DR as a spinning reservation in the operation of the power system [24].

Contrary to popular belief about the unreliability of responsive loads, studies show that the reliability of a set of responsive small loads is more than the reliability of large generators. For example, in a study conducted by the American Energy Regulatory Commission, in which the reliability of a set of small loads consisting of 1200 responsive small loads that each has a capacity of 500 kW and the reliability of each of these loads is 0.9, as compared with 6 generators that each has a capacity of

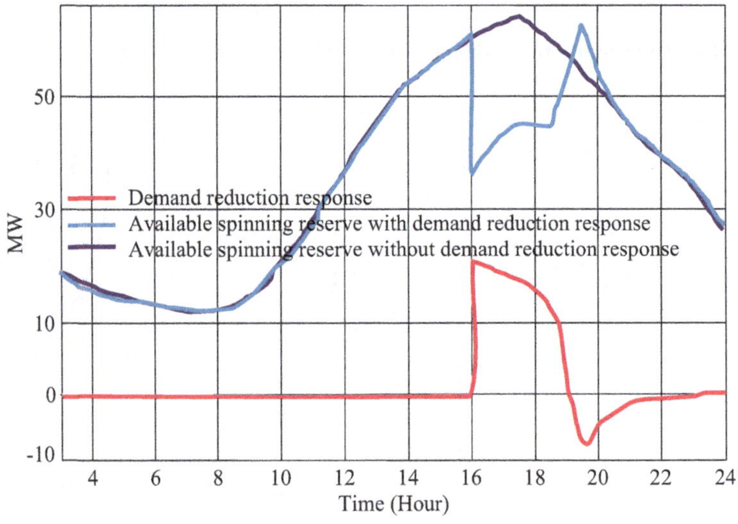

Fig. 18 Considerable spinning reserve ability leftovers even after demand decrease is in effect

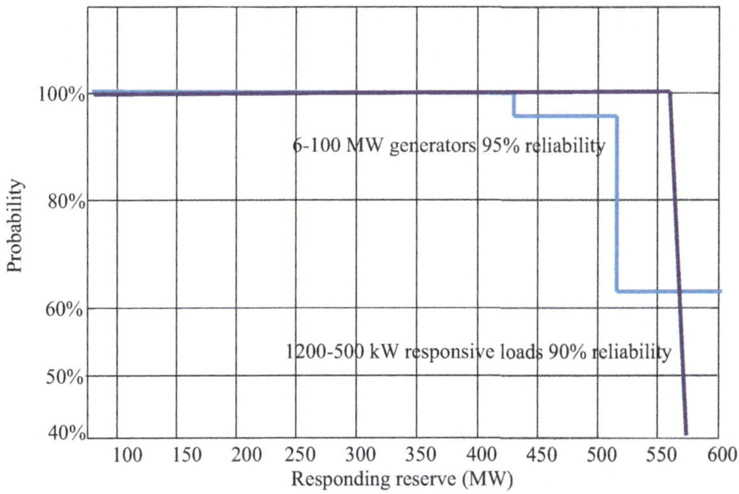

Fig. 19 Reliability of a set of responsive loads versus reliability of 6 large generators

100 MW and a reliability of 0.95 and the results of this study are shown in Fig. 19, which shows that the reliability of a set of responsive loads is significantly higher than the reliability of generators [20].

5.3 Other Benefits

The extra benefits that result from the realization of demand response plans on smart grid contain the following:

Robust retailers and retail market, provide new tools for customer load management, improve market operations and make the market more competitive, appropriate interaction of supply and demand in the market, also effective communication between retail and wholesale markets, activation of customers and retailers and the environmental benefits that will result from reduced fossil generator production [25].

6 Restrictions on the Use of Demand Response Programs

There are conditions and restrictions for the use of demand response resources that must be considered when using them. Some of these constraints are [26]:

- The number of times the load can be interrupted in a given period of time
- Maximum activity time
- Speed in responding to the requested need of the power grid.

7 Barriers to the Implementation of Demand Response Programs in the Conventional Non-smart Power System

The execution of demand response plans in the non-smart power system of different countries has been associated with some obstacles and problems that have been overcome by the emergence of smart grid in most of these failures. The most important of these obstacles are:

- The need to invest in the provision of intelligent measuring equipment.
- Lack of funding and rewards required by companies to upgrade DR programs.
- Variety in how incentives are paid to participants due to the implementation of DR programs, which makes it difficult to calculate and pay.
- The problem of how to measure the amount of load reduction caused by the implementation of DR programs.
- Due to the instability of the situation in the electricity markets, the situation of demand and generation resources in these markets is unknown. Sometimes there is a shortage of generation resources, and in this case, when using DR programs in a wide range of customers, DR activities are gradually forgotten and it is difficult to revive them.
- How to assess the amount of cost reduction that has occurred as a result of the implementation of DR programs is difficult.
- Delay in payment of bonuses and incentives for participants in DR programs.

8 The Effect of Smart Grid and Advanced Infrastructures in the Implementation of DR Programs

The full implementation of DSM needs communication structures and sensors, automatic measurement, specialized processors, and smart devices provided in the smart grid [27].

Figure 20 shows the classification of the technology needed to implement DR programs [28, 29].

The smart grid allows DR applications to communicate in a two-way so that it can record customer consumption on an hourly or shorter time and send that information to a data center at specific times through telecommunication systems. It is also able to receive energy price information from the electricity company and inform the customer. Figure 21 shows the energy management scheme in the infrastructure of a smart grid [26, 29].

Fig. 20 Categories of DR Technologies

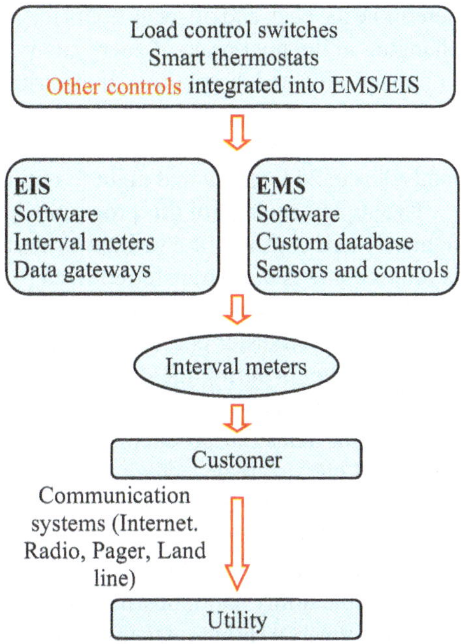

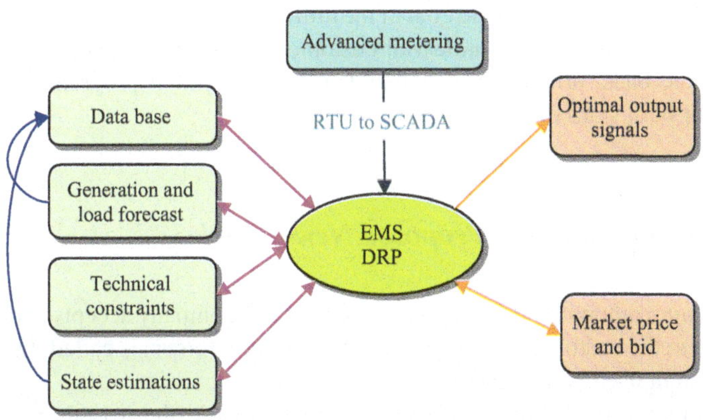

Fig. 21 The overall view of energy management in the smart grid

9 Presenting a New Method TOU in Smart Grid

As mentioned in previous sections, the TOU demand response program is one of the types of time-based programs that consider variable tariffs for consumers based on energy consumption hours. In this program, the energy price is calculated and received at least in two peak and non-peak states or three peak, mid-peak, and off-peak hours on different energy prices in each state. In this program, the utilities determine the peak and off-peak hours for each region based on regional load reviews, changing in the final price of energy, environmental limitations, and the limitations of transmission and distribution of electrical energy, so that the customers can match their consumption pattern with that. For this purpose, the price is adjusted in different periods of the day and night so that the customers can transfer their consumption loads during 24 h of day and night to optimize their costs.

The implementation of this program allows customers who can transfer their load to match their consumption with the price, and therefore, the peak load of the system is reduced and is often transferred from peak hours to other hours.

The TOU program threatens the interests of industrial subscribers because industrial customers transfer part of their electrical energy consumption from peak hours to off-peak hours during the day and night to reduce their costs and that eliminates some of the consumer loads that aren't transferable. It reduces costs but leads to a reduction in industrial production. This not only reduces the revenue of industrial customers but also reduces their desire to participate in the TOU program, which is not favorable for the utility and of course, it is not in line with the goals of the TOU program.

With the goal of increasing the efficiency of the TOU program for both sides of the program (the utility and industrial customers), in this part presents a new technique on how to run TOU program to increase the willingness of the customer to react to the TOU tariffs in a way that reduces the need for electrical energy consumption in the peak hours, which is the ultimate goal of the utility in implementing this program, and also maximizes the profit of industrial customers for being in this program, which is one of the basic principles of demand-side management programs (to provide benefits for both sides of the program).

9.1 Description of the Proposed New Model

The maximum amount of demand that the electricity company accepts and promises to provide to the customer (including the industrial customer) is called the "contract demand" (unit used is kVA or kW).

If the amount of demand consumed by customers in the period, exceeds their contract demand, they are subject to a penalty in the bill or a power outage (depending on what is stated in the contract with the electricity company). So, industrial consumers, by carefully calculating and considering the appropriate coefficient of simultaneity for their electrical equipment, select the contract demand, and buy it from the power company.

Industrial customers (and, of course, the electricity company) always make sure that their maximum demand does not exceed the contract demand.

The new method of implementation of TOU program is recommended, if the capacity of the power grid allows, when industrial customers reduce their energy consumption during peak hours, they are allowed to consume demand more than their contract demand during off-peak hours so that their production does not decrease.

9.2 Case Study

Figure 22 shows the region's load curve occurred in July (the warmest month of the year in the region). Off-peak, mid-peak, and peak hours of the area in this month, are in accordance with Table 1.

The proposed new model of TOU program was implemented for 3 big volunteered industrial customers that were covered by the Mazandaran and Golestan Regional Electric Company at the sub-transmission level. These industries were included: 1—pulp and paper manufacturer industry (with a contract demand of 35,000 kw), 2—cement manufacturer industry (with a contract demand of 35,000 kw), and 3—MDF manufacturer factory (with a contract demand of 6500 kw).

The electricity company signed a contract with these industries that according to it when the electricity company is facing the problem of peak electricity consumption if these industries reduce their load consumption during peak hours (from 20 to 23), they can consume more than their contract demand during off-peak hours (from 2 to 9) and this increase in consumption can be up to 4000 kW (for cement factory), 6000 kW (for paper mill) and 2000 kW (for MDF manufacturer factory) due to the limitations of the transmission and sub-transmission network.

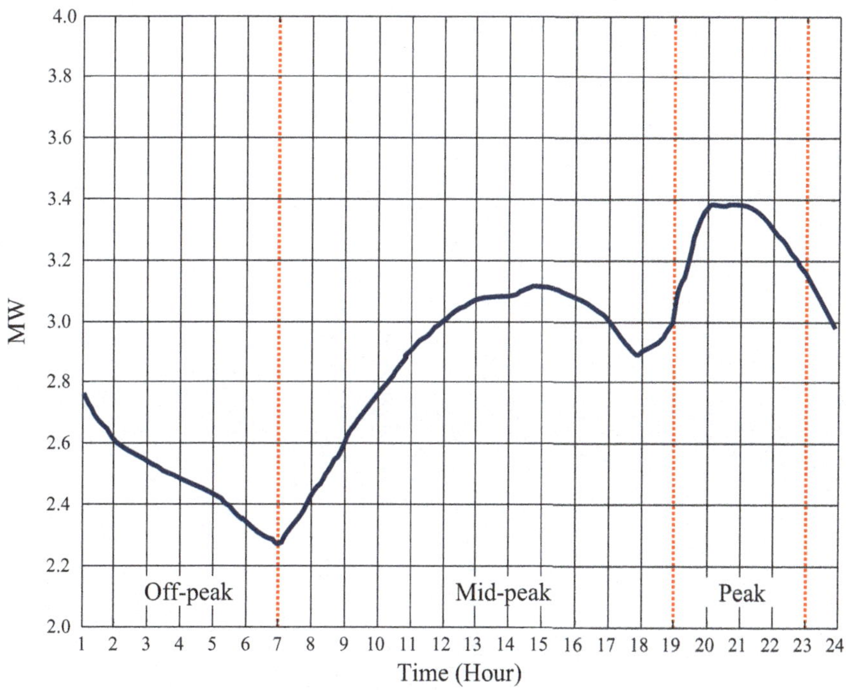

Fig. 22 Regional load curve in July

Table 1 Tariff hours in the region

Mid peak	Off peak									Mid peak									Peak				Mid peak
01:00	02:00	03:00	04:00	05:00	06:00	07:00	08:00	09:00	10:00	11:00	12:00	13:00	14:00	15:00	16:00	17:00	18:00	19:00	20:00	21:00	22:00	23:00	24:00

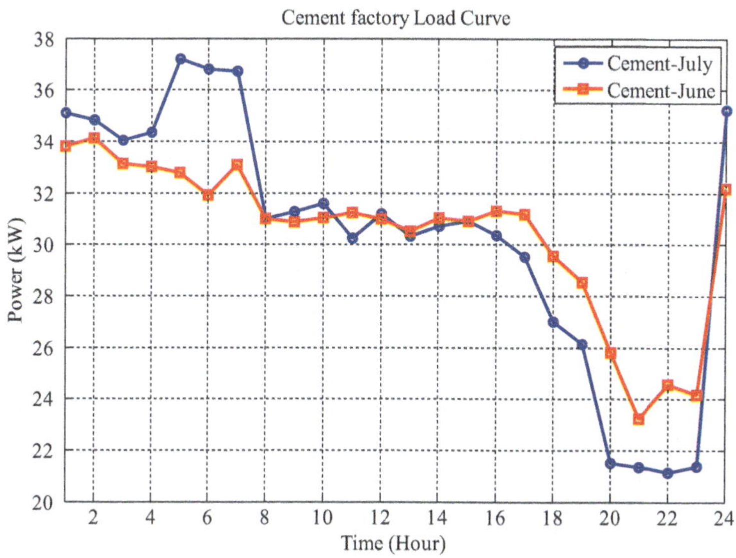

Fig. 23 Cement factory load curve in June and July

In June, TOU was launched without allowing the three factories to consume more load than contract demand during off-peak hours.

In July, the TOU was implemented in this way that allowed the three industrial customers to consume more electricity load during off-peak hours than contract demand if they reduced their consumption load during peak hours.

Figures 23, 24 and 25 show the load curve of the cement factory, the paper mill, and the MDF factory in July and June, respectively.

Tables 2, 3 and 4 shows a comparison between the effects of implementing the classic TOU program on the consumption load curve of the cement factory, paper mill, and MDF factory during June with the results of implementing the suggested model of TOU program during July, respectively.

These tables indicate that the new proposed model of TOU is able to decrease the consumption load during the peak hours down to 28.35% in cement factory, 4.4% in paper mill and 20.63% in MDF factory of the consumption load in mid-peak hours of a day, while for the classic TOU program, these values are 20.18% for cement, 0.58% for paper and 1.43% for MDF factories.

These tables also show that the suggested model TOU is able to increase the consumption load during the off-peak hours up to 15.82% in cement factory, 3.19% in paper mill and 14.13% in MDF factory of the consumption load mid-peak hours of a day, while for the conventional TOU program, these values are 6.0% for cement, −0.8% for paper and −0.6% for MDF factories.

Analysis of this case study shows if industrial customers are given the opportunity to compensate for the reduced energy consumption during peak hours on off-peak hours in such a way as mentioned above that the amount of production in the factories

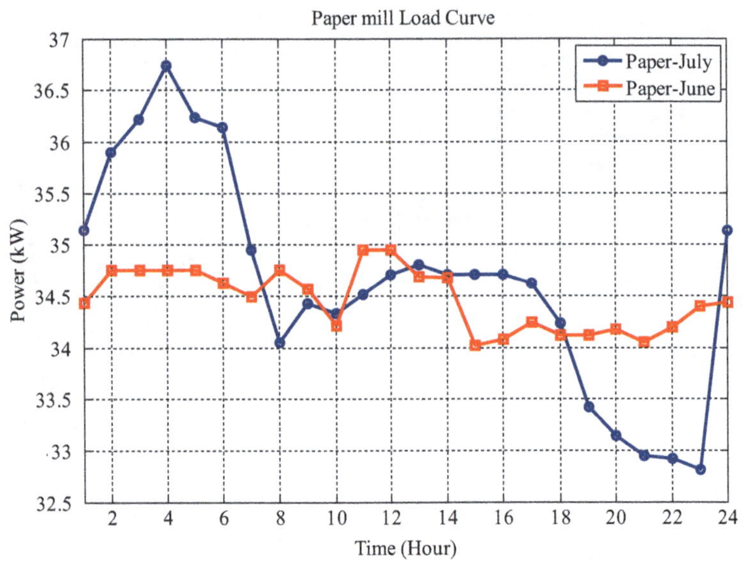

Fig. 24 Paper mill load curve in June and July

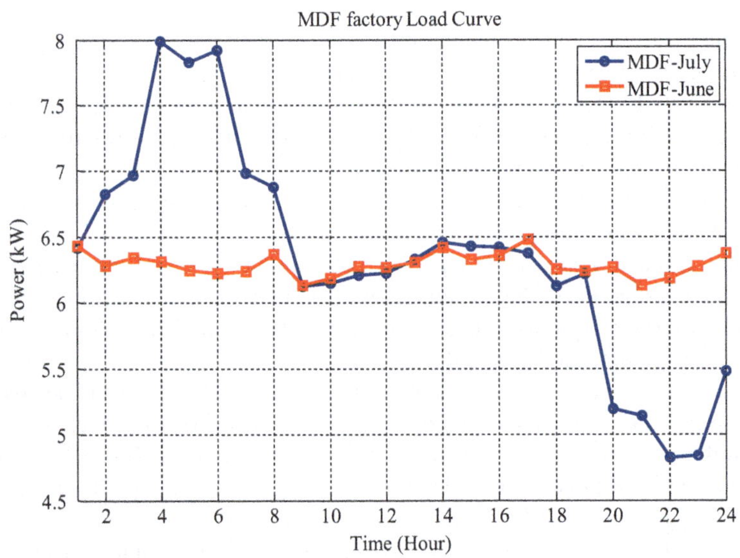

Fig. 25 MDF factory load curve in June and July

Table 2 The amount of increase in consumption load during off-peak hours and decrease in consumption load during peak hours in the Cement factory

Peak load decline% (compared to the mid-peak load)	Off peak load increase% (compared to the mid-peak load)	Ave. daily peak load (MW)	Ave. daily off-peak load (MW)	Ave. daily mid-peak load (MW)	Max. Daily off-peak load (MW)	Contract demand (MW)	Month	Scenario
20.18	6.0	24.49	32.52	30.68	34.15	35	June	Classical TOU-consumption up to the contract demand
28.35	15.82	21.38	34.56	29.84	37.25	35	July	New proposed model of TOU-consumption exceeds contract demand

Table 3 The amount of increase in consumption load during off-peak hours and decrease in consumption load during peak hours in the Paper mill

Peak load decline% (compared to the mid-peak load)	Off peak load increase% (compared to the mid-peak load)	Ave. daily peak load (MW)	Ave. Daily off-Peak load (MW)	Ave. daily mid-peak load (MW)	Max. daily off-peak load (MW)	Contract demand (MW)	Month	Scenario
0.58	-0.8	34.21	34.69	34.41	34.76	35	June	Classical TOU-consumption up to the contract demand
4.4	3.19	32.96	35.58	34.48	36.74	35	July	New proposed model of TOU-consumption exceeds contract demand

will remain constant, the customer's desire to participate in the TOU program will increase. Therefore, this program will be more successful in responding to demand.

Tables 5, 6 and 7 show the monthly production values of the industries present in the program, cement, paper, and MDF, respectively, during the months of June and July. As shown in these tables, the production of cement, paper, and MDF factories increased by 5.39%, 1.65%, and 0.61% in July, respectively. In fact, the new proposed method has led to positive (even small) changes in the production of plants in this program. Table 8 shows the price of electricity for industrial use in Iran in terms of Rials/kWh.

In Tables 9, 10 and 11, the values of energy consumptions of cement, paper, and MDF plants during June and July are listed with their electricity bills. As shown in these tables, the proposed model of TOU program has decreased the costs of energy consumption for the industrial customers present in the program by 877,540, 190,400, and 232,730 Rials in July compared to June.

The findings and results of new model suggested for TOU program can be reviewed from the following two aspects:

The first aspect is the effect of the implementation of the proposed program on the industrial load profile, which causes that the peak consumption of the industrial customers to transfer into the off-peak points of the area load curve more effectively than the classical TOU program.

And the second aspect is the economic profit for the industrial customers present in the new method of TOU program that guarantees the continuation of their participation in the TOU program.

Table 4 The amount of increase in consumption load during off-peak hours and decrease in consumption load during peak hours in the MDF factory

Peak load decline% (compared to the mid-peak load)	Off peak load increase% (compared to the mid-peak load)	Ave. Daily peak load (MW)	Ave. daily off-peak load (MW)	Ave. daily mid-peak load(MW)	Max. daily off-peak load (MW)	Contract demand (MW)	Month	Scenario
1.43	-0.6	6.22	6.27	6.31	7.99	6.5	June	Classical TOU-consumption up to the contract demand
20.63	14.13	5	7.19	6.3	6.37	6.5	July	New proposed model of TOU-consumption exceeds contract demand

Table 5 The monthly production values of the cement

Percentage of changes in the amount of production in July compared to June	Monthly production (ton)	Month	Scenario
–	14,092	June	Classical TOU
+5.39%	14,852	July	Proposed model of TOU

Table 6 The monthly production values of the Paper

Percentage of changes in the amount of production in July compared to June	Monthly production (ton)	Month	Scenario
–	13,465	June	Classical TOU
+1.65%	13,687	July	Proposed model of TOU

Table 7 The monthly production values of the MDF

Percentage of changes in the amount of production in July compared to June	Monthly production (m^3)	Month	Scenario
–	95,220	June	Classical TOU
+0.61%	95,803	July	Proposed model of TOU

Table 8 The price of electricity for industrial use in Iran Rials/kWh

Mid tariff	Peak tariff	Off-peak tariff
340	680	170

Table 9 The bills for cement factory during June and July

Bills (Rials)	Energy consumption (Kwh)			Scenario
	Off peak	Mid Peak	Peak	
66,389,080	64,512	138,702	12,152	Classical TOU (June)
65,511,540	68,574	137,184	10,605	Proposed model of TOU (July)
877,540	–	–	–	Underpayment (July compared to June)

Table 10 The bills for paper mill during June and July

Bills (Rials)	Energy consumption (Kwh)			Scenario
	Off peak	Mid peak	Peak	
75,808,440	68,826	154,619	16,967	Classical TOU (June)
75,618,040	70,598	154,415	16,346	Proposed model of TOU (July)
190,400	–	–	–	Underpayment (July compared to June)

Table 11 The bills for MDF factory during June and July

Bills (Rials)	Energy consumption (Kwh)			Scenario
	Off peak	Mid peak	Peak	
13,817,090	12,447	28,247	3084	Classical TOU (June)
13,584,360	14,272	27,854	2482	Proposed model of TOU (July)
232,730	–	–	–	Underpayment (July compared to June)

10 Conclusions

This chapter presents the following results:

At first, the issue of power demand side management is discussed that it has been mentioned in the strategic plan of the International Energy Agency during the years 2004 to 2009 as one of the main research topics. And then it was mentioned that the history of the formation of smart grid and its role in the development of consumer management programs.

Then, the types of demand response programs that can be implemented under the smart grid were discussed and introduced it was stated that the use of load management programs in smart power grids will be beneficial for all stakeholders.

Finally, a new method and technique of implementing one of the demand response programs called TOU were proposed in such a way as to ensure the interests of the electricity industry and customers simultaneously.

The proposed program was implemented among the volunteer industrial customers and by drawing the load curves of the industries present in the program during the two months of June and July, which are related to the months with the implementation of the TOU program in a classical way and the TOU program in the proposed method. The results of the implementation of the two programs were analyzed from two technical and economic aspects, which indicate the superiority of the results of the new proposed TOU over the traditional TOU program.

Eventually, this chapter of the book shows that if demand response programs are implemented with careful study of the region's load curve and also with constructive interaction between power companies and industries, it can both keep production in industrial customers constant and can help the electricity industry achieve optimal values of consumption management indicators.

References

1. Ghasemi, A., Shayeghi, H., Moradzadeh, M., Nooshyar, M.: A novel hybrid algorithm for electricity price and load forecasting in smart grids with demand-side management. Appl. Energy **177**, 40–59 (2016).
2. Boisvert, R.N., Cappers, P.A., Neenan, B.: The benefits of customer participation in wholesale electricity markets. Electr. J. **15**(3), 41–51 (2002)

3. Qdr, Q.J.U.D.E.: Benefits of demand response in electricity markets and recommendations for achieving them. US Dept. Energy, Washington, DC, USA, Tech. Rep (2006)
4. Newbery, D., Green, R., Neuhoff, K., Twomey, P.: A review of the monitoring of market power. Report prepared at the request of ETSO, www.etso-net.org, (2004).
5. Tang, Y., Song, H., Hu, F., Zou, Y.: Investigation on TOU pricing principles. In: 2005 IEEE/PES Transmission and Distribution Conference and Exposition: Asia and Pacific, pp. 1–9 (2005).
6. Kirschen, D.S.: Demand-side view of electricity markets. IEEE Trans. Power Syst. **18**(2), 520–527 (2003)
7. EPRI's IntelliGridSM Initiative. https://intelligrid.epri.com
8. Blumsack, S.A., Apt, J., Lave, L.B.: Lessons from the failure of US electricity restructuring. Electr. J. **19**(2), 15–32 (2006)
9. Violette, D., Freeman, R., Neil, C.: DRR Valuation and Market Analysis, Volume I: Overview, A report for the International Energy Agency (IEA) DRRTask XIII., (2006)
10. Violette, D.M., Freeman, R., Neil, C.: DRR valuation and market analysis volume II: Assessing the DRR benefits and costs. Report to the International Energy Agency Demandside Programme (2006)
11. Vehmas, J., Kaivo-oja, J., Luukkanen, J.: Energy efficiency as a driver of total primary energy supply in the EU-28 countries–incremental decomposition analysis. Heliyon **4**(10), e00878 (2018)
12. Qdr, Q.J.U.D.E.: Benefits of demand response in electricity markets and recommendations for achieving them. US Dept. Energy, Washington, DC, USA, Tech. Rep. (2006)
13. Mohajeryami, S., Moghaddam, I.N., Doostan, M., Vatani, B., Schwarz, P.: A novel economic model for price-based demand response. Electr. Power Syst. Res. **135**, 1–9 (2016)
14. Borenstein, S., Jaske, M., Rosenfeld, A.: Dynamic Pricing, Advanced Metering, and Demand Response in Electricity Markets (2002)
15. Kang, C., Jia, W.: Transition of tariff structure and distribution pricing in China. In: 2011 IEEE Power and Energy Society General Meeting, pp. 1–5 (2011)
16. "Entergy Arkansas, Inc.—Summer Advantage Program." [Online]. Available: https://www.entergyarkansas.com/your_home/save_money/EE/summer-advantage.aspx. Accessed: 01-Mar-2018
17. "Base Interruptible Program | San Diego Gas & Electric." [Online]. Available: https://www.sdge.com/business/bip. Accessed: 01- Mar-2018
18. Lawrence, D.J.: 2001 performance of New York ISO demand response programs. In: 2002 IEEE Power Engineering Society Winter Meeting. Conference Proceedings (Cat. No. 02CH37309), vol. 2, pp. 995–998 (2002)
19. Popovic, Z.N., Popovic, D.S.: Direct load control as a market-based program in deregulated power industries. In: 2003 IEEE Bologna Power Tech Conference Proceedings, vol. 3, pp. 4 (2003)
20. Paulson, E.J.: Demand response as ancillary services in the PJM RTO. In: IEEE Power Engineering Society General Meeting. IEEE (2005)
21. Misaghian, M.S., Saffari, M., Kia, M., Nazar, M.S., Heidari, A., Shafie-khah, M., Catalão, J.P.: Hierarchical framework for optimal operation of multiple microgrids considering demand response programs. Electr. Power Syst. Res. **165**, 199–213 (2018)
22. Celebi, E., Fuller, J.D.: A model for efficient consumer pricing schemes in electricity markets. IEEE Trans. Power Syst. **22**(1), 60–67 (2007)
23. Levy, R., Abbott, R., Hadden, S.: New Principles for Demand Response planning. Electric Power Research Institute (EPRI), Palo Alto, USA (2002)
24. Kirby, B.J.: Spinning Reserve from Responsive Loads. Department of Energy, United States (2003)
25. Jazayeri, P., Schellenberg, A., Rosehart, W.D., Doudna, J.A.D.J., Widergren, S.A.W.S., Lawrence, D., et al.: A survey of load control programs for price and system stability. IEEE Trans. Power Syst. **20**(3), 1504–1509 (2005)
26. Kim, D.H., Kim, J.O.: Determination of the optimal Incentives and amount of load reduction For a retailer to maximize profits considering demand response programs. In: 2008 IEEE 2nd International Power and Energy Conference, pp. 1290–1295 (2008)

27. Vos, A.: Effective business models for demand response under the smart grid paradigm. In: 2009 IEEE/PES Power Systems Conference and Exposition, pp. 1–1 (2009)
28. Siano, P.: Demand response and smart grids—a survey. Renew. Sustain. Energy Rev. **30**, 461–478 (2014)
29. California Public Utilities Commission: California long term energy efficiency strategic plan. California Public Utilities Commission, San Francisco, CA (2008)

A Smart Bidirectional Power Interface Between Smart Grid and Electric Vehicle

M. Nandhini Gayathri

1 Introduction

Climate change has become an important issue for governments all over the world. The transport sector is one of the major consumers of non-renewable resources and the primary constituent of declension of the environment through emission. One of the methods of reducing emissions is a shift toward electric vehicles (EV); almost 40% of the global CO_2 emissions pertain to energy production. Hence, electrification of automobiles will be futile unless the electricity charging them is generated by clean means. Renewable energy is a feasible option over the conventional fossil fuel sources (oil, gas, coal) for the electrical energy production. However, owing to their intermittent and unpredictable nature, reliable production remains a critical challenge which causes difficulty in integration with the power grid. It is vital to find a way to integrate clean energy sources and reduce the overload in the network to make EVs a viable option and to mass manufacture them. The recent technological progression in the electrical distribution system and load management termed as "smart grids" (SG) has paved way for the usage of EVs through the emergence of the concept of vehicle to grid (V2G) which denotes am infrastructure that supports communication between the EVs and the power grids and to inject electricity into the grid based on the demand. Factors such as accessibility, convenience and affordability of EVs help in facilitating energy storage between recharge cycles for optimal integration of renewable energy. The ability of EVs to replenish the energy back to the grid will be an added advantage to proprietor, power system and the environment.

The adversity of environmental problems around the world forces a paradigm shift to moderate the emissions of greenhouse gasses. The transport sector is solely

M. N. Gayathri (✉)
School of Electrical and Electronics Engineering, SASTRA Deemed To Be University, Thanjavur 613401, India
e-mail: nandhini.gayathri@eee.sastra.edu

B. Vinoth Kumar et al. (eds.), *Intelligent Paradigms for Smart Grid and Renewable Energy Systems*, Algorithms for Intelligent Systems,
https://doi.org/10.1007/978-981-15-9968-2_5

responsible for 26 and 13.1% of the ultimate energy consumption and the total CO_2 emissions, respectively, in the world. Hence, vehicle electrification, i.e., electric and hybrid, is considered as a potential solution to mitigate global CO_2 emissions in many research publications. Among the different designs of vehicle electrification, plug-in EVs are of particular interest due to the direct charge from grid. China leads globally in EVs sales. The USA, Japan, Norway and Germany follow China in the order. In 2018, the global sales rate of plug-in electric vehicles increased 64% and reaches 2.1 million units when compared with 2017. By the end of 2018, it reached a grand total of 5.4 million units.

A plug-in hybrid EV (PHEV) made up of an electric motor along with an internal combustion (IC) engine to utilize both rechargeable batteries and other energy sources. PHEVs operate on charge depleting mode and charge sustaining mode. In CD mode, the energy is produced from the on-board battery packs. In CS mode, the IC engine is used during the charge depletion of battery. The PHEVs categorized into three; series, parallel and series–parallel. The self-sufficient hybrid car with an IC engine recharges the on-board batteries while running from gasoline and thereby does not rely on the grid. The on-board battery can power the vehicle to run up to 40 miles before the gasoline IC engine is needed making it fully capable for both short and long trips.

A common harmonious objective of the new technologies which is reduction of greenhouse gas emissions helps in revolutionizing this paradigm shift. Apart from electric mobility, other technologies influence renewable sources along with battery backup, at the residential and industrial levels to support the electrical power grid. Renewable energy system [RES] has grown excessively in the past decade; this necessitates enhanced power management at both residential and industrial level. The intermittent weather-dependent quality of RES can be mitigated by the inclusion of energy storage system [ESS] technologies. This requires an even more sophisticated control strategy.

2 Vehicle to Grid

The growth of vehicle electrification technology integrates the two major sector, transportation sector and electric power sector. The conventional transportation sector is independent of power grid due to fossil fuels as the primary source. In modern electrified transportation sector, vehicle is made up of the combination of IC engine and electric motor or electric motor merely. So the batteries are used to energize the motor. Battery is charged either with the help of IC engines or power grid or renewable sources or the combination of all three based on the availability. The battery allows storing and dissipating the electrical energy through the corresponding power converters. The charging of battery from the power grid is known as grid to vehicle (G2V). In G2V mode, the power transfer takes place from grid to EV battery. Similarly, the energy available in the battery can be feedback to the power grid. In this mode, the power transfer takes place from EV battery to power grid is known as vehicle to grid (V2G).

V2G concept has many advantages with certain limitations. The major advantages are over load sharing, load equalization and voltage regulation. It improves the overall performance and ultimately profits. The limitation is a need of stable operation of power grid by modifying the conventional structure of power grid components and adopting public support and policy makers. The V2G concept increases the charging cycles of EV batteries and causes battery degradation. So there is a need of compromise between V2G energy transfer and battery degradation. The modification of power grid to support V2G needs upgrading of hardware and software. It involves major expenditure. In addition, bidirectional power converter with intelligent controller is required to charge and discharge the EV battery as per the requirement. The frequent energy transfer between battery and grid results in massive energy losses.

The two major sectors of any country are electric power sector and transportation. In the developed country like USA, the energy capacity of the largest vehicles group is more than the total capacity of electrical power plants. The larger energy produced from passenger vehicle group by internal combustion or electric is idle for 95% of a day almost. As the trend of the automobile industry shifting toward electric and hybrid vehicle manufacturing, there is a great focus on using the on-board batteries in vehicles as energy storage devices. Many studies have shown that the value of utility from tapping into vehicular electrical storage has exceeded the cost of a reduced battery life and double-way replacement.

A potential design of EV involved in V2G technologies is the user interface between the vehicle and grid to deliver power to smart grid from EV battery. The intelligent charging controllers may have multiple choices for the user to charge and discharge the EV at free or to impose a low price to sustain a reasonable fee for the driver to be able to pay a specific travel distance. That will make it possible for owners to contribute as much as their lifestyle requires. It is also hoped that an incentive-based system will be more beneficial to customers [1].

2.1 Smart Grid

Smart grid is slightly differing from traditional power grid in which there is a communication established between the utility and the users. The two-way communication protocol enhances the power system stability, overall performance and the sustainability. The sensors and smart meters are used in real-time data acquisition and then it is processed and tracked by intelligent and autonomous controller.

The renewable sources and loads present in the traditional power grid are always stochastic in nature and difficult to forecast. The real-time advanced metering technology and artificial forecasting technology helps to utilize the renewable sources effectively and at the same time recommends the consumers to use the energy economically. A range of generation sources should increase overall reliability and the chances of attacks and natural disasters.

2.1.1 Smart Charging/Discharging

The EV batteries charging current is more than the current consumed by the vehicle at same power level. Moreover, the current consumption increases with respect to size, count and the type of battery. Many vehicles are charged simultaneously in a particular power grid for limited period results in overload in local grid nodes. Smart charge management techniques used in smart grid to overcome the aforementioned drawbacks. It also supports the bidirectional power transfer between grid and vehicle based on the surplus energy and demand level.

The use of unregulated EV charging results in power grid overloading and voltage fluctuation. It is solved by smart charging systems. In smart charging systems, optimization algorithms used to charge and discharge of EVs into grid at reduced cost. In intelligent charging method, different kinds of communication methods used to connect with smart grid by EV user. They are RFID, Bluetooth, WIFI, etc. To maximize the profit, the specification parameters such as state of charge, transportation time schedule or options for V2G services are varied to the desired one. Many smart charging models use old or previous consumption figures to generate new information. This approach may involve the use of the GPS feature on the mobile device of the EV owners to help evaluate driving style [2]. The charging and discharging characteristics of EV battery are slightly nonlinear in real time. Due to non-ideal condition, the charging and discharging behavior of the battery varies with respect to internal battery resistance. In order to build smart charging strategies, these charging behaviors must be considered [3].

2.1.2 Smart Grid Communication and Control

The contact and control aspect are the crucial part of the smart grid. Many constraints are present in both smart grid and EV side. Essential constraints are observed through sensors and smart meters and exchanged to intelligent controller through s two-way communication network.

In the smart grid, power providers and aggregators can know the real-time delivery and load demands through the two-way communication. It helps to optimize the power supply utilization. It makes switching from non-EV to EV much more realistic. It helps to optimize power delivery, minimize degradation and increase the power quality [4]. With a smart meter grid, the electricity network can collect information on power production and usage in the region to help plan generation and delivery for location-based pricing. Wireless network recommended economically instead of fiber optics to perform the communication between smart meters and smart grid control hub [5]. The major advantage of wireless network is wide area coverage.

The smart meter network has more challenge when compared to conventional data networks. The more number of users and real-time data transfer speed determines the performance of the smart network. Another problem will be to take into account the features of the control systems for charging schedules. Traditional scheduling algorithms that optimize efficiency or decrease average delay cannot be true on a smart grid. Addressing these problems may require the implementation of marginal position pricing and the power load variance chart in the scheduling algorithm [6]. On EV side, the battery charging and power to grid are important suggestions. EV owner insisted by a warning signal when the battery level lies less than the threshold along with information like nearby charging stations, size, energy price, etc., Similarly, V2G compatible EV allows the driver to receive guidance on the charging/discharge process. All this knowledge sharing will be achieved by hall-effect current sensors.

The vehicle charging level, energy price, nearby charging stations and grid status are obtained and analyzed through smart charging system. It exchanges the data among EV, charging stations and smart grid. Smart charging system recommends the charging when the energy cost is lower in addition with other conditions like battery charging level and nearby charging station. If the transmitting ranges were not adequate, the data transfer will be done by manually or by a recommendation program [7].

3 EV Battery Chargers: Construction and Working of the Power Stages

Based on the location, EV battery charger (EVBC) named as either off-board EVBC or as on-board EVBC. Irrespective of location, a typical EVBC is made of power electronic converter with its controller unit as shown in Fig. 1. The conventional EVBC consists of two-stage power electronic converter. The first stage is known as front-end and it interfacing the electrical grid with current feedback control. It performs AC-DC power conversion. The second stage is known as back-end interfacing the EV battery with voltage/current feedback control. It performs DC-DC power conversion. Both the stages are connected through the DC link. To avoid the complexity and to provide smooth and precise control, the two-stage conversions preferred with closed-loop control. The battery voltage and current levels are observed and reference voltage or current signals are generated by the battery management system (BMS) on back-end DC-DC converter. The front-end AC-DC converter operation is determined by back-end power converter. Here, the source current requirement from the grid is completely dependent on battery charging current. So a global power theory employed to determine the front-end AC-DC converter current reference. After the generation of the reference signals, simplified or specialized control strategies applied for each power converter stages. The control strategy determines the switching devices and their duration over the period of operation. Apart from the

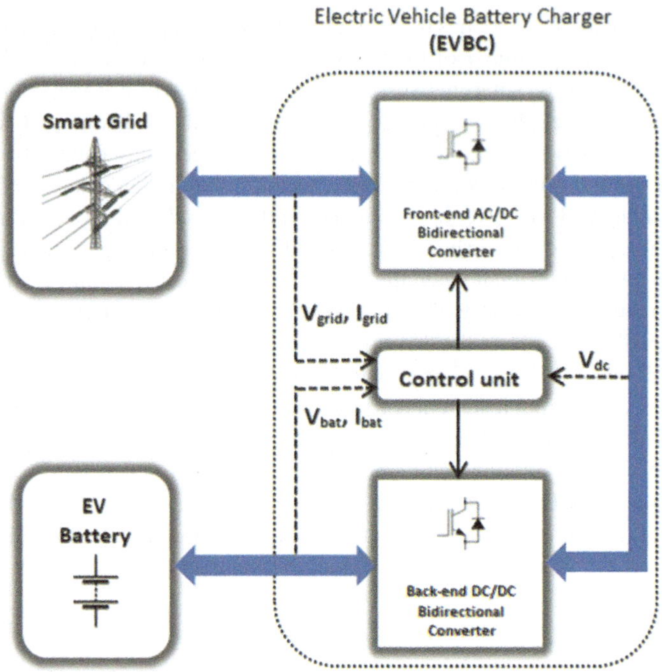

Fig. 1 A two-stage EVBC block diagram

main power flow, a bidirectional communication established for low-level control requirements.

Other than the on-board and off-board EVBC, the combination of both is also available. Based on the combination of on-board and off-board EVBC, many structures are made. Each structure as shown in Fig. 2a–f has its own distinguished benefits. The power transfer quality improved by reducing source harmonics, high input power factor and low ripple content with the help of different arrangements, for example, using multilevel structures, interleaved topologies, providing galvanic isolations. Specialized control algorithms used to perform the power management effectively in the smart grid/smart home. It helps to perform active power transfer between grid and EV. The EV consumer benefitted by financial incentives also through the G2V/V2G modes of EV.

4 Vehicle Electrification Through Wireless Charging Systems

Electric vehicles use a battery pack as the fuel tank to store the electrical energy that propels their motors. EVs are charged by plugging the vehicle to the energy source as

Fig. 2 **a** Combined structure without galvanic isolation. **b** Combined structure with galvanic isolation. **c** On-board/off-board EVBC without galvanic isolation configuration 1. **d** On-board/off-board EVBC without galvanic isolation configuration 2. **e** On-board/off-board EVBC with galvanic isolation configuration 1. **f** On-board/off-board EVBC with galvanic isolation configuration 2

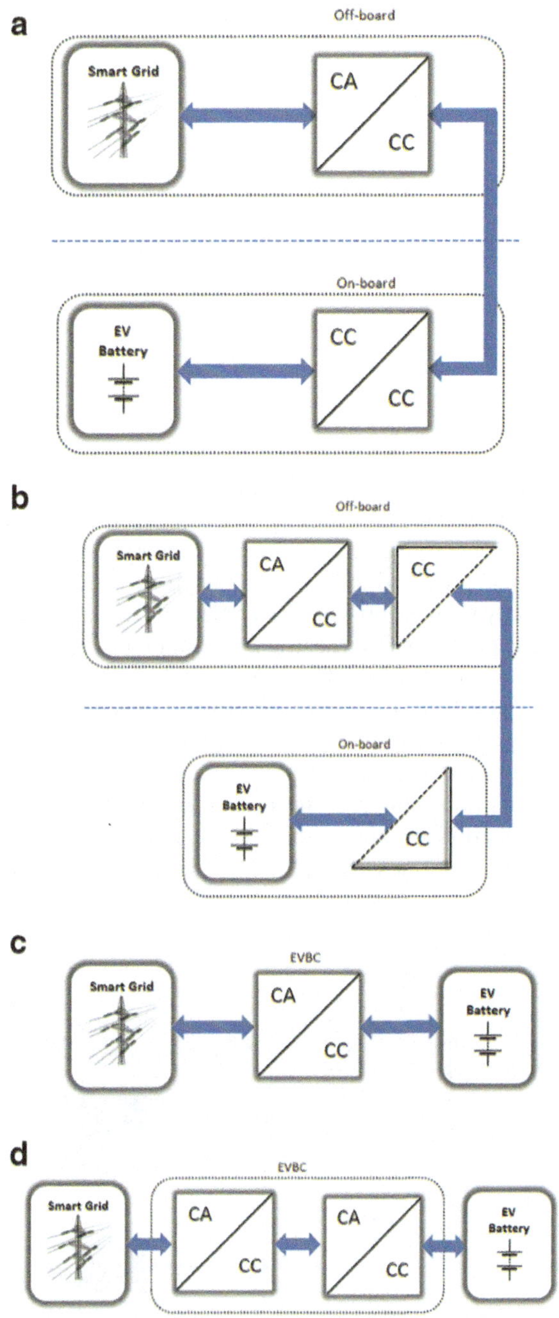

Fig. 2 (continued)

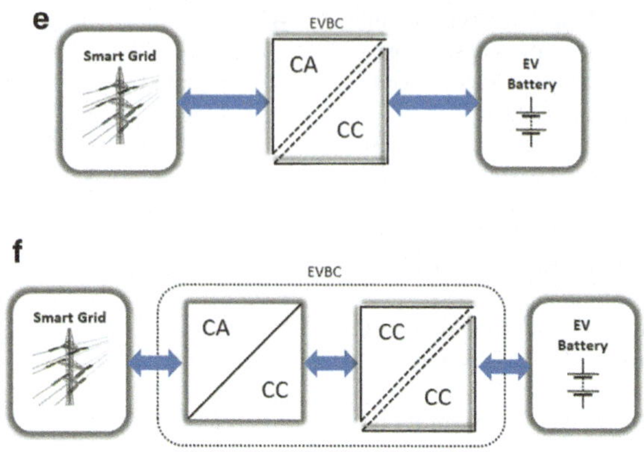

shown in Fig. 3; this source can be electric grid, or any other electric charge storing station and needs to be recharged periodically or by a continuous power supply from the grid connection. There are few ways by which this selection can be made, (i) when using a hybrid vehicle, the energy required to power the motors are obtained from the heat produced in the vehicle engine (ii) or the electric vehicle can be powered directly from the off-grid power plant or from a power source. This concept is shown in the picture below. Electric trains and trams are powered by the overhead wires or through electric rails. The power supply for the electric railway can be either AC or DC, with the former useful for longer distances and cheaper to install, while the latter being the one used for many years for the purpose of traction. Rail current collection comes in various forms, electric railways must be in alignment with the electrified rail to get the power for the traction and that is possible because of the railway tracks which avoid any kind of displacement of the vehicle. Electric vehicle will require an external separate electrification unit however self-contained it can be.

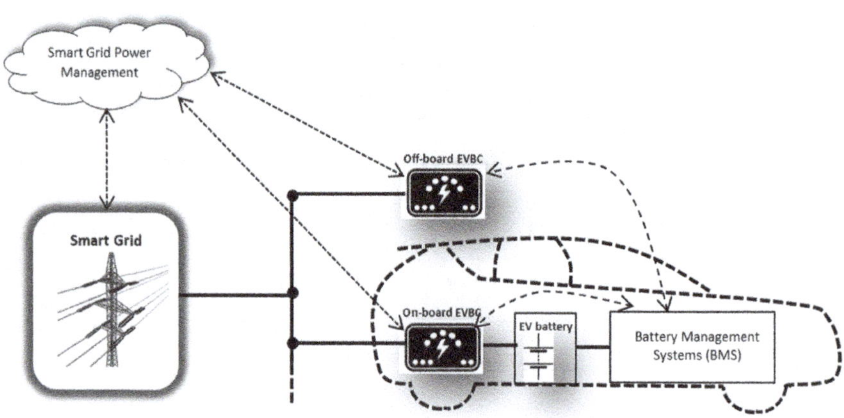

Fig. 3 Smart grid and EV interface

Galvanic contact was the only method to recharge the batteries in an EV, but now wireless power transfer technology is gaining attention and making it the newest way to quickly recharge the batteries. When using galvanic contact method, an isolating transformer was always installed for safety precaution and to prevent any transference of electric charge during contact. This discharge of charges during contact was completely removed by using WPT technology which uses the wireless mode when charging. This is now a commercial option and will have a greater reach in the upcoming years. An approach of WPT is displayed in this section with active power being 11 kW, misalignment between the primary coil (WPT) and secondary coil which is on-board the electric vehicle. A new validation with 10 kW WPT model and efficiency of 94% is designed for the EV. Various researches are held all over the world to help WPT reach greater audience when it comes to dynamic electrification. Existing prototypes for this method have a sustaining capacity of 20 kW for dynamic electrification of a vehicle moving in a 100 m road. Since it is a novel methodology in this industry, it is still awfully expensive for large scale purposes. The forthcoming lessons will talk about the use of WPT with focus on charging systems.

4.1 *Wireless Power Transfer*

Nicholas Tesla was the man who reported on the use of WPT technology. This is a two-phenomenon-based technology combining magnetism and electricity. Transformers were made smaller and lightweight compared to the previous ones by using Maxwell's equation with magnetically coupled coils as shown in Fig. 4a. This has paved way for effectively calculating the amount of transfer of energy in the circuit theory. Figure 4 shows this behavior with less frequency and negligible losses.

The figure shows the impedance compensation of the circuit added to both coils. The very first compensation circuit to be used was a pure capacitive compensation circuit. Based on the connection between coil and compensation network, it is classified into four. They are series–series (SS), series–parallel (SP), parallel–parallel (PP) and parallel–series (PS). The series–series compensation is shown in the diagram. Figure 5 displays no power loss in the winding. The losses in the windings of the

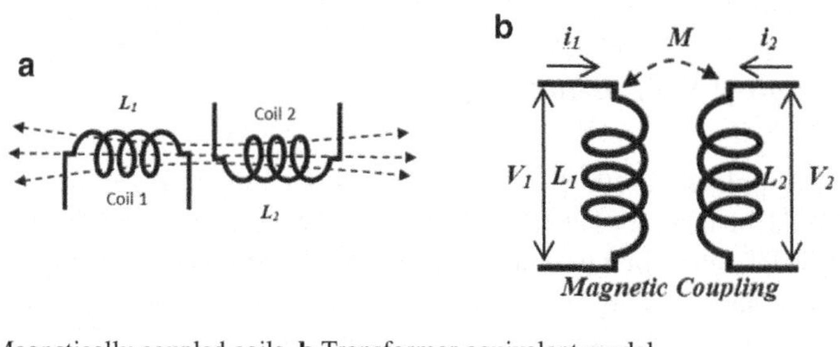

Fig. 4 **a** Magnetically coupled coils. **b** Transformer equivalent model

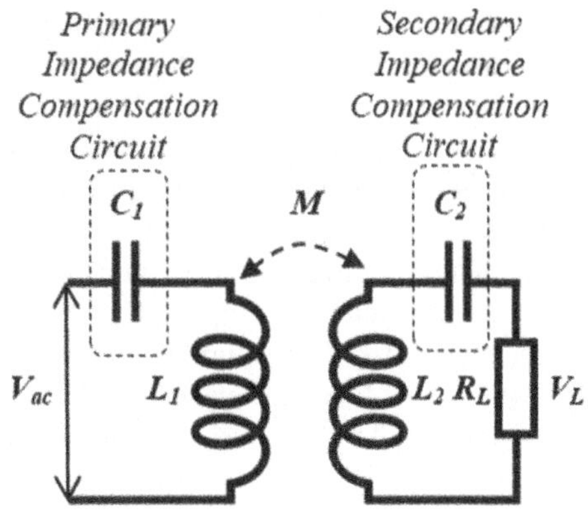

Fig. 5 The WPT basic configuration with compensation network

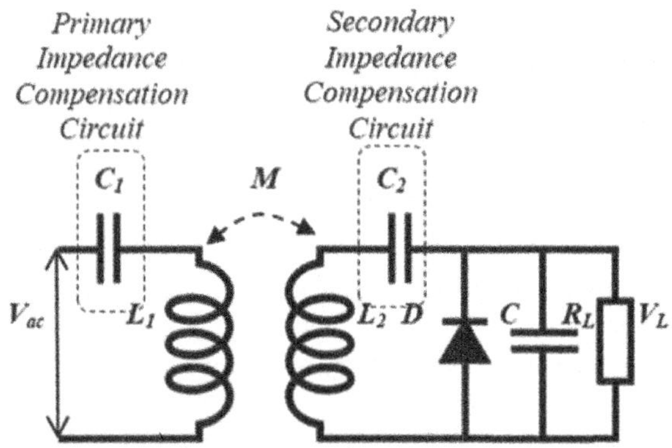

Fig. 6 The WPT basic configuration with voltage rectification

coils are clearly seen when SS compensation is implemented in the WPT model and it is represented in Fig. 6.

The equivalent circuit of the WPT represented by total series resistances of inductor and capacitor. The power transfer efficiency (η) of the power transfer analyzed by circuit analysis. It is maximum at the resonance condition (5). The amount of power transfer is increased by providing new compensation topology.

4.2 Stationary WPT Charging

In automotive applications, the Society of Automobile Engineers (SAE) recommends stationary WPT in the year of 2016. Figure 7 shows a cross-sectional view of a typical coil-to-coil WPT setup. In the setup, the primary coil and secondary coils are placed back to back as shown in Fig. 8. To increase the mutual inductance, ferrite plates are preferred and result in the maximum amount of power transfer (3). Semi-enclosed coils by parallel aluminum plates are used to avoid the magnetic field flux presents

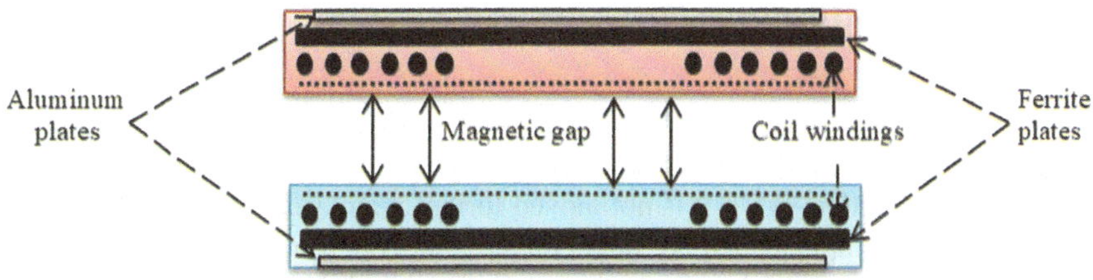

Fig. 7 Cross-sectional view of coil-to-coil WPT structure

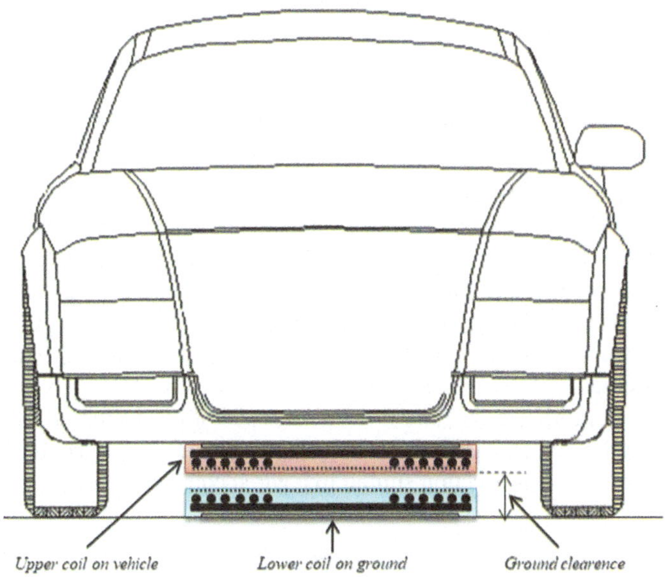

Fig. 8 Stationary WPT setup

outside the gap. The SAE J2954 predefines three different power levels. They are 3.7, 7.7 and 11 kW. The typical power transfer efficiency is about 85% achieved at full alignment. In the current situation, the maximum power transfer levels and its efficiency are limited by the maximum magnetic gap distance between the primary and secondary coils.

The SAE J2954 ground clearance range mentioned in terms of *Z*-classes. They are *Z*1, *Z*2 and *Z*3 and the corresponding ground clearance range is (100–150 mm), (140–210 mm) and (170–250), respectively.

4.3 Dynamic WPT Charging

The aim of the early stage of WPT is dynamic electrification of vehicles, especially to minimize battery size, not meant for stationary charging [1, 8]. But due to practical limitations of dynamic WPT, it is not moved further many years.

Few major attempts [9–13] are made on dynamic WPT, but they are abandoned due to poor efficiency and or high cost. An effective dynamic WPT is proposed using a homogeneous WPT technology. Experiments are done with 3 kW prototype on moving objects. In this dynamic WPT, primary coils in the ground side and secondary coils on inside of EV are proposed in [14].

4.4 WPT Control Through Field Exposure Control

High or medium power transfer at medium or low voltage causes high current flow in wired charging. So there is a need for safe architecture to charge the EV battery. Apart from safe architecture, careful handling required. Mishandling or failure in the EV battery architecture causes overheating of conductors, sparks and results in major accidents like damage to human or equipment.

These drawbacks are overcome by WPT systems as shown in Fig. 9 in which the power transfer takes place through a wireless medium. It provides galvanic isolation between the charging terminal and EV. The absence of conductors avoids the risks associated with wired chargers. Apart from the contactless interface, the rest of the system requires safe construction similar to a wired charger.

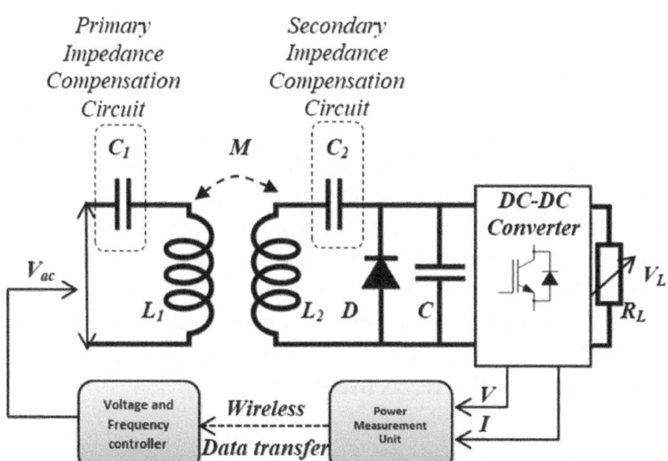

Fig. 9 Simplified dynamic WPT topology with rectification and stabilization

The drawback of WPT is the eddy current produced by strong EMF generated on the WPT coils. It causes heat in the nearby resistive and ferromagnetic materials and results in energy losses and fire. The presence of these objects within the field (between primary and secondary coils) is known as "foreign objects," which is an undesirable one. So foreign object detection methods used to turn off the live WPT and alert the user by alarm, when foreign objects are present. Recently, WPT enhanced by introducing compact dual band [155], maximum power transfer, maximum efficiency [157], bidirectional power flow between grid and vehicle [132], on-board EVBC [159], IoT [160] to strength the vehicle electrification in the market.

5 Vehicle-To-Vehicle Power Transfer

The EV interface to smart grid allows exchanging active power between G2V and V2G. Apart from this, there is a new recommendation of operation called vehicle to vehicle (V2V). The communication system between vehicles decides the V2V power transfer.

5.1 Power Transfer Using the Front-End Power Converters

In this mode, both EVs are connected together as shown in Figs. 10 and 11. The EVs voltage rating should be same as the smart grid to perform V2V power transfer. The power transfer done with or without the connection of smart grid. The one which supplies energy configured in V2G known as donor and another one receives energy configured in G2V known as recipient. So both are the combination of V2G and G2V and performs peer-to-peer power sharing. In this power transfer mode, all four converters employed and the overall efficiency reduced to 65% approximately. Apart from supplying energy to another recipient EV, it is possible to supply standalone electrical loads also. This mode of power transfer is known as vehicle to load (V2L). In V2V, donor EV supplies standalone electrical load.

5.2 V2V Power Transfer Using the Back-End Power Converters

In this mode, the power transfer done without AC link. In this mode, back-end converters employed to perform the power transfer as shown in Fig. 12. Both the front-end converters are disabled. Two-stage conversions improve the power transfer efficiency when compared with four-stage conversion. Here also, donor and recipient EVs configured in V2G and G2V, respectively. Power transfer allows between two

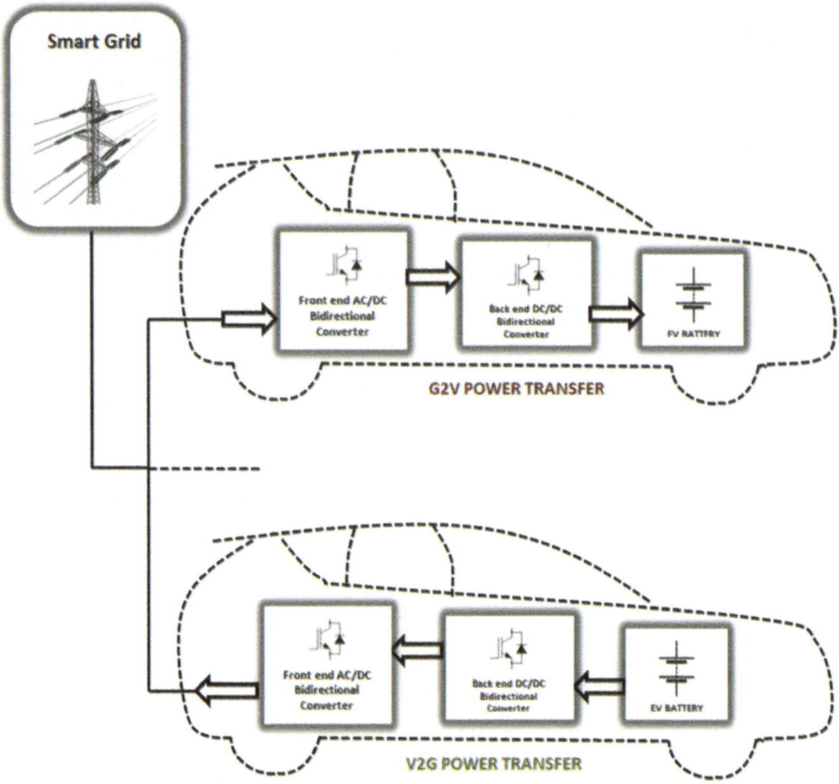

Fig. 10 Power transfer between two EVs through same electrical grid

EVs in remote area also as it is independent of electrical grid. This approach can help in powering an EV with completely depleted batteries restricted by the difficulty to move to a charging station or power outlet. A typical on-board EVBC is made up of a two-quadrant buck-boost converter and a four-quadrant full-bridge converter as the back-end and front-end, respectively. In this V2V configuration, the four-quadrant front-end converter disabled and reconfigured as split-pi buck-boost converter. It works similar to a four-quadrant DC-DC converter. So it allows the power transfer between two different voltage levels of EV batteries.

All the power transfer methods discussed above are performed through wireless mode also. The research is going on to improve the performance of wireless power transfer.

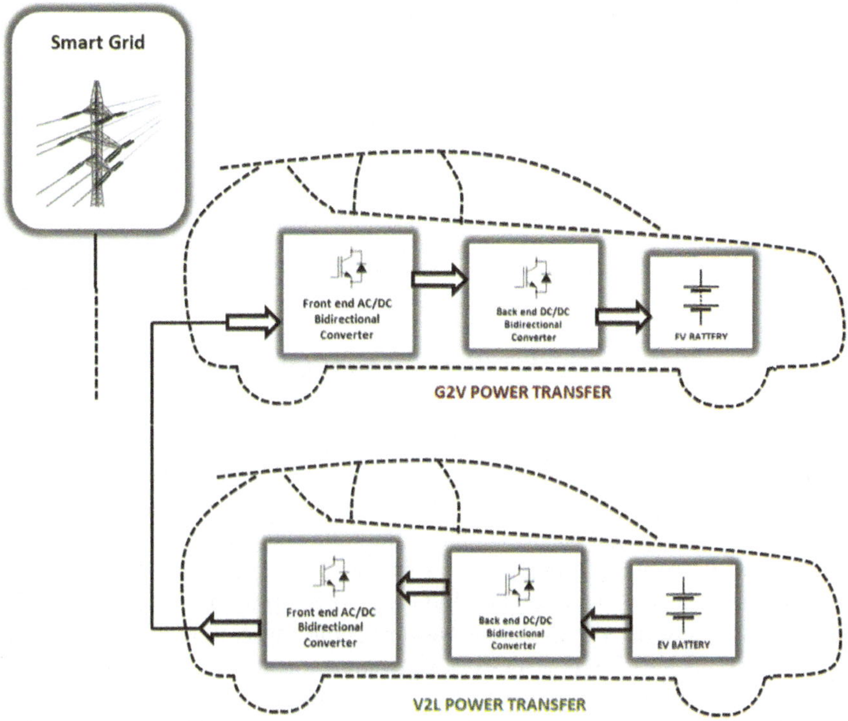

Fig. 11 Power transfer between two EV batteries without using electrical grid interface

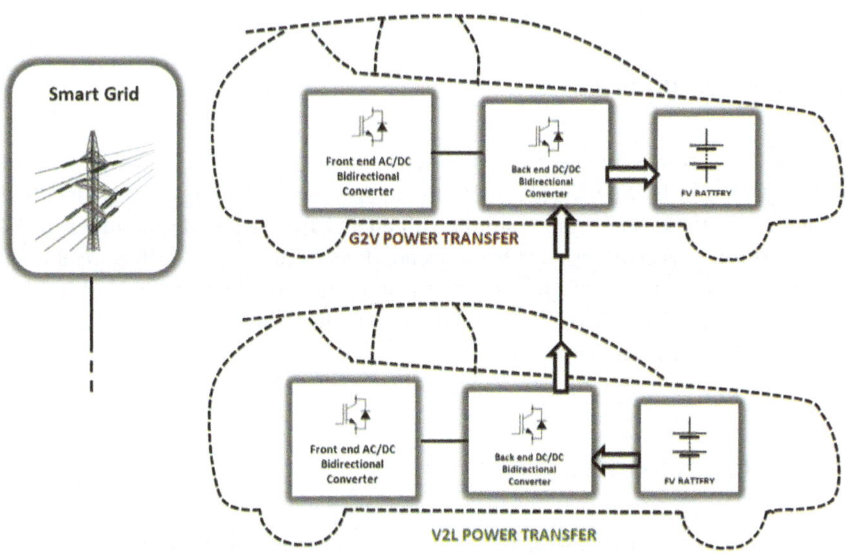

Fig. 12 Power transfer between two EV batteries using only the back-end power stages without using the electrical grid interface

6 Integration of Renewable Energy (RE) for V2G Services

The V2G setup offers a wide range of solutions varying from power utilities, grid operators and aggregators to EV owners and the community. The services provided include various ancillary facilities, reactive power compensation by voltage control, time shifting and active power support. Such facilities have become indispensable because of their alleviation of the growing instability and unreliability of the network due to renewable energy incorporation [15, 16].

6.1 Spinning Reserve: Ancillary or Auxiliary Service

This is a support service given to the power network to boost and preserve the stability and performance of the power grid, which also improves sustainability. Many ancillary services are needed for a variety of purposes, such as security, reliability and grid stability. Such systems comprise the following: reactive source, voltage control, management, operating reserve spinning, operating additional storage and re-establishing energy variance [17]. V2G technology provides auxiliary facilities to the power network via a rotating backup facility, where the energy stored in the grid-linked EVs is used as an extra generation potential to compensate for the losses due to power interruptions [18]. The rotating backup facility provided by V2G technology gives a plan to start failure recovery and also decreases the backup production capacity [19, 20].

6.2 Time Shifting

For time-shift utilities, reserve capacities and technologies are essential to necessitate and supply power for a period of 5–12 h. In this specific case, energy storage facilities are necessary to consume and assimilate all energy from RESs during low demand period. This absorbed energy can be replaced, if necessary, by cost-effective alternative power sources fetched from the network and sold during high power requirement periods, by alleviating the booting or updating of other common and more prevailing peak power production plants [21].

6.3 Active Power Assistance

EV provides a wide range of active power assistance methods. By means of bidirectional connection, the extra energy of EVs that would have been wasted can be returned to facilities and aggregators via the smart network through exclusive recharge stations, parking spaces, etc. Active power assistance aims to ease the market for power utilities. And the demand is fluctuating, in that demand highs and lows, with a reduction in midnights and significant needs during noon and early morning. The variation wears down the production power of the utilities. Furthermore, utility consumers will see power rates adjust in accordance with the requirements; they will have to compensate with a higher price for energy usage during high demand hours. Power systems are planned for the unmitigated scenarios, i.e., assuming utmost usage and need. Consequently, if the need is less than the utmost limit, the systems are underused. Working at full efficiency, the device often wears out over its lifetime. The EVs are capable of providing two forms of active power support, load leveling and peak savings, prolonging the lifespan of the power system and the budgetary burden on customers and EV buyers [22].

6.3.1 Load Balancing

Load balancing is the task of spreading the maximum demand curve during high usage hours, consequently reducing the operating burden on the machines. EVs operate as a combined delivery grid when at the V2G rate, to bring excess current back across the smart network to flatten the maximum load. The need for sole production and supplying felt by power services and aggregators is minimized by using intelligent supply networks, extending system usage period and reducing excessive maintenance or refurbishing costs. Using the power systems over an excessive period of time at a point below peak would result in lower total losses, extending efficiency and reducing burdening opportunities. Because of the unpredictable existence of weather-dependent renewable energy, the production power is unstable to reach the charge continuously. The use of supplying grids to store surplus energy, such as EVs, to serve as a buffer to supply power while renewable production rates are not in need would enable renewable energies to operate continuously throughout the year at a reduced dependency on ideal climatic conditions [19].

6.3.2 Peak Saving

By requiring power systems to not run at extremely bad peak rates, system failure is lessened and the system's overall life and generative capabilities are improved. This makes the delivery of power to longer and better quality. At high demand hours, the EV connected to the network raises the load on the low-voltage network. This raises immediate demand and, ultimately, the power requirement from average and high-voltage networks. The raised load would push more current from high- and medium-voltage networks down to low-voltage networks via transmission cables and transformers, which in effect raises transmission losses and thermal wear on components, reducing usability. This load is minimized by peak savings by organized EV charging, and distribution systems focused on EV by bidirectional infrastructure. The power supplied back to services and aggregators via V2G would reduce peak load, depletion of production capacity, distribution infrastructure and the high cost of electricity faced by EV buyers during peak demand hours by enabling the network to work at a reduced level [23].

6.3.3 Compensation for Reactive Power: Voltage Control

A constant challenge faced by power distributors and aggregators is guaranteeing that the potential difference and current transmitted across the grid are in phase. Nonetheless, a discrepancy between the two will occur with increasing load connected, leading to a reduction in the commercially viable power factor that needs corrective steps. Reactive power assistance is capable of providing the potential difference and current at the supply point to satisfy reactive load, which would have to be otherwise supplied by generators. Without support for reactive electricity, distribution voltages would drop under minimum rates and more current would be required to pass through power transmission lines, resulting in thermal wear and possible blackouts [24]. To lower the load felt at the utility level, modified capacitor stocks are used by utilities to provide local reactive power at the load bus. This complex, reactive volt-ampere (VAC) compensator banks are costly and difficult to update. Through using the DC-link condensers present in EV chargers, utilities may use the V2G distribution network as a reactive power support system through the bidirectional communication infrastructure as well as the active power support network. Since the reactive power is supplied by the DC-link condensers, no strain is put on the EV battery [25].

7 Optimization Techniques for V2G Power Transfer

The mathematical modeling of the systems used to analyze performance without real-time implementation. It saves money and time and also potential risks to the actual system. The mathematical modeling of the system is made up of all essential constraints and its performance parameters almost equal to the actual system. The presence of various constraints in mathematical modeling and its nonlinear behavior makes the system more complex to solve. So to solve the mathematical model and to find the suitable values, various optimization techniques used. The requirement of renewable energy integration in V2G services is maximized efficiency and minimized cost (Fig. 13).

Apart from maximization and minimization, any other desired criteria are also obtained by solving the mathematical model using optimization techniques. The different type of optimization techniques are,

i. Classical optimization techniques
ii. Metaheuristic optimization techniques
iii. Hybrid optimization techniques.

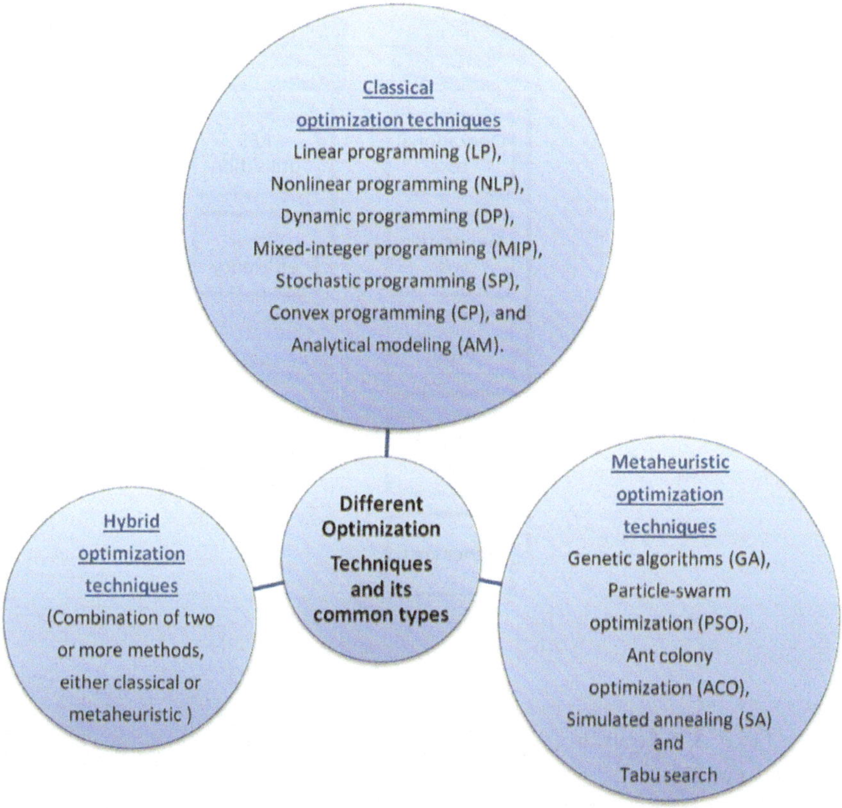

Fig. 13 Different optimization techniques and its common types

If the optimization functions are a continuous and/or differentiable, then classical optimization techniques used. The solutions are obtained through differential calculus. Metaheuristic optimization technique used to solve non-derivative, non-continuous objective functions. It provides the best solution faster than iterative or simple heuristics techniques. In hybrid optimization techniques, two or more of the previously described methods of either classical or metaheuristic are combined to get the solution. The common methods of each optimization technique are shown in Fig. 14.

The following constraints are considered during charging and discharging of the battery in V2G or V2A (vehicle to any) applications [26],

i. The constraint of battery state of charge (SOC)

$$\mathrm{Soc}_{\mathrm{EV,min}}^{i} \leq \mathrm{Soc}_{\mathrm{EV},t}^{i} \leq \mathrm{Soc}_{\mathrm{EV,max}}^{i} \leq \mathrm{Soc}_{\mathrm{EV}}^{i} \forall t \in ST, \forall i \in SE$$

$$\mathrm{Soc}_{\mathrm{EV},t}^{i} = \mathrm{Soc}_{\mathrm{EV},t_{\mathrm{start}}}^{i} + \frac{\eta_c^i P_{\mathrm{EV}}^i}{C_{\mathrm{EV}}^i} \sum_{t=t_{\mathrm{start}}}^{t} S_{\mathrm{EV},t}^{i,c} \quad (1)$$

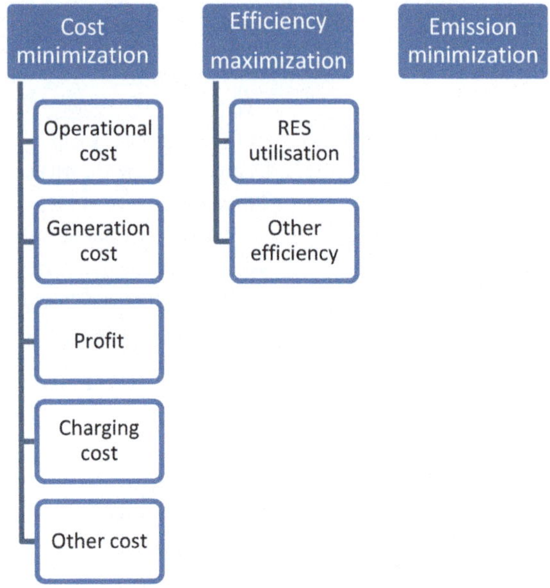

Fig. 14 Optimization objectives

$$-\frac{P^{i}_{\mathrm{EV}}}{\eta^{i}_{d}C^{i}_{\mathrm{EV}}}\sum_{t=t_{\mathrm{start}}}^{t}\left(S^{i,\mathrm{dV2}H}_{\mathrm{EV},t}+S^{i,\mathrm{dV2}B}_{\mathrm{EV},t}+S^{i,\mathrm{dV2}V}_{\mathrm{EV},t}+S^{i,\mathrm{dV2}G}_{\mathrm{EV},t}\right),\ s.t.t,\ t_{\mathrm{start}}\in T^{i}_{\mathrm{EV}} \tag{2}$$

where

$\mathrm{Soc}^{i}_{\mathrm{EV,min}}$ and $\mathrm{Soc}^{i}_{\mathrm{EV},t}$—minimum and maximum value of SOC
ST—Normally set to 24
SE—Usually set to 1 for a single household.

ii. Energy requirement from EV user

$$\mathrm{Soc}^{i}_{\mathrm{EV},t_{\mathrm{end}}}=\mathrm{Soc}^{i}_{\mathrm{EV},t_{\mathrm{start}}}+\frac{\eta^{i}_{c}P^{i}_{\mathrm{EV}}}{C^{i}_{\mathrm{EV}}}\sum_{t=t_{\mathrm{start}}}^{t}S^{i,c}_{\mathrm{EV},t}$$
$$-\frac{P^{i}_{\mathrm{EV}}}{\eta^{i}_{d}C^{i}_{\mathrm{EV}}}\sum_{t=t_{\mathrm{start}}}^{t_{\mathrm{end}}}\left(S^{i,\mathrm{dV2}H}_{\mathrm{EV},t}+S^{i,\mathrm{dV2}B}_{\mathrm{EV},t}+S^{i,\mathrm{dV2}V}_{\mathrm{EV},t}+S^{i,\mathrm{dV2}G}_{\mathrm{EV},t}\right),$$
$$s.t.t_{\mathrm{end}}\in T^{i}_{EV} \tag{3}$$

$$\left|\mathrm{Soc}^{i}_{\mathrm{EV},t_{\mathrm{end}}}-\mathrm{Soc}^{i}_{\mathrm{EV,exp}}\right|\leq\Delta,\ \Delta\to 0^{+} \tag{4}$$

iii. The constraint for EV charging statuses

$$S_{\mathrm{EV},t}^{i,c} + S_{\mathrm{EV},t}^{i,\mathrm{dV2}H} + S_{\mathrm{EV},t}^{i,\mathrm{dV2}B} + S_{\mathrm{EV},t}^{i,\mathrm{dV2}V} + S_{\mathrm{EV},t}^{i,\mathrm{dV2}G} = \begin{cases} 0 & \text{idling} \\ 1 & \text{either charging or discharging} \end{cases} \tag{5}$$

iv. Consideration of the lifetime of the EV battery

$$\Gamma_i = \left[\left(S_{\mathrm{EV},t+1}^{i,c} - S_{EV,t+1}^{i,\mathrm{dis}C}\right) - \left(S_{\mathrm{EV},t}^{i,c} - S_{\mathrm{EV},t}^{i,\mathrm{dis}C}\right)\right]^2 \tag{6}$$

$$S_{\mathrm{EV},t+1}^{i,\mathrm{dis}C} = S_{\mathrm{EV},t}^{i,\mathrm{dV2}H} + S_{\mathrm{EV},t}^{i,\mathrm{dV2}B} + S_{\mathrm{EV},t}^{i,\mathrm{dV2}V} + S_{\mathrm{EV},t}^{i,\mathrm{dV2}G}, \tag{7}$$

where $S_{\mathrm{EV},t}^{i,\mathrm{dis}C}$—discharging status of the ith EV.

v. Constraint for special charging conditions

$$\sum_{i \in \mathrm{SE}} S_{\mathrm{EV},t}^{i,c} - \sum_{i \in \mathrm{SE}} S_{\mathrm{EV},t}^{i,\mathrm{dV2}V} \geq 0. \tag{8}$$

The minimum and maximum values of EV battery are typically set to 20% and 90%, respectively, to prolong the battery. In addition to that charging, degradation should also be considered. The energy requirement of an EV user should satisfy the traveling purpose. At each time slot, EV should be idling, charging or in of the V2G or V2A discharging activities. Similarly, in V2V charging, the number of EVs transferring other vehicles must be equal to the number EVs receiving energy from other cars at each time slot.

The major optimizations of V2G application are [27],

i. Minimizing peak-valley difference of power grid

$$\min G_1 = \min \max_{1 \leq t \leq 24} (L_t + L_{t0}), \tag{9}$$

$$\min G_2 = \min\left[\max_{1 \leq t \leq 24} (L_t + L_{t0}) - \min_{1 \leq t \leq 24} (L_t - L_{t0})\right], \tag{10}$$

where,

$\min G_1$—smallest peak load
$\min G_2$—smallest peak-valley difference
L_{t0}—original daily load data.

ii. Minimizing power grid cost input

$$c_{\mathrm{e}} = c_{\mathrm{EL}} + \omega_{\mathrm{d}} + c_{\mathrm{s}} \tag{11}$$

$$c_{\mathrm{EL}} = \rho_{ij}^{\mathrm{down}}.\left(\frac{1}{\eta^{\mathrm{down}} \cdot \eta^{\mathrm{up}}} - 1\right). \tag{12}$$

$$\max \mathrm{Cost} = c_{\mathrm{e}} \cdot \left(\max_{1 \le t \le 24} L_t - \min_{1 \le t \le 24} L_t\right) - \min G_2\Bigg]. \tag{13}$$

where

c_{e}—unit power cost
c_{e}—conversion cost
c_{EL}—unit power loss cost.

iii. Maximizing EV user satisfaction

$$\max \theta = 1 - \frac{\sum_{j=1}^{24} |L_t - L_{t0}|}{\sum_{j=1}^{24} L_{t0}} \tag{14}$$

The primary goal of the power grid is to minimize the peak load, peak-valley difference and grid cost input in the system with customer satisfaction. The different optimization objective functions and its constraints with findings are discussed in the part.

7.1 Optimization Objective Function(s) for Renewable Energy Integration in V2G Service

The optimization includes one or more of the following in renewable energy integration in V2G service. They are cost minimization to consumer, profit maximization to the service provider, system efficiency maximization and greenhouse gases emission minimization to the environment, etc. The various minimization functions concerning the number of variables and constraints are listed below. The following functions apply to maximization also.

Minimize[f, x]
minimizes the function ff or single variable x (15)

Minimize[f, $\{x, y, \ldots\}$]
minimizes the function f for multiple variables $x, y \ldots$ (16)

Minimize[$\{f, \mathrm{cons}\}$, $\{x, y, \ldots\}$]

$$\text{minimizes the function } f \text{ subject to the constraints cons.} \tag{17}$$

$$\text{Minimize}\left[\ldots, x \in \text{reg}\right]$$
$$\text{minimizes the function } f \text{ subject to the constraints } x \text{ to be in the region} \tag{18}$$

$$\text{Minimize}[\ldots, \ldots, \text{dom}]$$
$$\text{minimizes the function } f \text{ subject to the constraints variables to the domain dom, typically Reals or Integers} \tag{19}$$

Most efficient and cost-effective EVs, maximizing V2G interactions achieved through optimization techniques. And also it improves smart grid technologies, power generation and distribution (Table 1).

Table 1 Summarization of objective functions, constraints and main findings

Objective	Constraints	Findings
Cost minimizations [20, 28–53]	Operating cost, schedule cost, start-up cost, shut-down cost, fuel cost, emission cost, ancillary services cost, number of renewable sources, boilers heat production cost, electricity cost, no-load cost, marginal cost, discharging price, load shedding, interruption time, active power output, market price, energy price, electricity price, online scheduler profit, offline scheduler profit, energy transfer between grid and battery, storage capacity, charging time, starting hour of charge/discharge, ending hour of charge/discharge, costs of upgrades, costs of losses, maintenance cost, penalty cost of renewable sources power imbalances, emissions	Minimizing costs and maximizing profits
Efficiency maximization [54–62]	Share of renewable energy sources, number of thermal units, optimal charging policy, imbalance cost	Maximizing the use RES through the smart grid during optimal times, scheduling of EV fleet charging
Emission minimization [63]	Optimization objective, gasoline consumption	Reduction in carbon dioxide emissions

8 A Smart Interface: Smart Grid, Smart Homes and EVs

A sustainable transportation future relies heavily on numerous factors and some of the available structures or possibilities implemented for an electric vehicle battery charger have been explained along with the technological advancement. With the help of EV flexibilities and a basic understanding of the above-mentioned sections, the upcoming topics explain the opening opportunities an EVBC has on the development of smart grids and smart homes.

8.1 Innovation in Smart Homes: Different Modes and Approaches: On-Board Mode—EVBC

The different operational modes for the use of an on-board EVBC are discussed alongside the illustrations. From Fig. 15, it is clearly seen that there are a considerable amount of restrictions and opportunities for on-board plug-in electric vehicles (PEV) to be used on a smart home. To implement this, a bidirectional communication needs to be established between electric vehicles (EV), utilities, smart home and other electrical appliances maintaining the power management under control. Power management is an important element that allows an interactive relationship between all the emerging smart grid technologies, EVBC, smart homes and consumer appliances. On the one hand, there is an interaction between power management at smart grids and power management at smart homes.

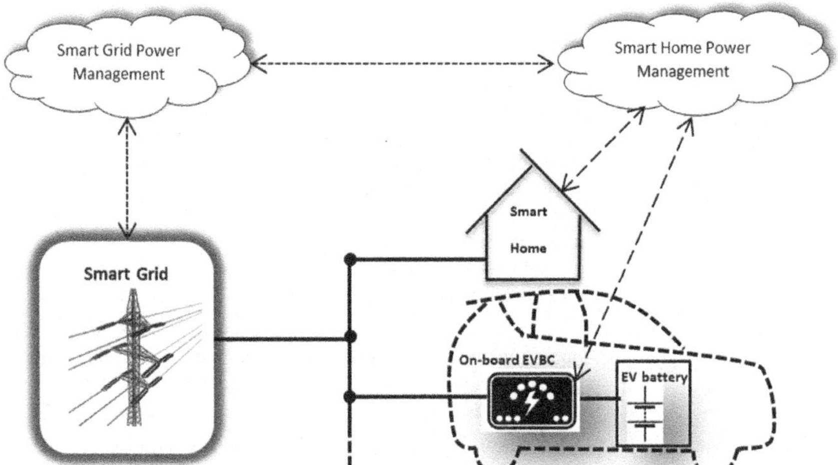

Fig. 15 An on-board EVBC and smart home interface

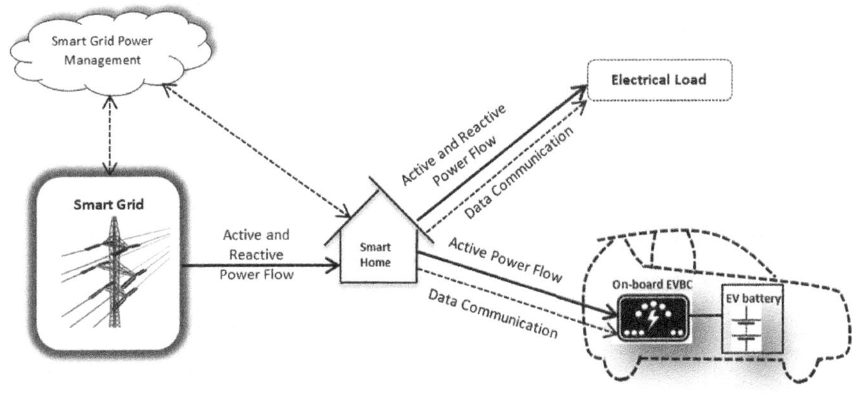

Fig. 16 On-board EVBC: G2V operation mode

On the other hand, there is a communication between the EVBC and consumer electrical appliances and its real need is to control the appliances and their communication according to the varying operational schedules.

8.1.1 Grid to Vehicle (G2V) Operation Mode

All the commercial vehicles nowadays prefer to use the G2V mode for EV battery charging. Figure 16 shows on-board plug-in EVBC connected to a smart home. Here, power flow is unidirectional with the energy grid and bidirectional to communicate other factors such as the status of charging and functional set points. In this mode, the current flowing on the grid side is separate from any other consumer electronics connected. The home switch breaker reduces the current flow if it exceeds the substantial current value by triggering the circuit breaker on. To avoid this, power management forcibly stops the G2V mode, heavily adding to its disadvantage. But comparably, the flexible G2V mode is the one that adjusts the charging power of an EV according to the operational statuses of the appliances connected to it. The charging power value is adjusted seeing the input supply from RES, leading to a balance in the power generation, distribution and consumption cycle.

8.1.2 Vehicle to Grid Interface in V2G Operation Mode

The power flow is bidirectional in V2G mode where the EV returns a considerable amount of the energy stored back to the grid. This mode works according to the smart grid energy management or smart home energy management and electric vehicle user, making it the most flexible ESS option in terms of the grid stability. Also, V2G maintains a communication link as shown in Fig. 17 with the smart grid to

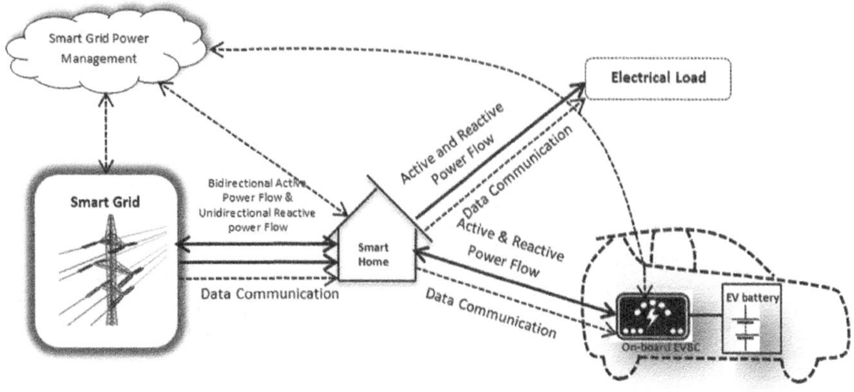

Fig. 17 On-board EVBC in V2G operation mode

work around the schedules for the successful operation of EVBC and addressing the power quantity that must be directed back to the grid.

8.1.3 Vehicle to Load (V2L) as Voltage Source Operation Mode

The previous two techniques had current feedback control applied and EVBC control performed based on the absorption/injected active power. This method acts as a voltage source for the loads (consumer electronics) when the EV is not attached to the grid. V2L uses voltage feedback control which implies that voltage waveform generated by the electric vehicle battery charger and current waveform by appliances. The reverse charger technology allows electricity to be sent to a specific electrical device. This mode is usually used when the electric vehicle is situated at an isolation from the power grid (instances such as catastrophic events when electrical grid is unavailable) and uses the charge from the battery in store as shown in Fig. 18. Hence, it is mandatory to manage an EV owner agreement addressing the important factors such as the battery charge status and about conserving the agreeable battery charge level for the next usage.

One of the real-life applications for this mode of operation is the "LEAF" project presented by Nissan requiring an electric vehicle power station outside the grid (system permanently installed), creating a disadvantage to this V2L mode.

8.1.4 Vehicle to Home (V2H)

As a result of V2L operational mode, EVBC became a viable option when dealing with an offline uninterruptible power supply as shown in Fig. 19. This is helpful in our case of smart homes, as EVBC begins operation almost instantly as a voltage source. Smart homes notify the EVBC about the outage in power and receive critical

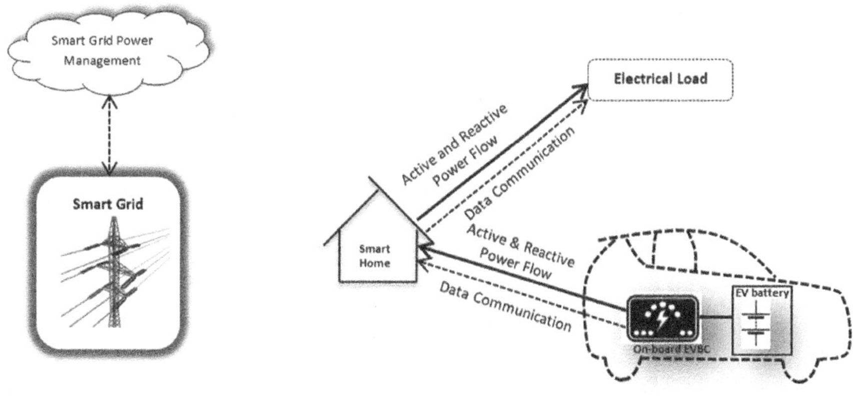

Fig. 18 On-board EVBC in V2L operation mode

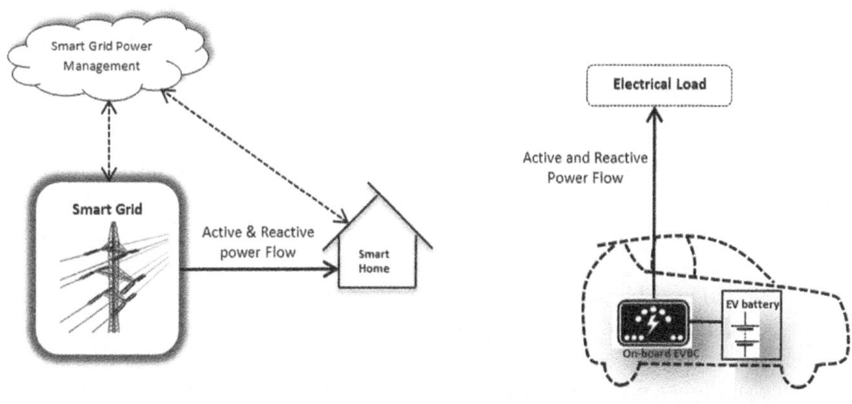

Fig. 19 On-board EVBC in V2H operation mode

information such as battery state of charge from the EVBC (for instance, getting permission to set up a control model based on the priority in the functioning of electrical devices). To identify the fault and failure in power in the electric grid, it is necessary to monitor and measure the grid voltage as the grid side converter is being controlled by a voltage feedback. In such cases, EVBC initiates the process as soon as the smart home disconnects from the energy grid and makes sure it identifies whenever voltage gets restored by starting the synchronization for the phase voltage's transition to normal mode. However, EVBC can be in standby or return to either G2V or V2G modes.

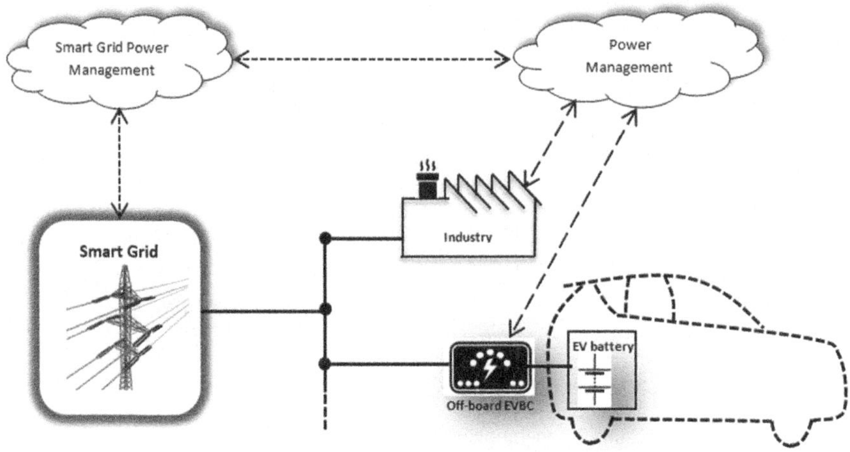

Fig. 20 An off-board EVBC and an EV plugged-in at an industry

8.2 Off-Board EV Battery Charger

The important technologies available for an off-board EVBC with respect to smart grids are discussed in this section. As the name implies, off-board EVBC has to keep a solid connection with the electrical grid even when not attached to an electric vehicle as shown in Fig. 20. In this method, smart grid contextualization is the need of the hour and uses bidirectional communication to send/receive information from the off-board EVBC and to transmit set points to the EVBC. The main difference of an off-board EVBC and an on-board EVBC is that the operating power is high (presence of higher power for a short span of time), while still enabling the use of G2V/V2G modes just like an on-board EVBC.

8.2.1 Operation Mode: Grid to Vehicle and Vehicle to Grid

Another interesting aspect of an off-board EVBC with V2G mode is shown in Fig. 21. Its use is very focused and particular to the application, meaning that when the battery is charging the goal is to charge as fast as possible without any interruption but with V2G mode, the chances of interruptions are high and thereby leading to longer hours of charging.

8.2.2 Operation Mode: Power Quality Compensator

After the vehicle completes charging in the previous method, an off-board EVBC goes out of operation for certain time periods until another EV enters for charging. This

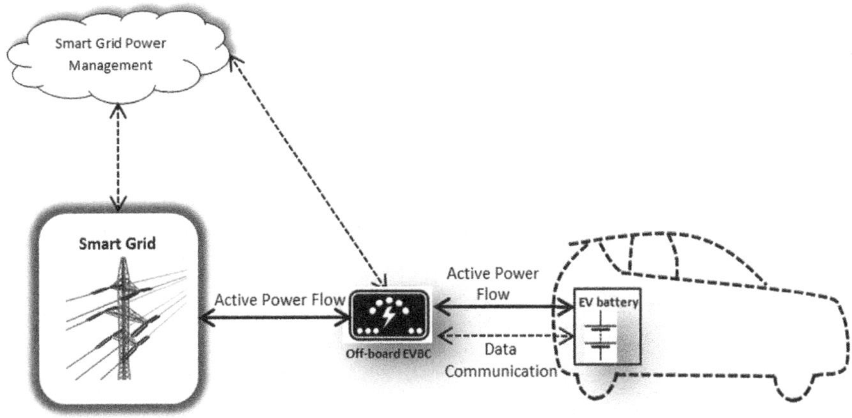

Fig. 21 G2V/V2G modes through off-board EVBC

new mode accomplishes its goal both when the EV is plugged-in and not plugged-in as shown in Fig. 22, without affecting the off-board EVBC, the stored charge in the EV battery and even performing the grid to vehicle or vehicle to grid modes. Therefore, the transmission of active power from the grid to the electric vehicle and vice versa is not needed. Additional hardware is also not required which is huge benefit in using this mode of operation. The linear and nonlinear appliances are the only means to determine the power quality problems.

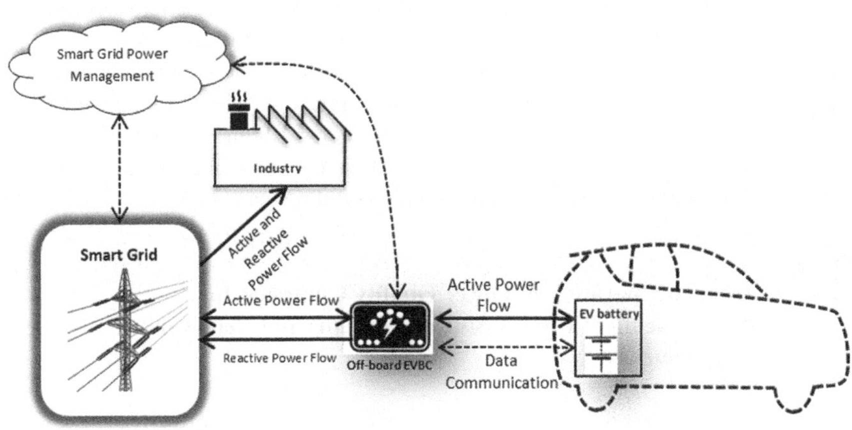

Fig. 22 Power quality compensator through off-board EVBC

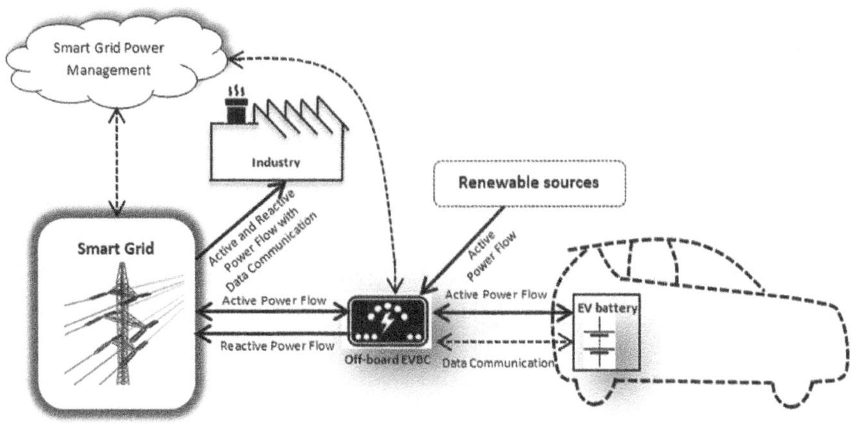

Fig. 23 Power quality compensator through off-board EVBC with an EV plugged-in into grid

8.2.3 Combined Operation of Power Quality Compensator and Renewable Energy Sources (RES)

Previous sections dealt with the compensation of power quality and possible exchanges of power with the energy grid. In order to fully utilize the renewable energy sources in hand for the smart grids and to reduce the dependency of power required from the electrical grid for EV, we are looking for new opportunities in this section. Solar photovoltaic panels being the abundant energy source available, this can be installed in the EV charging stations to improve the energy efficiency. However, with both off-board EVBC and RES requiring the same kind of front-end power stages, the need for unifying both the systems and creating an interface between the two with the electrical grid comes into the picture. The result is an increased efficiency as the established link between RES and EV for charging as shown in Fig. 23 is known to be the boost in efficiency compared to other solutions such as various front-end power stages and back-end power stages.

8.2.4 Combined Operation of Power Quality Compensator and Interfacing of Energy Storage Systems and Renewable Energy Sources

The existing models are redesigned to get an increase in the effectiveness of the technique. This is done by including a bidirectional DC link to the off-board EV battery charger. Based on the circumstances on the off-board EVBC, the interfacing of modes, grid to vehicle or vehicle to grid, the interfacing of RES and flexible ESS; a whole system is allocated and designed. Here in this mode as shown in Fig. 24, complications from the previous modes are taken into consideration and power is being directly injected from the RES to the EV and ESS to avoid getting into the

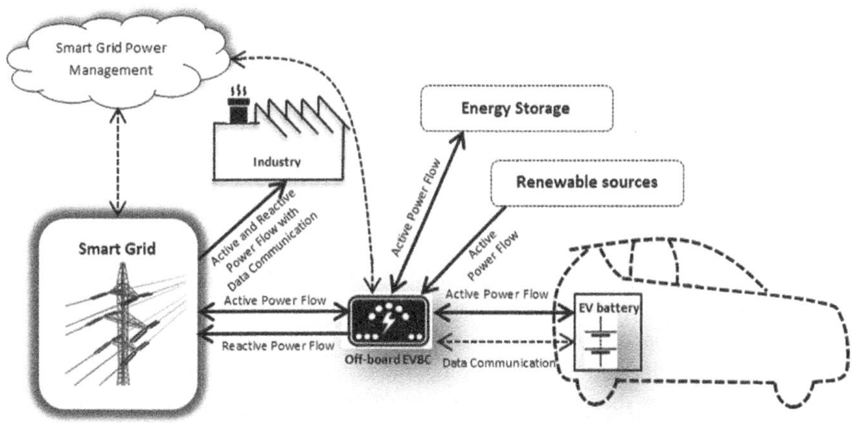

Fig. 24 Power quality compensator through off-board EVBC with an EV plugged-in into the electrical power grid: unified operation with RES, ESS

energy grid. The number of power stages is reduced to a greater extent to achieve better results.

Plugged-in EV has a very peculiar aspect of having high power spike in a shorter span of time, and ESS comes into rescue when we try to totally reduce the voltage flickering in the electrical grid side (power for the EV is supplied from the ESS). In cases where the EV is not plugged-in, integration of ESS and RES takes place allowing an operation like the load shift process. For convenience in the electrical installation, the power generated from the RES can be preserved in the ESS. Off-board EVBC can be operated even when the EV is not plugged-in, there is no production from RES, and no requirement for ESS to compensate to maintain the power quality.

9 Conclusion

This chapter of the book presents the technologies, challenges and a global perspective for vehicle electrification in smart networks. The new paradigm of moving the transportation industry to vehicle electrification, particularly plug-in electric vehicles (EVs), is fuelled by climatic concerns. Nonetheless, this latest approach also supports a range of emerging technologies, such as communication technologies, control electronics for on-board and off-board charging systems, two-way power transfer in vehicle-to-vehicle mode synchronized technology incorporating a battery charging device and a single-function engine driver, wireless power transfer for charging processes, and both on-board and off-board functioning modes of the EV in smart homes and grids. The value of these emerging technologies for vehicle electrification and the relationship between them is discussed in this chapter of the book. The defined EV battery charging modes can be carried out separately of the charging device configuration (i.e., the number and types of power stages for the on-board and off-board charging systems). Moreover, considering that certain modes of operation only need the front-end power stage (AC-DC converter), wireless power transfer technologies may also be considered. In the same way, standardized charging systems for batteries and motor drivers can also be regarded for the application of the operating modes offered. Vehicle optimization for grid for renewable energy implementation is also proposed. Moreover, the integrated technologies of wireless power transfer and integrated systems can also be used in the implementation of other operating modes. In view of the importance of these technologies in terms of power transmission, communication technologies are absolutely important for defining the operating modes, for forming a two-way link for transferring data and power control between the smart network or smart home, the consumer and the EV. Every chapter of the book discusses these developments, explaining the importance of automotive electrification not only as a new model for the transport industry but also as a pioneer of smart grids.

References

1. Davies, R.A.: Wireless' autos: a Russian dream. Maclean's Mag. **58**(13), 34 and 36 (1945)
2. EUROSAT.: Energystatistics—an overview. In: Eurostat Regional Yearbook, Edition (2018)
3. Boulanger, A.G., Chu, A.C., Maxx, S., Waltz, D.L.: Vehicle electrification: Status and issues. Proc. IEEE **99**(6), 1116–1138 (2011)
4. Bose, B.K.: Global warming—energy, environmental pollution, and the impact of power electronics. IEEE Ind. Electron. Mag. **4**(1), 6–17 (2010)
5. Ribeiro, B., Brito, F.P., Martins, J.A.: Survey on electric/hybrid vehicles. In: Transmission and Driveline, SAE International (2010)
6. Wencong, S., Rahimi-Eichi, H., Zeng, W., Chow, M.-Y.: A survey on the electrification of transportation in a smart grid environment. IEEE Trans. Ind. Electron. **8**(1), 1–10 (2012)
7. Oviedo, R.M., Fan, Z., Gormus, S., Kulkarni, P.: The reign of EVs? An economic analysis from consumer's perspective. IEEE Electr. Mag. **2**(2), 61–71 (2014)
8. Babat, G.I.: Patent GB657035A. High frequency electric transport system with contactless transmission of energy, **657**, 035 (1946)
9. Systems Control Technology Inc.: Roadway Powered Electric Vehicle Project Track Construction and Testing Program Phase 3D. PaloAlto, California (1994)
10. Lashkari, K., Shladover, S.E., Lechner, E.H.: Inductive power transfer to an electric vehicle. In: Proceedings of 8th International Electric Vehicle Symposium (1986)
11. Lechner, E.H., Shladover, S.E.: The roadway powered electric vehicle—an all-electric hybrid system. In: Proceedings of the 8th International Electric Vehicle Symposium (1986)
12. Boys, J.T., Covic, G.A.: The inductive power transfer story at the University of Auckland. IEEE Circ. Syst. Mag. **15**, 6–27 (2015)
13. Zhang, Z., Chau, K.T.: Homogeneous wireless power transfer for move-and- charge. IEEE Trans. Power Electron. **30**(11), 6213–6220 (2015)
14. Fujita, T., Yasuda, T., Akagi, H.: A dynamic wireless power transfer system applicable to a stationary system. IEEE Trans. Ind. Appl. **53**(4), 3748–3757 (2017)
15. Monterio, V., Pinto, J.G., Exposto, B., Goncalves, H., Ferreira, J.C., Couto, C., et al.: Assessment of a battery charger for electric vehicles with reactive power control. In: Proceedings of the IEEE IECON2012: 38th Annual Conference on Industrial Electronics Society, 25–28 Oct 2012, pp. 5142–5147 (2012)

16. Han, S., Han, S., Sezaki, K.: Development of an optimal vehicle-to-grid aggregator for frequency regulation. IEEE Trans. Smart Grid **1**(1), 65–72 (2010)
17. Mullan, J., Harries, D., Bräunl, T., Whitely, S.: The technical economic and commercial viability of the vehicle-to-grid concept. Energy Policy **48**, 394–406 (2012)
18. Sortomme, E., El-Sharkawi, M.A.: Optimal scheduling of vehicle-to-grid energy and ancillary services. IEEE Trans. Smart Grid. **3**(1), 351–359 (2012)
19. Ghanbarzadeh, T., Goleijani, S., Moghaddam, M.P.: Reliability constrained unit commitment with electric vehicle to grid using hybrid particle swarm optimization and ant colony optimization. In: Proceedings of the IEEE Power and Energy Society General Meeting, 24–29 July 2011, pp. 1–7 (2011)
20. Soares, J., Vale, Z., Canizes, B., Morais, H.: Multi-objective parallel particles warm optimization for day-ahead vehicle-to-grid scheduling. In: Proceedings of the IEEE CIASG 2013: Symposium on Computational Intelligence Applications in Smart Grid, 16–19 Apr 2013, pp. 138–145 (2013)
21. Díaz-González, F., Sumper, A., Gomis-Bellmunt, O., Villafáfila, R.: A review of energy storage technologies for windpower applications. Renew. Sustain. Energy Rev. **16**(4), 2154–2171 (2012)
22. Yong, J.Y., Ramachandaramurthy, V.K., Tan, K.M., Mithulananthan, N.: Are view on the state-of-the-art technologies of the electrical vehicle, its impacts and prospects. Renew. Sustain. Energy Rev. **49**, 365–385 (2015)
23. Leemput, N., Geth, F., Claessens, B., Van Roy, J., Ponnette, R., Driesen, J.: A case study of coordinated electric vehicle charging for peak shaving on a low voltage grid. IEEE Innovative Smart Grid Technologies (ISGT Europe) (2012)
24. Kesler, M., Kisacikoglu, M.C., Tolbert, L.M.: Vehicle-to-grid reactive power operation using plug-in electric vehicle bidirectional off-board charger. IEEE Trans. Industr. Electron. **61**(12), 6778–6784 (2014)
25. Yong, J.Y., Ramachandaramurthy, V.K., Tan, K.M., Mithulananthan, N.: Bi-directional electric vehicle fast charging station with novel reactive power compensation for voltage regulation. Int. J. Electr. Power Energy Syst. **64**, 300–310 (2015)
26. Mao, T., Zhang, X., Zhou, B.: Modeling and solving method for supporting 'vehicle-to-anything' ev charging mode. Appl. Sci. **8**, 1048 (2018)
27. Qu, Z., Song, J., Liu, Y., et al.: Optimization model of EV charging and discharging price considering vehicle owner response and power grid cost. J. Electr. Eng. Technol. **14**, 2251–2261 (2019)
28. Zhang, M., Chen, J.: The energy management and optimized operation of electric vehicles based on microgrid. IEEE Trans. Power Deliv. **29**, 1427–1435 (2014)
29. Chen, C., Duan, S.: Optimal integration of plugin hybrid electric vehicles in microgrids. IEEE Trans. Industr. Inf. **10**, 1917–1926 (2014)
30. Tan, Z., Yang, P., Nehorai, A.: An optimal and distributed demand response strategy with electric vehicles in the smartgrid. IEEE Trans. Smart Grid **5**, 861–869 (2014)
31. Aunedi, M., Strbac, G.: Efficient system integration of wind generation through smart charging of electric vehicles. In: Proceedings of 8th International Conference and Exhibition on Ecological Vehicles and Renewable Energies (EVER), pp. 1–12 (2013)
32. Derakhshandeh, S., Masoum, A.S., Deilami, S., Masoum, M.A., Hamedani, G.M.: Coordination of generation scheduling with PEVs charging in industrial microgrids. IEEE Trans. Power Syst. **28**, 3451–3461 (2013)
33. Saber, A.Y., Venayagamoorthy, G.K.: Resource scheduling under uncertainty in a smart grid with renewables and plugin vehicles. IEEE Syst. J. **6**, 103–109 (2012)
34. Saber, A.Y., Venayagamoorthy, G.K.: Plugin vehicles and renewable energy sources for cost and emission reductions. IEEE Trans. Industr. Electron. **58**, 1229–1238 (2011)
35. Khodayar, M.E., Wu, L., Shahidehpour, M.: Hourly coordination of electric vehicle operation and volatile wind power generation in SCUC. IEEE Trans. Smart Grid **3**, 1271–1279 (2012)
36. Li, C.T., Ahn, C., Peng, H., Sun, J.: Synergistic control of plugin vehicle charging and wind power scheduling. IEEE Trans. Power Syst. **28**, 1113–1121 (2013)

37. Vayá, M.G., Andersson, G.: Integrating renewable energy forecast uncertainty in smart charging approaches for plug-in electric vehicles. In: IEEE Grenoble PowerTech (POWERTECH), pp. 1–6 (2013)
38. Zhao, J., Wen, F., Dong, Z.Y., Xue, Y., Wong, K.P.: Optimal dispatch of electric vehicles and wind power using enhanced particle swarm optimization. IEEE Trans. Ind. Inf. **8**, 889–899 (2012)
39. Li, C.T., Ahn, C., Peng, H., Sun, J.: Integration of plug-in electric vehicle charging and wind energy scheduling on electricity grid. In: IEEE Innovative Smart Grid Technologies (ISGT), pp. 1–7 (2011)
40. Zhou, L., Li, F., Gu, C., Hu, Z., Blond, S.: Cost/benefit assessment of a smart distribution system with intelligent electric vehicle charging. IEEE Trans. Smart Grid **5**, 839–847 (2014)
41. Vasirani, M., Kota, R., Cavalcante, R.L., Ossowski, S., Jennings, N.R.: An agent-based approach to virtual powerplants of wind power generators and electric vehicles. IEEE Trans. Smart Grid **4**, 1314–1322 (2013)
42. Goonewardena, M., Le, L.B.: Charging of electric vehicles utilizing random wind: a stochastic optimization approach. In: IEEE Globecom Workshops (GCWkshps), pp. 1520–1525 (2012)
43. Chen, S., Tong, L.: IEMS for large scale charging of electric vehicles: architecture and optimal online scheduling. In: Proceedings of IEEE 3rd International Conference on Smart Grid Communications (SmartGridComm), pp. 629–634 (2012)
44. Sortomme, E., El-Sharkawi, M.A.: Optimal charging strategies for unidirectional vehicle-to-grid. IEEE Trans. Smart Grid **2**(1), 131–138 (2011)
45. Verzijlbergh, R.A., de Vries, L.J., Lukszo, Z.: Renewable energy sources and responsive demand. Do we need congestion management in the distribution grid? IEEE Trans. Power Syst. **29**(5), 2119–2128 (2014)
46. Jin, C., Sheng, X., Ghosh, P.: Energy efficient algorithms for electric vehicle charging with intermittent renewable energy sources. In: IEEE Power and Energy Society General Meeting (PES), pp. 1–5 (2013)
47. Hu, W., Su, C., Chen, Z., Bak-Jensen, B.: Optimal operation of plug-in electric vehicles in power systems with high wind power penetrations. IEEE Trans. Sustain. Energy **4**, 577–585 (2013)
48. Kristoffersen, T.K., Capion, K., Meibom, P.: Optimal charging of electric drive vehicles in a market environment. Appl. Energy 88, 1940–1948 (2011)
49. Shaaban, M.F., El-Saadany, E.: Accommodating high penetrations of PEV sand renewable DG considering uncertainties in distribution systems. IEEE Trans. Power Syst. **29**, 259–270 (2014)
50. Gunter, S.J., Afridi, K.K., Perreault, D.J.: Optimal design of grid connected PEV charging systems with integrated distributed resources. IEEE Trans. Smart Grid, **4**, 956–967 (2013)
51. Ghofrani, M., Arabali, A., Etezadi-Amoli, M.: Electric drive vehicle to grid synergies with large scale wind resources. In: IEEE Power and Energy Society General Meeting, pp. 1–6 (2012)
52. Ghofrani, M., Arabali, A., Etezadi-Amoli, M., Fadali, M.S.: Smart scheduling and cost-benefit analysis of grid-enabled electric vehicles for wind power integration. IEEE Trans. Smart Grid **5**(5), 2306–2313 (2014)
53. Ghofrani, M., Arabali, A., Ghayekhloo, M.: Optimal charging/discharging of grid-enabled electric vehicles for predictability enhancement of PV generation. Electr. Power Syst. Res. **117**, 134–142 (2014)
54. Schuller, A., Hoeffer, J.: Assessing the impact of EV mobility patterns on renewable energy oriented charging strategies. Energy Procedia **46**, 32–39 (2014)
55. Gottwalt, S., Schuller, A., Flath, C., Schmeck, H., Weinhardt, C.: Assessing load flexibility in smart grids: electric vehicles for renewable energy integration. In: IEEE Powerand Energy Society General Meeting (PES), pp. 1–5 (2013)
56. Strnad, I., Skrlec, D., Tomisa, T.: A model for the efficient use of electricity produced from renewable energy sources for electric vehicle charging. In: Proceedings of 2013 4th International Youth Conference on Energy (IYCE), pp. 1–8 (2013)

57. Zhu, L., Yu, F.R., Ning, B., Tang, T.: Optimal charging control for electric vehicles in smart microgrids with renewable energy sources. In: Proceedings of IEEE 75th Conference on Vehicular Technology Conference (VTC Spring), pp. 1–5 (2012)
58. Mets, K., De Turck, F., Develder, C.: Distributed smart charging of electric vehicles for balancingwindenergy. In: Proceedings of IEEE 3rd International Conference on Smart Grid Communications (SmartGridComm), pp. 133–138 (2012)
59. Pantoš, M.: Stochastic optimal charging of electric-drive vehicles with renewable energy. Energy **36**, 6567–6576 (2011)
60. Tushar, M.H.K., Assi, C., Maier, M., Uddin, M.F.: Smart microgrids: Optimal joint scheduling for electric vehicles and home appliances. IEEE Trans. Smart Grid **5**, 239–250 (2014)
61. Fazelpour, F., Vafaeipour, M., Rahbari, O., Rosen, M.A.: Intelligent optimization to integrate a plug-in hybrid electric vehicle smart parking lot with renewable energy resources and enhance grid characteristics. Energy Conv. Manag. **77**, 250–261 (2014)
62. Battistelli, C., Baringo, L., Conejo, A.: Optimal energy management of small electric energy systems including V2G facilities and renewable energy sources. Electr. Power Syst. Res. **92**, 50–59 (2012)
63. Patil, R., Kelly, J.C., Filipi, Z., Fathy, H.: A framework for the integrated optimization of charging and power management in plug-in hybrid electric vehicles. In: Proceedings of 2012 American Control Conference (ACC), pp. 1327–1334 (2012)

Nature-Inspired Optimization Algorithms for Renewable Energy Generation, Distribution and Management—A Comprehensive Review

Vamsi Krishna Reddy Aala Kalananda and Venkata Lakshmi Narayana Komanapalli

1 Introduction

Meta-heuristics are widely employed in several engineering fields to achieve optimality while satisfying various constraints. The incorporation of meta-heuristics in the renewable energy has taken a huge leap forward in last two decades and has been proven to be advantageous with a greater potential to successful optimize the energy needs of our times. The need for renewable energy is growing rapidly as the governments around the world are adopting a clean and green energy generation strategy in order to cope up with the climate change and ensure sustainability. The imperative shift towards renewables is reinforced by the fact that the world energy requirement grows rapidly, while coal-based conventional energy sources dwindle over time.

The growth in renewable energy sector has taken a big leap with more and more governments of various countries adopting to a clean and green generation to curb the carbon emissions to control pollution and global warming. According to the International Energy Agency (iea.org), the energy produced through all means of generations amounts to 23,696 TWh (Terawatt-hour) in the year 2017 with the major power generated from coal standing at 9863.33 TWh followed by natural gas at 5882.82 TWh and nuclear power at 2636.03 TWh, respectively. Hydro power dominates the renewable energy sector with a generation of 4197.29 TWh followed by wind power at 1127 TWh. The contribution of other renewables is quite small compared to the conventional sources with the biofuel at 481.52 TWh followed by

V. K. R. Aala Kalananda · V. L. N. Komanapalli (✉)
School of Electrical Engineering, Vellore Institute of Technology, Vellore, Tamil Nadu 632014, India
e-mail: vlnarayana.k@vit.ac.in

V. K. R. Aala Kalananda
e-mail: vamsikrishna.ak@vit.ac.in

B. Vinoth Kumar et al. (eds.), *Intelligent Paradigms for Smart Grid and Renewable Energy Systems*, Algorithms for Intelligent Systems,
https://doi.org/10.1007/978-981-15-9968-2_6

solar PV at 443.55 TWh geothermal at 85.34 TWh. The rise in the contribution of renewables towards world electricity generation for the past three decades is represented in Table 1 and illustrated in Fig. 1 (data retrieved from International Energy Agency—iea.org).

It is clear from the data in Table 1 that the growth in the hydro power generation was gradual compared to wind power generation with a substantial growth after the year 2005. Solar PV power on the other hand saw its growth after the year 2010 with the development of PV panels with better efficiency. Biofuel power generation had a moderate growth through the years with geothermal having the least growth. The

Table 1 Variation of power generation through renewables from 1990 to 2017

Year	Hydro (TWh)	Wind (TWh)	Biofuel (TWh)	Solar PV (TWh)	Geothermal (TWh)
1990	2191.67	3.88	105.47	0.09	36.42
1991	2268.63	4.19	72.13	0.1	37.39
1992	2267.16	4.63	83.02	0.12	39.3
1993	2397.67	5.61	85.8	0.15	40.23
1994	2419.73	7.31	90.02	0.17	41.05
1995	2545.96	7.95	95.24	0.19	39.89
1996	2583.18	9.45	94.79	0.22	42.18
1997	2614.54	12.08	101.32	0.27	42.38
1998	2628.63	16.07	102	0.35	45.35
1999	2636.26	21.52	108.42	0.61	48.66
2000	2695.85	31.34	114.4	0.99	51.98
2001	2638.2	38.45	113.94	1.32	51.57
2002	2711.12	52.85	126.73	1.58	52.29
2003	2726.33	64.23	136.57	2.01	54.09
2004	2894.22	84.43	149.97	2.66	56.5
2005	3019.5	103.92	170.87	3.92	58.28
2006	3124.34	133.05	183.16	5.52	59.61
2007	3165.71	170.83	201.07	7.47	62.29
2008	3285.59	221.05	218.97	11.92	64.91
2009	3338.82	277.44	240.4	20.04	67.03
2010	3530.72	341.38	280.55	32.22	68.1
2011	3603.78	436.01	298.04	63.76	69.29
2012	3765.96	524.04	325.97	99.03	70.27
2013	3898.28	646.36	357.79	139.64	71.74
2014	3976.01	718.07	391.54	190.25	77.52
2015	3989.82	838.31	417.8	250.57	80.56
2016	4162.26	958.15	462.01	329.14	82.18
2017	4197.29	1127.31	481.59	443.55	85.34

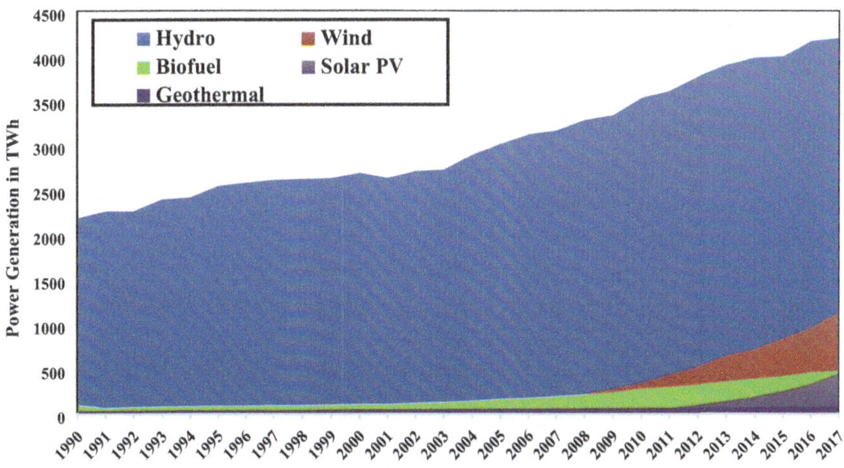

Fig. 1 Growth of power generation through renewables over the last three decades

demographic breakdown with the top 20 countries leading in the renewable energy field for the year 2018 is presented in Table 2.

The data from 2018/2017 makes it clear that the developed countries are switching to renewables with China topping the charts for every category. The *Three Gorges Dam* spanning by the Yangtze River is a major contributor to the hydro power in the country. In the USA, the wind power generation is majorly contributed by the *Alta Wind Energy Centre* in California followed by Oregon and Indiana. In India, the major contribution to the renewables is through hydro power and wind with no geothermal generation capacity. Apart from hydro power, solar and wind have seen major improvements in these countries, while geothermal has seen a little to no major improvement. The contribution of the renewables to the world's power generations is still insignificant considering the fact that most of our energy is produced through fossil fuels like coal and natural gas. The breakdown of the contribution of the renewables compared to conventional sources for the year 2017 is presented in Table 3 and illustrated in the pie charts in Figs. 2, 3 and 4 (data retrieved from International Energy Agency—iea.org).

The data from Fig. 2 makes it clear that nearly 75% of the energy is generated through the conventional sources, and it is quite essential that our dependence on fossil fuels is lowered down for a sustainable future. The renewables must be developed to act as the main stream of power generation and not just as a back to the conventional sources. In fact, the opposite, renewables as the main source with fossil fuel-based generation acting as the backup or reserve should be the motto of the developing nations. The advancements in the renewables are quite essential, and one such way to do so is by optimization and careful planning and utilization of the renewables.

The flagship report "*World Energy Outlook 2019*" released in November 2019 called for an urgent shift towards the renewables towards tacking the climate change.

Table 2 Comparison of renewable power generation of the world's top 20 countries

Country	Hydro (TWh)	Wind (TWh)	Biofuel (TWh)	Solar PV (TWh)	Geothermal (TWh)	Total (TWh)
China[a]	1189.84	295.02	79.43	130.65	0.125	1819.94
USA	315.619	277.91	58.95	87.18	18.96	758.619
Brazil[a]	370.9	42.37	52.25	0.83	0	466.35
Canada	383.48	29.65	7.12	3.84	0	424.09
India[a]	141.8	51.06	43.76	26.03	0	262.65
Germany	24.17	111.59	45.1	46.16	0.16	227.18
Russia[a]	187.13	0.14	0.08	0.55	0.43	188.33
Japan	90.67	7.63	19.009	67.6	2.44	187.349
France	70.13	28.5	5.87	10.19	0.13	114.82
Italy	50.92	17.49	16.85	22.65	6.08	113.99
UK	7.96	57.11	32.08	12.92	0	110.07
Spain	36.74	50.81	5.48	7.51	0	100.54
Turkey	59.75	19.88	2.63	7.47	6.9	96.63
Mexico	32.52	13.07	1.88	3.19	5.36	56.02
Australia	16.02	15.17	3.51	9.92	0	44.62
Indonesia[a]	18.63	0.0006	0.23	0.02	12.76	31.6406
Thailand[a]	9.52	1.1	15.38	4.54	0.0001	30.5401
South Korea	7.28	2.46	6.82	8.76	0	25.32
Iran[a]	15.05	0.3	0.02	0.08	0	15.45
Taiwan[a]	8.87	1.73	0.11	1.69	0	12.4

[a]Represents that the data is collected for the year 2017 as the data for 2018 was not available

Table 3 Contribution of non-renewables and renewables to the world's power demand

Non-renewables		Renewables	
Mode of generation	Contribution (TWh)	Mode of generation	Contribution (TWh)
Coal	9863.33	Hydro	4197.29
Natural gas	5882.82	Wind	1127.31
Nuclear	2636.03	Biofuel	481.52
Oil	841.87	Solar PV	443.55
Waste	114.04	Geothermal	85.34
Municipal wastes	74.05	Renewable waste	37.36
Others	36.02	Solar thermal	10.84
		Tidal	1.04
Total	19,448.16	Total	6384.25

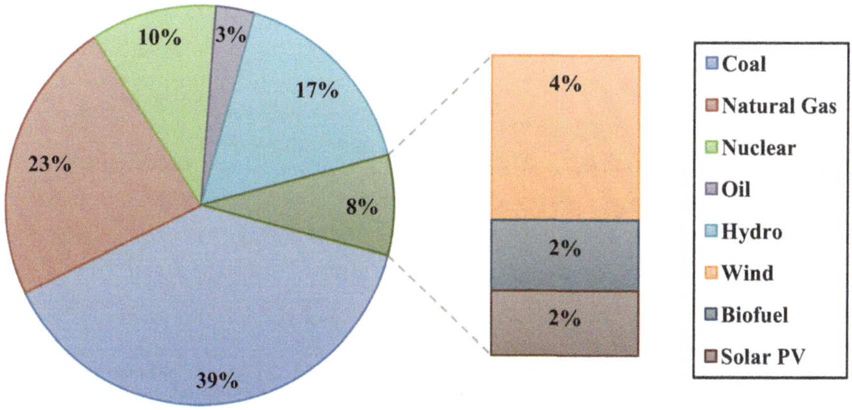

Fig. 2 Share of power generation through renewables and non-renewable means

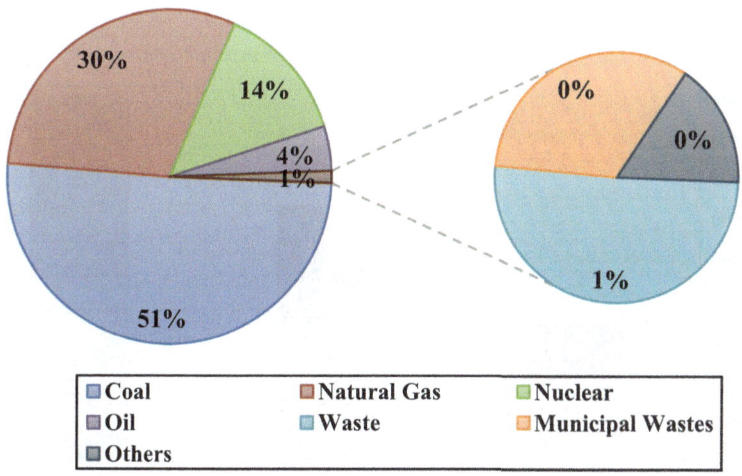

Fig. 3 Breakdown of various sources of non-renewable generation

It proposed various schemes towards the development of renewables for sustainable development. It stated that wind and solar PV are the two key areas of focus with potential to reach the goal of supplying nearly 40% of the world's energy by the year 2040. The report also focused on biogas which could replace the natural gas demand by up to 20% globally. The report emphasizes on the point that tackling all the problems and limitations in the world of renewables by the year 2030 is the pathway to slow down climate change and limit the pollution and our dependence on fossil fuels. The report predicts the growth of renewables for the next two decades which is represented in Fig. 5 (data retrieved from International Energy Agency—iea.org).

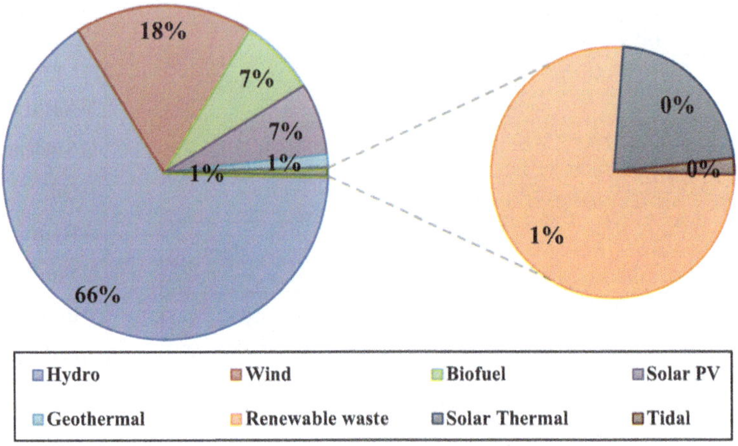

Fig. 4 Breakdown of various sources of renewable generation

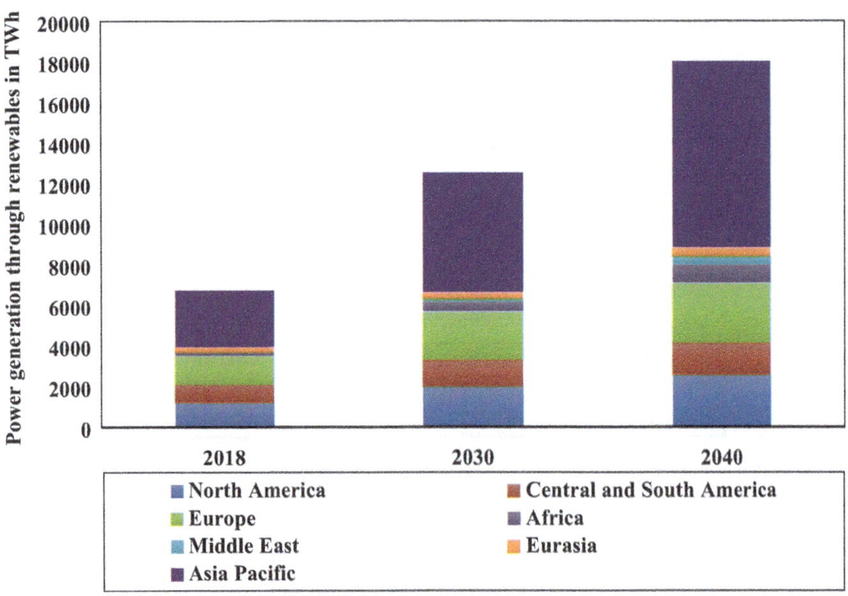

Fig. 5 Prediction of power generation through renewables for the next two decades

The operation and management of renewable energy sources is completely different in contrast to that of the conventional energy sources. The control, operation and the management of conventional sources are mostly demand driven with the variations in the load demand requiring the augmentation of generation so as to match this demand. They could be operated through the day and throughout the year with the seasonal variations and demographics of the region causing little to no restriction. For example, the coal and diesel power plants are capable of being operated at any conditions provided that the availability of the fossil fuels is fulfilled. The generation of these plants could be ramped up or lowered down to match the incoming demand with no need for any sort of storage requirement. These had the flexibility to transfer the additional energy generated to other deficit locations in a grid-connected system. The only major problem these conventional sources faced is that of economic load dispatch with the addressing of losses in the transmission and distribution. The study of literature presents a plethora of optimization techniques and tools to tackle the problem of economic load dispatch successfully. When it comes to renewables, the operation and management are quite complex with various restrictions and constraints holding back the full potential of power generation and management. Hence, the use of various meta-heuristics and optimization tools/algorithms has been deployed over time to enable optimal results while dealing with the constraints and restrictions.

This book chapter is catalogued as follows. Section 2 deals with the various problems and constraints holding back the development and integration of renewables for both grid-connected and standalone systems with their mathematical formulations. Section 3 describes the role of nature-inspired algorithms towards solving these challenges and problems holding back the potential of the renewables. Section 4 provides a descriptive and comparative analysis of the various meta-heuristics and other optimization techniques for renewable power optimization and a generalized optimization model with various objective functions and constraint functions handled by various renewables followed by the conclusion in Sect. 5.

2 Challenges in the Renewable Power Industry

There are various hurdles and restrictions that are encountered in the renewable power industry while some being minor, demographic oriented, seasonal, efficiency-related, operational, management-based, control-sided, resource limitations, grid-synchronizing issues, etc. This section provides an insight of various challenges encountered by the renewables categorizing them based on the type of production. To accomplish this, a literature survey of the most cited articles from a wide catalogue of journals across various platforms like Elsevier, Springer, IEEE, De-Gruyter, Taylor and Francis, Wiley and several other international journals are studied and analysed. The impact of the issue concerning the renewable is identified, and the conflicting factors coinciding with it are provided. A comparison table is framed based on the impact of the issue and final generalized framework comprising of all the challenges and restrictions is prepared. The important optimization problems are listed in the tables corresponding each domain and the most cited problems have been explained with their mathematical modelling in detail.

2.1 *Hydro Power*

Hydro power is undoubtedly the leading contributor to the renewables with about 66 and 17% to the total energy production. China tops the charts followed by Canada, Brazil, USA and India in the top five for hydro power generation as shown in Fig. 6. Hydro power is demographic oriented with the rivers playing the major role towards harnessing it. Apart from the demographic constraint, the seasonal factors rainfall and the environmental factors like landscape most likely influence the energy harnessing through it. Besides these factors, there are several other factors and problems that need to optimized to utilize the hydro power to its fullest extent. The major problems are shown in Table 4.

2.1.1 Optimal Dispatch of Cascade Hydropower Stations

The main objective here is to maximize or improve the annual power production of the cascade hydropower stations. This objective is achieved through the optimal selection of the operation periods with respect to the changing water levels to obtain an optimal curve of the water level. The change in outflow of the reservoir, hydraulic head and the turbine release are the important factors that could affect the power production. The decision variables are the power coefficient of the hydro power station, the average hydraulic head of the station, water outflow of the station and the dispatch period. This scheme would return the optimized inflow and outflow rates to optimize the water curve which in turn maximizes the power output. This is given by the following equation.

$$\text{OF(AE)} = max \sum_{a=1}^{HS_{\text{num}}} \sum_{b=1}^{S_d} O_{a,b} \Delta dp \quad (1)$$

and

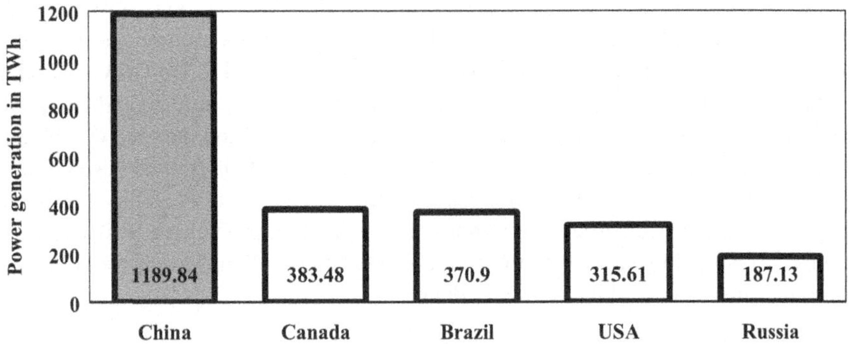

Fig. 6 Top five countries in hydro power generation

Table 4 Major problems of hydro power generation with objective function and constraints

S. No.	Problem description	Objective	Constraints
1.	Optimization of load dispatch of cascade hydropower plants [1]	Maximize annual energy production	Water level/storage volume, power output, outflow, water balance equation, hydraulic connection equation, boundary constraints
2.	Combined load frequency control with automatic voltage regulation through optimization [2]	Minimize integral square error	Minimum possible deviation for both frequency and voltage
3.	Multi-reservoir systems-optimal control and operation [3]	Maximize the efficiency of multi-reservoir system	Release volume, storage capacity of reservoir
4.	Optimal operation rules for reservoirs [4]	Minimize the total squared deviation	Water release volumes, inflow to the reservoir, storage capacity of reservoir
5.	Water resources combinatorial optimization problems [5]	Minimize the total annual costs	Total demand discharge, Pumping head
6. (a)	Short-term hydro-thermal scheduling [6]	Thermal power cost minimization and maximization of energy production	Generation limits, generating unit ramp rate limits
6. (b)	Optimization of fixed-head hydro-thermal power system [7]	Thermal power cost minimization	Transmission losses, generation limits, power balance constraints, water availability constraints, generating unit ramp rate limits
6. (c)	Optimal wind integrated hydro-thermal power generating scheduling [8]	Minimization of emissions and maximize the renewable power production	Generating limits of hydro and thermal plants, reservoir storage limits, water balance constraints, operation time period, power balance equation
6. (d)	Optimization through self-scheduling of hydro-thermal Genco in smart grids [9]	Maximization of profit margin in an electricity market	Hydro and thermal constraints, additional emission allowance

$$O_{a,b} = C_a h_{a,b} Q_{a,b} \tag{2}$$

where

OF is the objective function, AE is the yearly energy generation of all the cascaded hydro power plants in KWh (Kilowatt-hour), HS_{num} is the total number of hydropower plants or units, S_d is the size of the total dispatch period, a stands for the current time period, b expresses the current or bth station, $O_{a,b}$ is the output hydroelectric power of the station a for the period b in KW, Δdp is the dispatch period, C_a is the power coefficient of station a, $h_{a,b}$ implies the average/mean hydraulic head of the unit a for the time period b in metres, $Q_{a,b}$ represents the outflow discharge of the station a for the period b in (m^3/sec).

The key constraints to be handled are the water level or storage volume constraint, power output constraint, outflow constraint, water balance equation, hydraulic constraint and other boundary constraints. It is worth mentioning that these constraints are subject to change based on the demographics of the region. Various schemes and scenarios are possible with a combination of these constraints or all of these constraints combined. Each of the constraints is defined as follows.

The water level/storage volume constraint is given below.

$$\underline{W}_{a,b} \leq W_{a,b} \leq \overline{W}_{a,b} \tag{3}$$

where

$W_{a,b}$ corresponds to the water level of the reservoir a for the period b in metres, $\underline{W}_{a,b}$ and $\overline{W}_{a,b}$ are the lower and upper limits of the water level of the reservoir a for the time period b in metres jointly.

The power output restriction constraint for one unit/station is given below.

$$\underline{O}_{a,b} \leq O_{a,b} \leq \overline{O}_{a,b} \tag{4}$$

The power output restriction constraint for one entire system is given below.

$$\underline{O}_b \leq \sum_{a=1}^{HS_{\text{num}}} O_{a,b} \leq \overline{O}_b \tag{5}$$

where

$O_{a,b}$ is the power output of the station a for the period b in KW, $\underline{O}_{a,b}$ and $\overline{O}_{a,b}$ are the lower and upper limits of the power outputs for the station a for the period b in KW, respectively, and $\underline{O}_b$ and $\overline{O}_b$ are the lower and upper limits of the power outputs of all the power stations for the period b in KW, respectively.

The outflow constraint is as follows.

$$\underline{Q}_{a,b} \leq Q_{a,b} \leq \overline{Q}_{a,b} \tag{6}$$

where

$Q_{a,b}$ represents the discharge outflow rate of the reservoir a for the time period b in (m^3/sec), $\underline{Q}_{a,b}$ and $\overline{Q}_{a,b}$ are the lower and upper limits of the discharge outflow rate of the reservoir a for the time period b in (m^3/sec), respectively.

The water balance equation is as follows.

$$B_{a,b+1} = B_{a,b} + \left(V_{a,b} - Q_{a,b}\right).\Delta dp \tag{7}$$

where

$Q_{a,b}$ expresses the outflow rate of the reservoir a for the time period b in (m^3/sec), $B_{a,b}$ corresponds the storage volume of the reservoir a at the initial period b in m^3, $B_{a,b+1}$ is the storage volume of the reservoir a at the final period b in m^3, $V_{a,b}$ is the inflow rate to the reservoir a for the period b in (m^3/sec).

The hydraulic connection equation is as follows.

$$V_{a+1,b} = Q_{a,b} + I_{a,b} \tag{8}$$

where

$Q_{a,b}$ is the outflow of the reservoir a for the period b in (m^3/sec), $V_{a+1,b}$ is the inflow to the reservoir $a + 1$ for the period b in (m^3/sec), $I_{a,b}$ is the interval inflow rate to the reservoir $a + 1$ for the time period b in (m^3/sec).

The boundary constraint on the water levels of the reservoir is given as follows.

$$\begin{aligned} L_{a,1} &= L_{a,i} \\ L_{a,S_a+1} &= L_{a,f} \end{aligned} \tag{9}$$

where

$L_{a,i}$ denotes the initial water level of the reservoir a in metres and $L_{a,f}$ denotes the final water level of the reservoir a in metres.

2.1.2 Combined Load Frequency Control with Automatic Voltage Regulation Through Optimization

This problem requires the optimization/optimal tuning of the controller for the combined load frequency control (LFC) and automatic voltage regulation (AVR) problem for two or more interconnected hydro-electric power units with other power systems. The controller to be optimized could be a I (integral), PI (proportional-integral) controller, PID (proportional-integral-derivative) controller, PD (proportional-derivative) controller or others (fuzzy logic controller, neural network controller, neuro-fuzzy etc.) in order to dynamically regulate the frequency and voltage with respect to the changes in load. The desired control is achieved by optimizing the different gains of the controller with respect to a performance index which is to either maximized or minimized. A best example of a performance index would be the integral square error (ISE) which is to be minimized as given by the following quadratic equation.

$$\text{OF(ISE)} = \min\left(\int_{0}^{\tau} \left(\Delta F_1^2 + \Delta F_2^2 + \Delta V_1^2 + \Delta V_2^2 + \Delta \text{Power}_{\text{tie - line}(1,2)}^2\right)\tau.d\tau\right) \tag{10}$$

and

$$\Delta \text{Power}_{\text{tie - line}(1,2)} = \frac{2\pi T}{s}(\Delta F_1 - \Delta F_2) \tag{11}$$

where

OF is the objective function, *ISE* is the integral square error, τ is the total operating time, ΔF_1 is the change in frequency for the power station 1, ΔF_2 is the change in frequency for the power station 2, ΔV_1 is the variation in voltage profile for the power generating station 1, ΔV_2 is the variation in voltage profile for the power generating station 2, $\Delta \text{Power}_{\text{tie - line}(1,2)}$ is the variation in the tie-line power flow between the interconnected stations 1 and 2, T is the hydro governor constant and s represents the pole in the transfer function.

The implementation of this system eliminates the disturbances on the interconnected power systems enabling non-interrupted power flow caused by load perturbation. Constraints can be laid down on the minimum possible deviation for both frequency and voltage.

2.1.3 Multi-reservoir Systems-Optimal Control and Operation

This system requires the optimal operation of a reservoir or a group of reservoirs for the management of surface water resources. This optimization can be either done through continuous-time variables or discrete time variables. The multi-reservoir operation optimization is done usually for four-reservoir benchmark system (FRBS) and ten-reservoir benchmark system (TRBS) with the data for the optimization being the minimum and maximum water storage capacities of the reservoirs, time series (either continuous-time or discrete time) inflow rate into the reservoirs, minimum and maximum release volumes of all the reservoirs in the system, benefit coefficients of the reservoirs. The objective function is formulated as follows.

$$\text{OF}(MRS) = max\left(\sum_{x=1}^{R}\sum_{y=1}^{P} Bc_x(y) \times Vr_x(y)\right) \tag{12}$$

where

OF is the objective function, *MRS* stands for multi-reservoir system, $x = 1, 2, 3, \ldots, R$ where R is the total number of reservoirs, $y = 1, 2, 3, \ldots, P$ where

P stands for the total operating period, $Bc_x(y)$ is the benefit coefficient of reservoir x at time y, $\mathrm{Vr}_x(y)$ is the release volume of reservoir x at time y.

This optimization problem searches for the optimal release volumes of the reservoirs which are the decision variables to maximize the benefits during the operation while satisfying the constraints on storage volumes of the reservoirs and limits imposed on the release volumes from the reservoirs. These are given by the following equation.

$$\begin{aligned} \mathrm{Vr}_x^{\min} &\le \mathrm{Vr}_x(y) \le \mathrm{Vr}_x^{\max} \\ \mathrm{Sc}_x^{\min} &\le \mathrm{Sc}_x(y) \le \mathrm{Sc}_x^{\max} \\ \mathrm{Sc}_x^{P+1} &= S_x(I) \end{aligned} \tag{13}$$

where

$\mathrm{Vr}_x^{\min}$ and $\mathrm{Vr}_x^{\max}$ are the minimum and maximum values of release volumes from reservoir x, $\mathrm{Sc}_x^{\min}$ and $\mathrm{Sc}_x^{\max}$ are the minimum and maximum values of storage capacity of reservoir x, Sc_x^{P+1} is the storage volume of the reservoir x at the end of the operating time period and $S_x(I)$ stands for the initial storage volume in reservoir x.

2.1.4 Scheduling of Hydro-Thermal Units

(a) **Short-term hydro-thermal scheduling**

Shot-term scheduling refers to the hourly operation of hydropower stations for either standalone operation or in coordination with others like hydro-thermal operation. This operation is applicable to multi-reservoir cascaded operation of hydropower stations. The optimality is achieved in two aspects, i.e. the minimization of fuel costs of the thermal power plant and maximizing the usage of the hydro power plant through optimal scheduling. The objective function is to optimally dispatch of both hydro and thermal power plants which is done by minimizing the fuel cost of the thermal unit and the difference between the power generated to the actual generating potential of the hydro unit and maximizing the power output. The objective function is given by Eq. (1).

The various constraints on generation limits, unit ramp rate limits and the hydraulic network constraints are drafted as given below.

The generation limits are given as follows.

$$P_{h\,z}^{\min} \le P_{h\,zx} \le P_{h\,z}^{\max}$$

$$P_{t\,y}^{\min} \le P_{t\,yx} \le P_{t\,y}^{\max} \tag{14}$$

where

$P_{h\,z}^{\min}$ and $P_{h\,z}^{\max}$ are the minimum and maximum generation limits of the zth hydro power plant, $P_{t\,y}^{\min}$ and $P_{t\,y}^{\max}$ are the minimum and maximum generation limits of the yth thermal power plant.

The generating unit ramp rate limits are specified as follows.

$$\begin{aligned} P_{t\,yx} - P_{t\,y(x-1)} &\leq \mathrm{RU}_y \\ P_{t\,y(x-1)} - P_{t\,yx} &\leq \mathrm{RD}_y \end{aligned} \tag{15}$$

where

RU_y is the ramp up rate limit for the yth thermal unit and RD_y is the ramp down rate limit for the yth thermal unit.

The hydraulic constraints are specified as follows.

$$\begin{aligned} \mathrm{SV}_{h\,z}^{\min} &\leq \mathrm{SV}_{h\,zx} \leq \mathrm{SV}_{h\,z}^{\max} \\ \mathrm{DR}_{h\,z}^{\min} &\leq \mathrm{DR}_{h\,zx} \leq \mathrm{DR}_{h\,z}^{\max} \end{aligned} \tag{16}$$

where

$\mathrm{SV}_{h\,z}^{\min}$ and $\mathrm{SV}_{h\,z}^{\max}$ are the minimum and maximum storage volumes of the zth hydro unit, $\mathrm{DR}_{h\,z}^{\min}$ and $\mathrm{DR}_{h\,z}^{\max}$ are the minimum and maximum discharge rates of the zth hydro unit.

2.2 Wind Power

Wind power is the second largest contributor in the field of renewables with a total contribution of 18%. Although wind power is an intermittent source of energy, the combined use of storage systems and other renewables makes it feasible for load dispatch. The power output of wind farms is variable seasonally and demographically, and weather forecasting is essential to estimate or predict the variations in the power generated. Hence, several optimization techniques to overcome the various problems and limitations are formulated to maximize its potential and utilize it effectively. A comparison of the top five countries with the largest wind power generating capacities is given in Fig. 7. It is obvious that China and USA are the major producers followed by Germany and UK. The fact is to do with the availability of both onshore and offshore wind generation that put them in top. The major problems are listed out with their corresponding objective and constraints in Table 5.

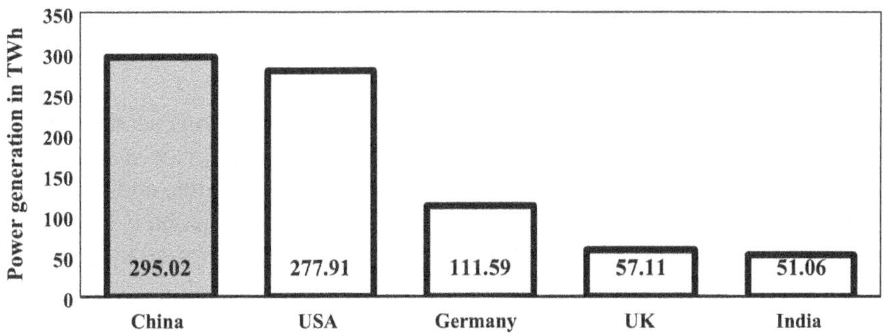

Fig. 7 Top five countries in wind power generation

Table 5 Major problems of wind power generation with objective function and constraints

S. No.	Problem description	Objective	Constraints
01.	Optimal control strategy of doubly-fed induction generation-based wind turbine [10]	Minimize the speed, torque and rotor flux loop errors	Control constraints
02.	Optimization of maximum power point tracking (MPPT) of variable-speed wind generators (VS-WG) [11]	Maximize the maximum power point tracking (MPPT) and fault ride-through (FRT)	Control and error limiting constraints
03.	Economic dispatch of wind power generators [12]	Minimize the various costs associated with wind power generation	Power generation limits, losses
04.	Optimal unit commitment with wind power and hydro power with pumped storage [13]	Maximize the benefits from both hydro and wind power systems	Pumped storage and generation limits, wind reserve inventory, Energy absorbed
05.	Optimal sizing and energy management of standalone hybrid photovoltaic/wind system [14]	Minimize the total net present cost (TNPC) of the system	Loss of expected load, loss of expected energy, storage limits, power generation limits
06.	Optimal placement of wind turbines [15]	Minimize the cost associated with the installation of turbines	Placement constraints
07.	Optimal wind farm layout problem [16]	Minimize the cost associated with the installation of turbines	Feasible zone, distance between any two wind turbines

2.2.1 Optimal Control Strategy of Doubly-Fed Induction Generation-Based Wind Turbine

In this problem, the optimal control of the doubly-fed induction generator (DFIG) for wind turbine is required. The optimization aims at improving the speed control through the appropriate tuning of controller of the wind turbine while minimizing the rotor flux and electromagnetic torque ripples. The objective function is formulated as function of the integral square error (ISE) which is to be minimized to reduce the speed, torque and rotor flux loop errors.

The objective function is formulated as follows.

$$\mathrm{OF} = min\left(\int_{o}^{i} \left(w_s e_{s_r}^2(i) + w_T e_{T_e}^2(i) + w_f e_{f_r}^2(i)\right) d(i)\right) \tag{17}$$

where

OF is the objective function,
w_s, w_T and w_f are the weighing factors for the speed, torque and flux of the DFIG,
$e_{s_r}(i)$, $e_{T_e}(i)$ and $e_{f_r}(i)$ are the errors corresponding to the rotor speed, electromagnetic torque and rotor flux, respectively.

Various performance indices like the settling time, percentage of the start-up overshoot, percentage of the step-up overshoot, torque ripples, percentage of droop at step-down are measured and analysed to determine the efficiency of the optimization process.

2.2.2 Optimization of Maximum Power Point Tracking (MPPT) of Variable-Speed Wind Generators

The maximum power point tracking (MPPT) in wind generators is a prominent aspect towards the extraction of maximum power through the wind turbines. Often, this problem is coupled with the enhancement of the fault ride-through (FRT) in the grid-coupled wind farms. MPPT is simply aimed at obtaining the optimal turbine speed to match the wind speed so as to obtain the maximum power output.

This optimization procedure requires the mathematical modelling of the wind turbine, permanent motor synchronous generator (PMSG), controller which are then formulated into a high-dimensional multi-objective cost function whose ISE (integral square error) is to be minimized. This mathematical formulation is as follows.

$$\mathrm{OF} = max(\mathrm{MPPT}) + max(\mathrm{FRT}) \tag{18}$$

and

$$\text{MPPT}(dv_1, dv_2, dv_3, \ldots\ldots dv_{76}) = \int_0^i (P_{\text{MAX}} - P_{\text{act}})^2 di + \int_0^i (Q' - Q)^2 di \tag{19}$$

$$\text{FRT}(dv_{77}, dv_{78}, dv_{79}, \ldots\ldots dv_{152}) = \int_0^i (V'_{\text{DC}} - V_{\text{DC}})^2 di + \int_0^i (V'_{\text{rms}} - V_{\text{rms}})^2 di \tag{20}$$

where

OF is the objective function which is divided into the MPPT and FRT part,
$dv_1, dv_2, dv_3, \ldots\ldots dv_{76}$ are the decision variables for the MPPT part in the considered problem,
$dv_{77}, dv_{78}, dv_{79}, \ldots\ldots dv_{152}$ are the decision variables for the FRT part in the considered problem,
i is the time index or the period of operation,
P_{MAX} is the reference maximum power that is capable of being generated,
Q', V'_{DC}, V'_{rms} are the reference values of reactive power, DC voltage, root mean square reference voltage of the grid,
P_{act}, Q, V_{DC}, V_{rms} are the actual values of real power, reactive power, DC voltage, root mean square actual voltage of the grid.

This problem from [11] has 76 dimensions for the MPPT part and 74 dimensions for the FRT part. The number of variables is subject to change depending on the factors considered and the type of the problem. All the control parameters and decision variables are defined within the fitness function, and constraints are not added. Additional constraints may or may not be defined based on the requirement of the scheme.

2.2.3 Economic Dispatch of Wind Power Generators

Although the generation of wind power requires no additional operational costs apart from the initial investment and capital expenditures, the availability of wind power and load dispatch can have a considerable effect on the optimal operation of the wind generators. This problem deals with the matching of load with the generated power with low losses by an optimal economic dispatch strategy. Additionally, the wind speed profile and prediction is made through fuzzy set theory and added to the objective function. The objective function here is a multi-objective strategy with the cost functions of all wind generators along with the penalty cost function and reserve cost function, while the role of the optimization algorithm would be to minimize all these objectives.

It is formulated as follows.

$$\text{OF} = min\{(\text{CC}) + (\text{CW}) + (\text{PF}) + (\text{RF})\} \tag{21}$$

and

$$\begin{aligned}
\mathrm{CC} &= \sum_{x}^{X} C_x(P_x) \\
\mathrm{CW} &= \sum_{y}^{Y} C_y(W_y) \\
\mathrm{PF} &= \sum_{y}^{Y} C_{P,W,y}(\mathrm{AW}_y - W_y) \\
\mathrm{RF} &= \sum_{y}^{Y} C_{R,y}(W_y - \mathrm{AW}_y)
\end{aligned} \tag{22}$$

and

$$\begin{aligned}
&C_x(P_x) = \frac{a_x}{2} P_x^2 + b_x P_x + c_x \\
&C_y(W_y) = D_y W_y \\
&C_{P,W,y}(\mathrm{AW}_y - W_y) = K_{P,W,y}(\mathrm{AW}_y - W_y) \\
&C_{R,y}(W_y - \mathrm{AW}_y) = K_{R,y}(W_y - AW_y)
\end{aligned} \tag{23}$$

and

$$(\mathrm{AW}_y - W_y) = \int_{W_y}^{W_{r,y}} (W - W_y)\mathrm{fW}(W)dW \tag{24}$$

$$(W_y - \mathrm{AW}_y) = \int_{0}^{W_y} (W_y - W)\mathrm{fW}(W)dW \tag{25}$$

where

OF is the objective function, CC is the cost function associated with the conventional/thermal generator, CW cost function of the wind power generator, PF is the penalty function and RF is the reserve function.
C_x is the cost function of the xth conventional/thermal generator, C_y is the cost function of the yth wind generator, $C_{P,W,y}$ is penalty cost function of the yth wind generator for under-utilization of the available wind power, $C_{R,y}$ is the reserve function of the yth wind generator associated with the uncertainty of the wind power. P_x is the power generated from the xth conventional/thermal generator, W_y is the power generated from the yth wind power generator, AW_y is the available power from the yth wind generator, a_x, b_x and c_x are the cost coefficients associated with the xth conventional power generation, D_y is the direct cost coefficient for

the yth wind power generator, $K_{P,W,y}$ is the penalty cost coefficient for the yth wind power generator, fW(W) is the wind energy conversion system (WECS) wind power probability distribution function (PDF) and $K_{R,y}$ is the reserve cost coefficient for the yth wind power generator.

The constraints levied on the power generation limits, losses are formulated as follows.

$$P_x^{\min} \leq P_x \leq P_x^{\max}$$
$$0 \leq W_y \leq W_{r,y} 0 \leq W_y \leq W_{r,y}$$
$$\sum_x^X P_x + \sum_y^Y W_y = L \quad (26)$$

where

$P_x^{\min}$ and $P_x^{\max}$ represent the minimum and maximum allowable power generated from the xth conventional generator, L represents the system load and losses.

2.2.4 Optimal Placement of Wind Turbines

The optimal placement of wind turbines is necessary in order to maximize the production capacity while limiting the need for any additional wind turbines. The objective function is formulated in order to decrement the cost associated with the installation of turbines while maximizing the total power extracted from the turbines. This is formulated as follows.

$$\text{OF} = min\left(\frac{cost}{(P_T)}\right) \quad (27)$$

and

$$cost = T\left(\frac{2}{3} + \frac{1}{3}e^{-0.00174T^2}\right) \quad (28)$$

where

OF is the objective function, *cost* denotes the cost function, P_T denotes the total power from T number of turbines.

The determination of the downstream velocity of the wind is an essential factor to determine the power output. This is formulated as follows.

$$v_d = v_0\left[1 - \sqrt{\sum_{t=1}^{T}\left(1 - \frac{v}{v_0}\right)^2}\right] \tag{29}$$

and

$$v = v_0\left[1 - \frac{2 \times A_i}{1 + E(d/r_1)}\right] \tag{30}$$

and

$$r_1 = r\sqrt{\frac{1 - A_i}{1 - 2A_i}} \tag{31}$$

$$E = \frac{0.5}{ln\left(\frac{h}{s}\right)} \tag{32}$$

The turbine thrust coefficient C_{th} is given by the following equation.

$$C_{\text{th}} = 4A_i(1 - A_i) \tag{33}$$

where

v_d denotes the velocity of the wind downstream for all the turbines, v_0 denotes the mean velocity of the wind downstream, v denotes the velocity of the wind downstream, A_i denotes the axial induction factor,
E denotes the entrainment constant, d is the distance from the downstream of the turbine,
r_1 is the downstream radius, r denotes the rotor radius, h is the hub height of the wind turbine, s is the surface roughness and $t = 1, 2, \ldots T$ denotes the number of wind mills.

2.3 Solar PV

Solar photovoltaic power has steadily grown with time, and it promises to be a reliable and cheap alternative to the conventional power sources. Solar power is naturally available for free of cost requires theoretically no operational cost. Solar is becoming more affordable with several industries, organizations working to improve its penetration of usage and enhance its efficiency. Although solar is available to be harvested throughout the year and is available worldwide, the storage and dispatch is a major problem. These problems need to be optimized based on the requirement such that the usage is maximum with little to no losses. This section provides an overview of various problems available for optimization in the field of solar PV with various constraints and limitations. The various objectives and the constraints are

Table 6 Major problems of solar power generation with objective function and constraints

S. No.	Problem description	Objective	Constraints
1.	MPPT for PV with reduced steady-state oscillation [17]	Enhancing maximum power point tracking (MPPT)	Oscillation limits
2.	Optimal control strategies of PV-diesel systems [18]	Minimization of operational costs of the diesel unit	State of charge (SOC) of batteries, minimum time of operation
3.	Optimization approach for energy management in PV [19]	Minimization of total cost of the system	Energy balance equation, storage capacity, power ratings
4.	Optimization of an off-grid hybrid PV–wind diesel system with different battery technologies [20]	Minimization of the total system cost	Power loss, battery costs, power loss in the converters
5.	Optimization-based solar PV parameter estimation [21]	Minimize the root mean square error (RMSE)	No constraints are defined
6.	Short-term PV power forecasting [22]	Training optimization of support vector machine (SVM)	Power generation limits

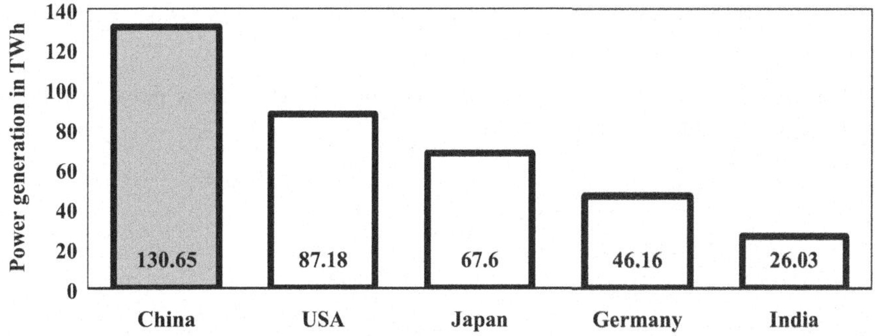

Fig. 8 Top five countries in solar power generation

shown in Table 6. The top five countries of solar power generation are detailed in the Fig. 8.

2.3.1 MPPT for PV with Reduced Steady-State Oscillation

The maximum power point tracking (MPPT) is one major area of research and has seen great contributions through various techniques and methods. One key contribution to this area is the implementation of the optimization techniques to further improve its efficiency. One major advantage of optimization algorithms is to do with the fact that these optimization algorithms require no prior training and large training data. The algorithms are sufficiently fast enough to track down the MPP with very low oscillations. The formulations of fitness function are such that the algorithm always tries to update the MPP on a continuous basis based on the sensor data.

The generalized fitness function is defined as follows.

$$\mathrm{OF} = \wp\left(D_i^n\right) > \wp\left(D_i^{n-1}\right) \tag{34}$$

where

OF is the objective function, $\wp$ is the power output of the solar array, D_i represents the duty ratio of the ith cycle, n *and* $n-1$ represents the current and previous iterations.

The duty ratio is decided with respect to the power output as follows.

$$D_{\text{new}} = \begin{cases} D_{\text{old}} + \theta \; if \; \wp > \wp_{\text{old}} \\ D_{\text{old}} - \theta \; if \; \wp < \wp_{\text{old}} \end{cases} \tag{35}$$

where

D_{new} and D_{old} are the new and old duty cycles of the MPPT controller, $\wp$ and $\wp_{\text{old}}$ are the current and previous power outputs from the solar array, θ is the fixed step size.

The approximate linear correlation between the change in array power and duty cycle is given by the following relation.

$$D_{\text{new}} = D_{\text{old}} - \frac{1}{S}\left(\wp_{\text{old,MPP}} - \wp_{\text{MPP}}\right) \tag{36}$$

and

$$S = \frac{\wp_{\text{MPP}}}{\Delta D} \tag{37}$$

where

S is the slope of the linear segment, $\wp_{\text{old,MPP}}$ is the power output at the previous maximum power point, $\wp_{\text{MPP}}$ is the power output at the current maximum power point, ΔD is the difference between the old and new duty cycles.

Constraints may or may not be featured in this type of problems. The problem studied here has a constraint to limit the oscillations given as follows.

The slope of the line segment is modified according to the changes in the power output to limit the oscillations given equation.

$$S_{\text{new}} = \begin{cases} S_{\text{old}}\ if\ \Delta\wp > 0 \\ \frac{S_{\text{old}}}{2}\ if\ \Delta\wp < 0 \end{cases} \tag{38}$$

and

$$\Delta\wp = \wp - \wp_{\text{old}} \tag{39}$$

where

S_{new} and S_{old} represents the new and old slopes of the linear segment varied as per the duty ratio, $\Delta\wp$ represents the change in output power.

2.3.2 Optimization Approach for Energy Management in PV

This problem deals with the optimal energy management of PV-wind interconnected system with focus towards the optimal load dispatch with an effective storage system. The load shifting, state of charge (SOC) of the batteries, maximum storage capacity of the batteries and the excess energy management are vital aspects in this optimization process.

The generalized fitness function is formulated as follows.

$$\text{OF} = min\left(\sum_{c}^{C} W_F \times \mathbb{C}_c\right) \tag{40}$$

and

$$\mathbb{C}_c = (\alpha_{\text{PV}}.C_{\text{PV}}) + (\alpha_{\text{wi}}.C_{\text{wi}}) + (Ms.C_{\text{Ms}}) + (Pr.C_{\text{Pr}}) \tag{41}$$

where

OF is the objective function, $c = 1, 2, \ldots, C$ represents the total number of clusters, W_F is the weighing factor based on the number of days within the cluster, $\mathbb{C}_c$ is the cost function,
$C_{\text{PV}}, C_{\text{wi}}, C_{\text{MS}}$ and C_{Pr} are the costs corresponding to the installation of PV panels, wind generators, storage systems and power rating of the system, respectively,
$\alpha_{\text{PV}}, \alpha_{\text{wi}}$, Ms and Pr are the installed capacitates of solar PV, wind, storage systems and power rating of the system, respectively.

The various constraints are formulated as follows.

The energy balance equation is modified as the following inequality constraint.

$$\sum_{i=1}^{I} P_{\text{PV},i} + \sum_{i=1}^{I} P_{w,i} \geq \sum_{i=1}^{I} L_i \tag{42}$$

where

$i = 1, 2 \dots I$ represents the time index, $P_{\mathrm{PV},i}$ represents the power generated form the PV panels for the ith period, $P_{w,i}$ represents the power extracted from the wind turbines for the ith period, L_i is the instantaneous load for the ith period.

The limits on the minimum and maximum storage capacity are formulated as the constraint given below.

$$SC_{\min} \leq SC_i \leq SC_{\max} \tag{43}$$

where

$SC_{\min}$ and $SC_{\max}$ denote the minimum and maximum storage capacities of the batteries and SC_i represents the storage capacity at the ith time period.

The maximum allowable power rating of the storage system is given by the following equation.

$$max(P_r) = \left|P_{\mathrm{PV},i} + P_{w,i} - L_i + L_{S,i}\right| \tag{44}$$

where

P_r is the power rating of the storage system, $P_{\mathrm{PV},i}$ represents the power generated form the PV panels for the ith period, $P_{w,i}$ represents the power extracted from the wind turbines for the ith period, L_i is the instantaneous load for the ith period, $L_{S,i}$ is the shifted load for the ith period for economic load dispatch.

2.3.3 Optimization-Based Solar PV Parameter Estimation

This problem is intended towards the development of a highly accurate PV simulation technique in order to maximize the overall performance prior to the instalment. The extraction of the optimal parameters is highly essential for the further analysis and development of simulation models.

The objective function is to minimize the root mean square error (RMSE) which corresponds to the error functions of the single- and double-diode models. It is formulated as follows.

$$\mathrm{OF} = \min(\mathrm{RMSE}) \tag{45}$$

and

$$\mathrm{RMSE} = \sqrt{\frac{1}{R}\sum_{r}^{R} E(V_o, I_o, X)^2} \tag{46}$$

where

OF is the objective function, RMSE is the root mean square error, $r = 1, 2, \ldots R$ denotes the number of readings, E stands for the error function, V_o, I_o are the output voltage and current, X is the vector of the solution.

The error function of the double-diode model is as follows.

$$
\begin{aligned}
E(V_o, I_o, X) = I_{\mathrm{PV}} - I_{\mathrm{RS1}}\left[exp\left(\frac{q(V_o + I_o R_s)}{n_1 kT}\right) - 1\right] \\
- I_{\mathrm{RS2}}\left[exp\left(\frac{q(V_o + I_o R_s)}{n_2 kT}\right) - 1\right] \\
- \frac{(V_o + I_o R_s)}{R_{\mathrm{Sh}}} - I_o
\end{aligned}
\tag{47}
$$

where

I_{PV} is the photogenerated current, I_{RS1} is the reverse saturation current of the first diode, I_{RS2} is the reverse saturation current of the second diode q is the electric charge, $n_1 and\ n_2$ denotes the ideality factors for the diodes 1 and 2, k is the Boltzmann's constant, T is the absolute solar cell temperature in Kelvin, R_s and R_{sh} are the series and shunt resistance of the solar cell.

The error function of the single-diode model is as follows.

$$
\begin{aligned}
E(V_o, I_o, X) = I_{\mathrm{PV}} - I_{\mathrm{RS}}\left[exp\left(\frac{q(V_o + I_o R_s)}{nkT}\right) - 1\right] \\
- \frac{(V_o + I_o R_s)}{R_{\mathrm{Sh}}} - I_o
\end{aligned}
\tag{48}
$$

where

I_{PV} is the photogenerated current, I_{RS} is the reverse saturation current, q is the electric charge, n denotes the ideality factor, k is the Boltzmann's constant, T is the absolute solar cell temperature in Kelvin, R_s and R_{sh} are the series and shunt resistance of the solar cell.

2.4 *Geothermal Power*

Geothermal power falls behind the power generated through solar by a greater margin and the reason being the non-availability of geothermal power extraction points (mostly tectonic plate boundaries) in of the other countries. Hot springs are the key points of geothermal power extraction. USA leads the other countries followed by Indonesia. Geothermal power plants theoretically require no operational costs but on the other hand require higher capital investment costs. Even though the power output from the geothermal plants is smaller in comparison to the other renewables, it could

Table 7 Major problems of geothermal power generation with objective function and constraints

S. No.	Problem description	Objective	Constraints
1.	Optimization of double-flash geothermal power plant [23]	Maximize the specific work output of the geothermal plant	Silica saturation index (SSI), quality of steam, temperature of the flash vessel
2.	Thermodynamic analysis and optimization of a geothermal Kalina cycle system [24]	Maximize the efficiency of the power generation	Power generation limits
3.	Performance analysis and optimization for maximum exergy efficiency of a geothermal power plant [25]	Minimization of total costs of the system	Exergy balance equation

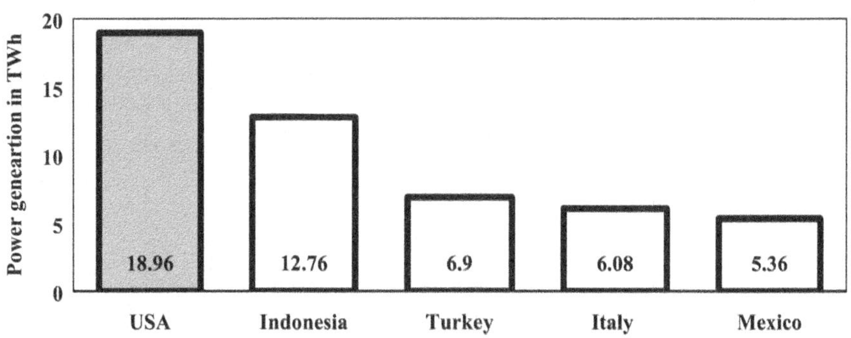

Fig. 9 Top five countries in geothermal power generation

act as a backup medium as the heat is available to generate the power throughout the day without a lot of seasonal variations. The various problems with their optimization objective and the possible constraints are organized in Table 7. The power generated from the geothermal means for various countries is shown in Fig. 9.

2.4.1 Optimization of Double-Flash Geothermal Power Plant

A double-flash geothermal power plant extracts the heat from the geothermal brine with a high-pressure and a low-pressure turbine to generate power. The working is similar to that of a conventional thermal power plant except that the need for a heating mechanism is replaced by the geyser or the hot spring. The optimization is done based on the maximization of power output from both the turbines by optimally varying the temperature of the separator between the two turbines and the operating temperature of the flash vessel.

The generalized objective function is as follows.

$$\text{OF} = max(\psi) \tag{49}$$

and

$$\psi = \frac{P_{\text{TOTAL}}}{\mathcal{G}} \tag{50}$$

and

$$P_{\text{TOTAL}} = P_{\text{LP}} + P_{\text{HP}} \tag{51}$$

where

OF is the objective function, ψ indicates the specific work output of the geothermal plant in Joule per Kilogram (J/Kg), P_{TOTAL} is the total power output from the geothermal power plant, $\mathcal{G}$ is the geothermal brine entering the plant in Kilograms (Kg), P_{LP} is the power output from the low-pressure turbine, P_{HP} is the power output from the high-pressure turbine.

The silica saturation index (SSI) limits the quality of steam to reduce the turbine damage; temperature limits of the temperature of the flash vessel are the possible constraints in this optimization to be handled.

These constraints are modelled as follows.

$$\text{SSI} = \frac{\text{ASC}}{\text{ASC}_e} \tag{52}$$

where

SSI is the silica saturation index, ASC is the amorphous silica concentration, ASC_e is the equilibrium amorphous silica concentration.

The SSI is limited to a value of 1.2 to reduce the damage to the turbines due to silica deposition on the blades.

$$\text{SSI} \leq 1.2 \tag{53}$$

Additionally, the quality of steam (q_s) has to be over 0.85 to prevent excessive damage to the turbine.

$$q_s \geq 0.85 \tag{54}$$

The temperature of the flash vessel (T_F) should not drop below 102 °C so as to reinject the brine solution without pumping it back.

$$T_F \geq 102\,^{\circ}\text{C} \tag{55}$$

2.4.2 Thermodynamic Analysis and Optimization of a Geothermal Kalina Cycle System

In a Kalina cycle system, there is an indirect heat transfer from the hot water obtained from the geothermal source to ammonia-water which is the working liquid. The heated ammonia-water drives the turbines to generate power. The major advantage with ammonia-water compared to steam is the larger net power compared to the regular stem technique. A detailed information regarding the mass flow rates and specific enthalpies are presented in [24].

The objective function is framed considering the energy and exergy of the ammonia-water with the efficiency of the power generation to be maximized. This is modelled as follows.

$$\mathrm{OF} = max\left(\mathrm{TH}_{\eta}\right) + max\left(\mathrm{EX}_{\eta}\right) \tag{56}$$

and

$$\mathrm{TH}_{\eta} = \frac{P_{\mathrm{Tu}} - P_{\mathrm{Pu}}}{\mathrm{HT}_{ev}} \tag{57}$$

and

$$\mathrm{EX}_{\eta} = 1 - \frac{\mathrm{EX}_{\mathrm{Td}}}{\mathrm{EX}_{i} - \mathrm{EX}_{d}} \tag{58}$$

where

OF is the objective function, TH_{η} is the thermal efficiency of the geothermal plant, P_{Tu} is the power output from the ammonia-water turbine, P_{Pu} is the power consumed for the pumping of the working solution, $\mathrm{HT}_{\mathrm{ev}}$ denotes the heat-transfer capacity of the evaporator.

EX_{η} denotes the exergy efficiency, $\mathrm{EX}_{\mathrm{Td}}$ is the total exergy destroyed, EX_{i} represents the exergy input, and EX_{d} is the exergy destroyed at one particular instant.

2.4.3 Performance Analysis and Optimization for Maximum Exergy Efficiency of a Geothermal Power Plant

The performance of a geothermal plant is improved through optimization of the Rankine cycle and its system components. The optimization is done through the annualized cost method considering the costs of input and output flows, energy pricing, product demand, etc. with the main objective is to minimize the costs.

A generalized formulation of the fitness function is described as follows.

$$\mathrm{OF} = \min(\mathrm{OF}_1) + \min(\mathrm{OF}_2)$$

$$\begin{aligned} \mathrm{OF}_1 &= \sum M_{\mathrm{IP}} - \sum M_{\mathrm{OP}} \\ \mathrm{OF}_2 &= H_{\mathrm{net}}^{\mathrm{IP}} - w_{\mathrm{net}}^{\mathrm{OP}} + \sum \xi_{\mathrm{IP}} M_{\mathrm{IP}} - \sum \xi_{\mathrm{OP}} M_{\mathrm{OP}} \end{aligned} \quad (59)$$

where

OF is the objective function,
M_{IP} and M_{IP} are the input and output mass flow rates, $H_{\mathrm{net}}^{\mathrm{IP}}$ is the net heat input rate, $w_{\mathrm{net}}^{\mathrm{OP}}$ is the net work output rate,
ξ_{IP} and ξ_{IP} are the input and output specific enthalpies.

The various constraints are defined as follows.
The exergy balance equation is formulated as follows.

$$\sum \left[1 - \frac{T_{\mathrm{Ref}}}{T_i} \right] Q_i - w + \sum M_{\mathrm{IP}} \zeta_{\mathrm{IN}} - \sum M_{\mathrm{OP}} \zeta_{\mathrm{OP}} - \mathrm{EX}_D = 0 \quad (60)$$

where

T_{Ref} is the reference temperature in $\left({}^{\circ}\mathrm{C}\ or\ K\right)$, T_{Ref} is the temperature at the ith location in $\left({}^{\circ}\mathrm{C}\ or\ K\right)$,
Q_i is the heat-transfer rate in KW, w is the work rate in KW, ζ_{IN} is the input specific flow exergy in KJ/Kg, ζ_{OP} is the output specific flow exergy in KJ/Kg, EX_D is the rate of exergy destruction in KJ/sec or KW.

2.5 *Biofuel*

Biofuel is the fuel generated from biomass through either fermentation or transesterification. Biodiesel and bioethanol are the two major types of biofuels. Biofuel earned a great reputation serving as an alternative to the fossil fuels with no emissions and renewability. Biofuels can be used in diesel power plants with various optimizations added such that the power generation is maximized and the operational costs are minimized. The various problems with their optimization objective and the possible constraints are organized in Table 8. The power generated from the geothermal means for various countries is shown in Fig. 10.

2.5.1 Modelling of Cetane Number of Biodiesels from Fatty Acid Methyl Ester (FAME) Information

The estimation of the cetane number (CN) is essential for the description of the ignition characteristics of the biofuel and estimate the motor power quality. Support vector machine (SVM) approach coupled with the optimization algorithms has been extensively used to analyse the various fuel samples in order to obtain an accurate estimation of the CN.

Table 8 Major problems of biofuel power generation with objective function and constraints

S. No.	Problem description	Objective	Constraints
01.	Modelling of cetane number of biodiesel from fatty acid methyl ester (FAME) information [26]	Minimization of modelling errors	No constraints are defined
02.	Optimal configurations of combined power plants for small-scale cogeneration from biomass [27]	Maximization of thermal and electrical efficiency	Power generation limits

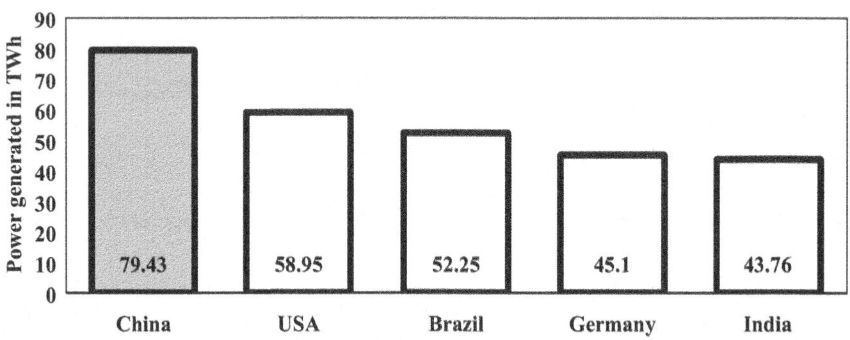

Fig. 10 Top five countries in biofuel power generation

In the example considered for the study, least square SVM or LSSVM approach is used in combination with optimization algorithms to determine the mean squared errors (MSE), the coefficient of determination (R^2), mean relative errors (MRE) and the standard deviations (STD) for an accurate estimate of the CN.

The generalized objective function is formulated as follows.

The relation between the CN and the input variables (fatty acid methyl esters profile) is denoted as follows.

$$f(C) = W^T\varphi + B \tag{61}$$

and

$$\mathrm{OF}_1 = min_{W,B,E}(F(W, E)) \tag{62}$$

and

$$F(W, E) = \frac{1}{2}\|W\|^2 + \frac{1}{2}\mathrm{M}\sum_{i=1}^{N} E_i^2 \tag{63}$$

where

$f(C)$ is the function relating to the connection of the output variables, W is the weight vector and W^T represents its transpose, φ is the mapping function, E_i denotes the error of the error of the ith sample data, M denotes a margin parameter, $i = 1, 2, \ldots, N$ denotes the number of samples.

The regression function $F(r)$ of the LVSSM is formulated as follows.

$$F(r) = \sum_{r=1}^{N} \sigma_r R(C, C_r) + b \tag{64}$$

and

$$R(C, C_r) = exp\left(-\frac{\|C_r - C\|^2}{\phi^2}\right) \tag{65}$$

where

$R(C, C_r)$ is the radial basis function (RBF), σ_r is a weighing factor, ϕ^2 is the squared bandwidth to be optimized by the optimization algorithm.

The mean square error (MSE) which is to be minimized forms the second objective function is formulated as follows.

$$\mathrm{OF}_2 = min(\mathrm{MSE}) \tag{66}$$

$$\mathrm{MSE} = \frac{\sum_{d=1}^{D}[\mathrm{CN}_P - \mathrm{CN}_E]}{D} \tag{67}$$

where

$d = 1, 2, \ldots, D$ are the number of data points used, CN_P and CN_P are the predicted and experimented values of the cetane numbers, respectively.

2.5.2 Optimal Configurations of Combined Power Plants for Small-Scale Cogeneration from Biomass

The small-scale biomass reserves can be effectively optimized through small combined cycles to generate heat which can be used to generate electricity for remote rural areas. The various thermal components utilized in this scheduling are to be optimally selected to increase the thermal efficiency and maximize the electrical power generation.

There are five design parameters to be optimized. These include the mass flow rate of air, compressor pressure ratio, efficiency of the heat exchanger, temperature and the pressure of the steam at the inlet of the steam turbine.

The objective function is formulated as a two-part function to maximize the thermal and electrical power efficiencies. It is given as follows.

$$\begin{aligned} \mathrm{OF} &= max(\mathrm{OF}_1 + \mathrm{OF}_2) \\ \mathrm{OF}_1 &= \mathrm{Th}_\eta = \frac{P_{\mathrm{Th}}}{M_f L_{\mathrm{HV}}} = \frac{M_s\left(se_o^S - se_i^S\right)}{M_f L_{\mathrm{HV}}} \\ \mathrm{OF}_2 &= \mathrm{El}_\eta = \frac{P_{\mathrm{El}}}{M_f L_{\mathrm{HV}}} = \frac{P_{El}^{top} - P_{El}^{bot}}{M_f L_{HV}} \end{aligned} \tag{68}$$

and

$$\begin{aligned} P_{\mathrm{El}}^{\mathrm{top}} &= M_a\left[\mathrm{Tur}_\eta^{\mathrm{Mech}}\left(\mathrm{se}_o^G - \mathrm{se}_i^G\right) - \frac{\left(\mathrm{se}_o^C - \mathrm{se}_i^C\right)}{\mathrm{Com}_\eta^{\mathrm{Mech}}}\right] \\ P_{\mathrm{El}}^{\mathrm{bot}} &= \mathrm{Exp}_\eta^{\mathrm{Mech}} M_s\left(\mathrm{se}_o^S - \mathrm{se}_i^S\right) \end{aligned} \tag{69}$$

where

OF is the objective function, Th_η stands for thermal efficiency, El_η stands for electrical efficiency,
P_{Th} stands for the thermal power in KW, P_{El} stands for electrical power output in KW, M_f is the fuel mass flow rate in Kg/sec, M_s is the steam mass flow rate in Kg/sec, M_a is the air mass flow rate in Kg/sec, L_{HV} stands for the lower heating value, $P_{\mathrm{El}}^{\mathrm{top}}$ and $P_{\mathrm{El}}^{\mathrm{bot}}$ stand for the electrical power output in KW for the topping and bottoming cycles, respectively,
se_o^S and se_i^S are the specific enthalpies in J/Kg at the outlet and the inlet of the steam expander,
se_o^C and se_i^C are the specific enthalpies in J/Kg at the outlet and the inlet of the compressor,
se_o^G and se_i^G are the specific enthalpies in J/Kg at the outlet and the inlet of the gas turbine,
$\mathrm{Tur}_\eta^{\mathrm{Mech}}$, $\mathrm{Com}_\eta^{\mathrm{Mech}}$ and $\mathrm{Exp}_\eta^{\mathrm{Mech}}$ are the mechanical efficiencies of the turbine, compressor and the expander, respectively.

2.6 Integrated and Hybrid Optimization of Renewables

The integration of the renewables with the grid helps ease off the burden on the fossil fuel-based generation, thereby reducing the carbon emissions. This helps for a coordinated load dispatch with equal distribution of load among the renewable and non-renewable sources. With a good energy storage infrastructure, the maintenance and operation cost of the conventional sources can be cut down significantly while overcoming the intermittencies associated with the renewables. Hybrid renewable systems are the scope for a sustainable future as their applications to the industrial and domestic sector could reduce the stress caused by the peak demand on the non-renewables sources of power. Hybrid PV-wind systems have become affordable and available for domestic users leading to lowered demand for the conventionally generated power. The surplus power from these hybrid systems can be injected to the grid or stored for later usage. While the power system has been made flexible with the integration of renewables and hybrid systems, it is important to address the various limitations and constraints and optimize them in order to reap the full-scale benefits of such systems. The various problems with their optimization objective and the possible constraints are organized in Table 9.

2.6.1 Optimal Scheduling of Renewable Generation in a Microgrid Under Load Uncertainty

Microgrids are the go-to way for effective small-scale localized renewable distributed generation with a low-cost operation and management. These microgrids provide efficient power generation at a lower cost while improving the stability of the larger regional power grid. The scheduling and dispatch of the load is a major challenge with constraints like intermittency and storage limitations. Hence, it is essential to optimize the costs incurred on power generation, operation and management with respect to the varying demand and the uncertainty associated with the renewables.

The generalized objective function is formulated as follows.

$$\begin{aligned} \mathrm{OF} = min(\sum_{r=1}^{R}[(\rho_r \times P_r^2) + (\sigma_r \times P_r) + \tau_r] \\ + K \times [P_W + P_{\mathrm{PV}} + P_{\mathrm{CHP}} - P_{\mathrm{LOAD}} - P_{\mathrm{LOSS}}]) \end{aligned} \quad (70)$$

where

OF is the objective function,$r = 1, 2, \ldots, R$ represents the total number of renewable plants,

ρ_r, σ_r and τ_r are the coefficients of the renewable energy systems,

P_r is the power generated through the renewable energy system, K is a penalty factor,

P_W, P_{PV} and P_{CHP} represent the power generated through wind, solar and combined heat and power energy systems, P_{LOAD} is the load demand and P_{LOSS} is the power loss in the system.

The load uncertainty is modelled through a probability distribution function (PDF) as follows.

$$Load(\mathrm{PDF}) = \frac{1}{\sqrt{2\pi} \times \mathrm{SD}} exp\left(-\frac{(P_{\mathrm{LOAD}} - M)^2}{2 \times \mathrm{SD}^2}\right) \quad (71)$$

where

Table 9 Major problems of integrated and hybrid renewable power generation with objective function and constraints

S. No.	Problem description	Objective	Constraints
01.	Optimal scheduling of renewable generation in a microgrid under load uncertainty [28]	Minimization of operational costs of the renewable sources	Power generation limits, power balance equation
02.	Optimization of a biomass-integrated renewable energy microgrid [29]	Minimization of cost of energy (COE)	Power balance equation, power generation and storage capacities
03.	Optimal energy management of a renewable-based isolated microgrid with pumped storage unit [30]	Minimization of start-up, operational and installation costs	Reservoir volume, water charge and discharge
04.	Optimized economic load dispatch and frequency regulation in reserve capacity integrated renewable smart grids [31]	Optimal economic load dispatch	Bus voltages, apparent power flow
05.	Optimal integrated renewable energy model with battery storage for remote areas [32]	Minimization of cost of energy (COE)	Total net present cost
06.	Modelling and optimization of an off-grid hybrid renewable energy system [33]	Minimization of total net present cost of the system (TNPC)	Battery storage, bounds constraints and power reliability
07.	Modelling and optimization of a hybrid PV-wind turbine pumped hydro storage energy system for mini-grid [34]	Minimization of total life cycle costs	Number of components
08.	Optimal power management of a small autonomous hybrid power system [35]	Minimization of total cost of equipment	Design constraints, state of charge (SOC) and number of the equipment

SD stands for standard deviation and M for the mean value of the power demand.

The constraints are formulated as follows.

The power generation limits on each of the renewable energy system are formulated by the following constraint.

$$P_r^{\mathrm{MIN}} \leq P_r \leq P_r^{\mathrm{MAX}} \tag{72}$$

where

P_r^{MIN} and P_r^{MAX} are the lower and upper limits on the permissible power to be generated by the renewable energy systems, respectively.

The power balance equation is formulated as follows.

$$\sum_{r=1}^{R} P_r = P_{\mathrm{LOAD}} \tag{73}$$

$r = 1, 2, \ldots, R$ represents the total number of renewable plants, P_r is the total power generated through renewables.

2.6.2 Optimization of a Biomass-Integrated Renewable Energy Microgrid

Biofuels are the next alternative to the fossil fuels with their low emissions and renewability. Integration of a biofuel-based power generation into a renewable microgrid eliminates the need for diesel-based power generation reducing the fuel costs and pollution. The biomass plant serves as backup power source for the peak load hours and when the other renewables fail.

The objective function is to minimize the cost of energy while considering the capital and operational costs and the energy flows between the system components.

$$\mathrm{OF(COE)} = \min\left(\frac{C_{\mathrm{NES}}}{\sum_{i=1}^{T} E_i^d + H_i^d}\right) \tag{74}$$

where

OF is the objective function, COE stands for the cost of energy, C_{NES} is the net energy supply cost,
$i = 1, 2, \ldots, T$ represents the time period of operation, E_i^d and H_i^d are the electrical and heating load demands for the ith period.

The various constraints are formulated as follows.
The power balance equation is formulated as follows.

$$\sum_{r=1}^{R} P_r = E_i^d + H_i^d \tag{75}$$

$r = 1, 2, \ldots, R$ represents the total number of renewable plants, P_r is the total power generated through renewables.

The limiting constraints on the power generation and storage capacities of the batteries are as follows.

$$P_{\mathrm{MIN}} \leq P_t \leq P_{\mathrm{MAX}}$$

$$B_{\mathrm{MIN}} \leq B_t \leq B_{\mathrm{MAX}} \tag{76}$$

where

P_{MIN} and P_{MIN} represent the maximum and minimum allowable power generation,
B_{MIN} and B_{MIN} represent the maximum and minimum storage capacities of the batteries,

2.6.3 Optimized Economic Load Dispatch and Frequency Regulation in Reserve Capacity Integrated Renewable Smart Grids

The coordination among various storages and reserves is essential to decrease the operational costs and optimally schedule load with the varying load demand. The dynamic variations in the load result in the frequency variations of the grid leading to the loss of synchronism. Hence, the output power is to be optimally scheduled with the varying load profile while achieving economic load dispatch from various renewable sources.

The objective function for the frequency regulation reserve capacity (FRRC) and economic load dispatch (ED) is formulated as follows.

$$\mathrm{OF} = min\left(\sum_{p=1}^{P}\left[\sum_{i}\mathrm{Ec}_{p,i}^{\mathrm{GEN}} + \sum_{i}\mathrm{Rc}_{p,i}^{\mathrm{GEN}} + \sum_{j}\mathrm{Rc}_{p,i}^{\mathrm{DR}} + \sum_{j}\mathrm{Rc}_{p,i}^{\mathrm{ESS}}\right]\right) \tag{77}$$

where

OF is the objective function, p indicates time period, i indicates the number of generators (GEN), j indicates the number of demand response (DR) aggregator or energy storage systems (ESS),
$\mathrm{Ec}_{p,i}^{\mathrm{GEN}}$ is the cost of energy generation through generator for the ith unit for the pth period,
$\mathrm{Rc}_{p,i}^{\mathrm{GEN}}$ is the reserve cost of energy generation through generator for the ith unit for the pth period,
$\mathrm{Rc}_{p,i}^{\mathrm{DR}}$ is the reserve cost of energy generation through demand response (DR) aggregator for the ith unit for the pth period,
$\mathrm{Rc}_{p,i}^{\mathrm{GEN}}$ is the reserve cost of energy generation through energy storage systems (ESS) for the ith unit for the pth period.

The constraints for the security-constrained unit commitment (SCUC) are as follows.

The constraint on the bus voltage is given as follows.

$$V_i^{\mathrm{MIN}} \leq V_i \leq V_i^{\mathrm{MAX}} \tag{78}$$

where

V_i^{MIN} and V_i^{MIN} are the minimum and maximum magnitudes of the bus voltages for the ith bus.

The maximum allowable apparent power flow through the transmission lines is expressed as follows.

$$|S_i \leq |S_i^{\text{MAX}}|| \tag{79}$$

where

S_i is the apparent power of the ith transmission line,
S_i^{MAX} is the maximum allowable apparent power of the ith transmission line.

2.6.4 Optimal Integrated Renewable Energy Model with Battery Storage for Remote Areas

The optimization of an integrated system for remote areas circles around the minimization of the total net present cost (TNPC) and the cost of energy (COE) in order to optimally schedule the power such that the demand is met with the existing resources at any point of time. Storage through battery systems is a viable option for rural remote areas because of their low maintenance and operational costs.

The objective function is formulated as follows.

$$\text{OF(COE)} = min\left[\frac{C_{\text{TNP}} \times C_{\text{RF}}}{\sum_{t=1}^{8760} Ae_{\text{gen}}(t)}\right] \tag{80}$$

and

$$C_{\text{RF}} = \frac{I(1+I)^{\text{LS}}}{(1+I)^{\text{LS}} - 1} \tag{81}$$

where

OF is the objective function, $t = 1, 2, \ldots, 0.8760$ indicates time period of operation for one year, C_{TNP} is the total net present cost, C_{RF} is the capital recovery factor, Ae_{gen} is the annual energy generated, I is the annual rate of interest, LS represents the life span of the equipment.

The constraints relating to the capital costs, operational and maintenance costs are formulated as follows.

The equality constraint concerning the total net present cost (C_{TNP}) *to be satisfied is formulated as follows.*

$$C_{\text{TNP}} = C_{\text{NPC}} + C_{\text{NPR}} + C_{\text{NP}\ O\&M} + C_{\text{NP FC}} \tag{82}$$

where

C_{NPC} is the net present capital cost, C_{NPR} is the net present replacement cost, $C_{\text{NP }O\&M}$ is the net present operation and management cost, $C_{\text{NP FC}}$ is the net present fuel cost.

2.6.5 Optimal Sizing and Power Management of a Small Autonomous Hybrid Power System

The small autonomous hybrid power systems or SAHPS in short are designed for the electrification of the rural communities. The optimal sizing of the variable renewable components is essential to predict the load demand and optimally dispatch the load.

The objective is to minimize the cost with respect to the design parameters with respect to various constraints. The objective function is as follows.

$$\text{OF} = min\left(C_{\text{Wind}} + C_{\text{Solar}} + C_{\text{Hydro}} + C_{\text{Batteries}} + C_{\text{DG}} + C_{\text{Others}}\right) \tag{83}$$

and

$$C_{\text{Wind}} = \sum_{w=1}^{W}\left[U_w P_w \frac{I(1+I)^{LF}}{(1+I)^{\text{LF}}-1} + {}^{M}_{O} P_w + \mathfrak{r}(P_w)\right] \tag{84}$$

$$C_{\text{Solar}} = \sum_{s=1}^{S}\left[U_s P_s \frac{I(1+I)^{LF}}{(1+I)^{\text{LF}}-1} + {}^{M}_{O} P_s + \mathfrak{r}(P_s)\right] \tag{85}$$

$$C_{\text{Hydro}} = \sum_{h=1}^{H}\left[U_h P_h \frac{I(1+I)^{\text{LF}}}{(1+I)^{\text{LF}}-1} + {}^{M}_{O} P_h + \mathfrak{r}(P_h)\right] \tag{86}$$

$$C_{\text{Batteries}} = \sum_{b=1}^{B}\left[U_b P_b \frac{I(1+I)^{\text{LF}}}{(1+I)^{\text{LF}}-1} + {}^{M}_{O} P_b + \mathfrak{r}(P_b)\right] \tag{87}$$

$$C_{\text{DG}} = \sum_{d=1}^{D}\left[U_d P_d \frac{I(1+I)^{\text{LF}}}{(1+I)^{\text{LF}}-1} + {}^{M}_{O} P_d + \mathfrak{r}(P_d)\right] \tag{88}$$

$$C_{\text{Others}} = C_{\text{co}} \times EE_{\text{NS}} \tag{89}$$

where

OF is the objective function,

C_{Wind}, C_{Solar}, C_{Hydro}, $C_{\text{Batteries}}$, C_{DG} and C_{Others} are the total costs of wind turbines, PV panels, pico-hydro plants, batteries, diesel generators and other equipment, respectively,

W, S, H, B and D are the total number of wind turbines, PV panels, pico-hydro plants, batteries and diesel generators, respectively,

U_w, U_s, U_h, U_b and U_d are the unit cost of power of wind, PV, pico-hydro, batteries and diesel generators in Rs/KW, respectively,
I is the rate of interest, LF represents the life span of the project,
P_w, P_s, P_h, P_b and P_d are the power capacities of wind, PV, pico-hydro, batteries and diesel generators,
$O^M P_w$, $O^M P_s$, $O^M P_h$, $O^M b$ and $O^M P_d$ are the maintenance and operating costs associated with wind turbines, PV panels, pico-hydro plants, batteries and diesel generators, respectively,
$\mathfrak{r}(P_w)$, $\mathfrak{r}(P_s)$, $\mathfrak{r}(P_h)$, $\mathfrak{r}(P_b)$ and $\mathfrak{r}(P_d)$ are the replacement costs associated with wind turbines, PV panels, pico-hydro plants, batteries and diesel generators, respectively,
C_{co} is the compensation coefficient and EE_{NS} is the expected energy not served.

The various constraints are formulated as follows.
The design constraints are formulated as follows.

$$P_{Wind}(i) + P_{Solar}(i) + P_{Hydro}(i) + P_{Battery}(i) + P_{DG}(i) \geq (1 - \mathfrak{r})P_{LOAD}(i) \quad (90)$$

$$P_{\text{Wind}}(i) + P_{\text{Solar}}(i) + P_{\text{Hydro}}(i) + P_{\text{Battery}}(i) + P_{\text{DG}}(i) - P_{\text{dump}}(i) \leq P_{\text{LOAD}}(i) \quad (91)$$

where

i indicates the time period, $\mathfrak{r}$is the maximum allowable ratio of the total unmet power to the total load demand for every time period
$P_{\text{Wind}}(i)$, $P_{\text{Solar}}(i)$, $P_{\text{Hydro}}(i)$, $P_{\text{Battery}}(i)$, $P_{\text{DG}}(i)$ and $P_{\text{dump}}(i)$ are the wind, solar, hydro, battery, diesel generator and dumped power, respectively, for the ith period,
$P_{\text{LOAD}}(i)$ is the load demand for the ith period.

The constraints on the state of charge (SOC) of the batteries are expressed as follows.

$$\begin{aligned} &\text{SOC}_{\min} \leq \text{SOC} \leq \text{SOCmax} \\ &0 \leq S_{\text{cap}} \leq S_{\text{cap,max}} \\ &H_c \leq HI_{\text{cap, max}} \end{aligned} \quad (92)$$

where

$\text{SOC}_{\min}$ and $\text{SOC}_{\min}$ are the allowable minimum and maximum state of charge (SOC) of the batteries,
S_{cap} and $S_{\text{cap,max}}$ are the storage capacity of the battery and maximum storage capacity of the battery, respectively,
H_c and $\text{HI}_{\text{cap,max}}$ are the hourly charge and discharge power and hourly inverter capacity, respectively.

The constraints on the number of the equipment are as follows.

$$\begin{aligned} 0 &\le N_{\text{PV}} \le N_{\text{PV,max}} \\ 0 &\le N_W \le N_{W,\max} \\ 0 &\le N_B \le N_{B,\max} \\ 0 &\le N_H \le N_{H,\max} \end{aligned} \quad (93)$$

where

$N_{\text{PV,max}}$, $N_{\text{PV,max}}$, $N_{\text{PV,max}}$ and $N_{\text{PV,max}}$ are the maximum allowable numbers of PV panels, wind turbines, batteries and pico-hydro plants, respectively.

3 Role of Nature-Inspired Algorithms for Renewable Power Optimization

Nature-inspired meta-heuristic optimization algorithms are a class of optimization algorithms which are based on the hunting or foraging techniques of various creatures found in the nature. These hunting techniques are unique to each creature, and they adopt a wide variety of hunting techniques to search for their prey. The hunting techniques modelled into the algorithms correspond to a group of animals/birds (known as swarm behaviour) rather than individual techniques. The group hunting/collective foraging is preferred so as to allow the algorithm to effectively explore the search landscape thereby avoiding local solutions and achieve the best possible global optimal solution. Some of the famous paradigms in this class are the particle swarm optimization (PSO), genetic algorithm (GA), artificial bee colony (ABC) algorithm, whale optimization algorithm (WOA), squirrel search algorithm (SSA), etc.

3.1 Nature-Inspired Algorithms—A Brief Overview

With many novel nature-inspired algorithms being proposed every year, the identification and deployment of the best from the vast collection is the key to the success for any kind of optimization task/problem at hand. While the nature-inspired algorithms are all known to perform well with an adequate degree of success, it is worth noting that not every algorithm perfectly optimizes each and every optimization problem. This is often referred to as "*No Free Lunch theorem*" which states that the degree of optimization for various algorithms varies from one problem to the other with one algorithm outperforming the other. Hence, variations or modifications are made to the algorithm such that the given optimization task at hand is optimized perfectly with little to no scope for further optimization. The performance of meta-heuristics with respect to randomized schemes and specialized schemes is represented in Fig. 11 (data retrieved from SCOPUS® database).

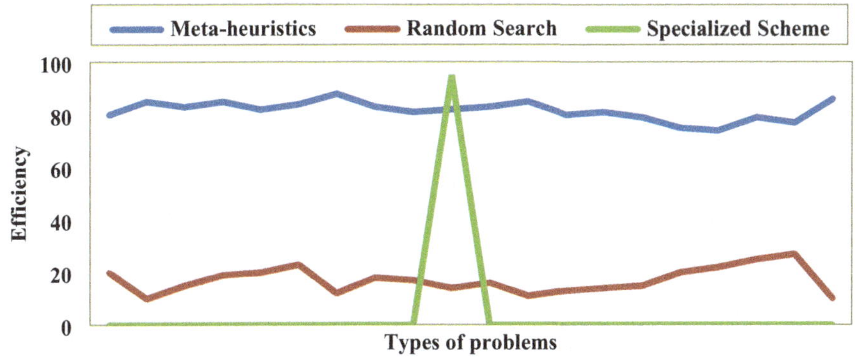

Fig. 11 Performance disparity with respect to different problems for meta-heuristics, random and specialized schemes

It is obvious that meta-heuristics have an upper hand in generating optimal solutions for almost any kind of problem even when limited information is known about the problem or when computational power is limited. The meta-heuristics/nature-inspired algorithms (NIAs) perform adequately, while the random searching schemes may not provide that kind of efficiency. Specialized schemes designed for a particular application-oriented task might have the upper hand but come higher computational cost and require complete information pertaining to the problem. The growth of the applications of meta-heuristics over the last three decades has been phenomenal with over 500,000 research articles and books indexed in the Scopus database alone. The application areas include mathematics, material science, business accounting and management, computer science, energy, social sciences, environmental sciences, physics and astronomy, decision science, various engineering fields, etc. The year-wise growth of the application of meta-heuristics over the last three decades is represented in the Fig. 12 (data retrieved from SCOPUS® database).

Nature-inspired/bio-inspired optimization algorithms have a fair share of applications with over 100,000 article and books in the Scopus database. The leaders within this class are the particle swarm optimization (PSO) and genetic algorithm (GA) with 6.1 and 3.1% of their total applications related to energy. The other notable NIAs are the artificial bee colony optimization (ABC), firefly algorithm (FA), bat algorithm (BA), whale optimization algorithm (WOA), etc. The hybridization of existing nature-inspired algorithms with other evolutionary/metaphor-based algorithms is prevalent among various fields intended towards the further improvement of the optimization. Improved, chaotic, enhanced, advanced versions of the existing NIAs have surface over time. The graph depicting the number of novel NIA's over the last two decades is presented in Fig. 13.

The application of NIAs to the renewable energy optimization has grown rapidly with the NIAs capable of handling multiple objectives with higher dimensions and numerous constraints while still attaining better optimality compared to other techniques with a low computational cost. A survey into the various bibliographical

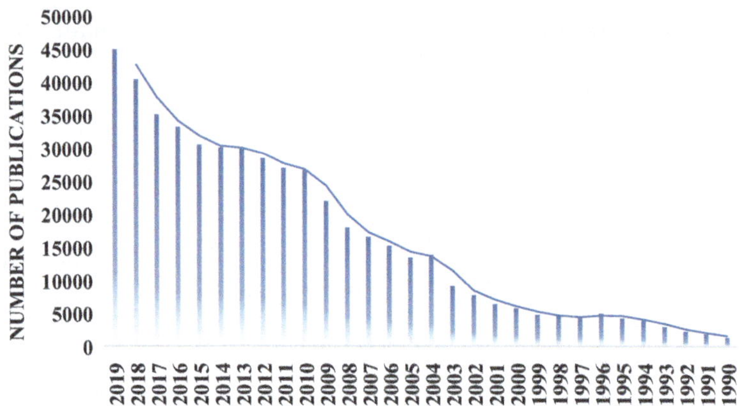

Fig. 12 Annual growth in the application of meta-heuristics for various domains

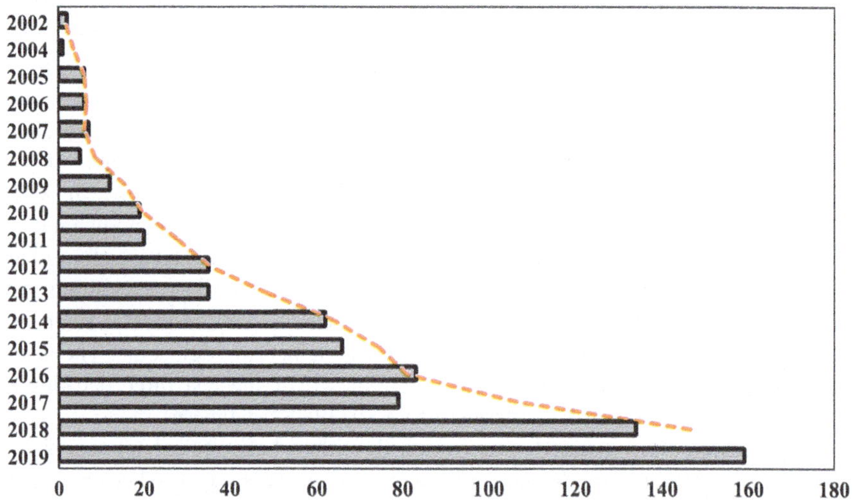

Fig. 13 Growth chart for the number of novel NIA's published over the last two decades

databases reveals that more than 25,000 documents have contributed to the renewable energy optimization since 2001. A comparative analysis of the various sectors with optimization applied for the time period 2019–2005 is presented in the Fig. 14 and Table 10 (data retrieved from SCOPUS® database).

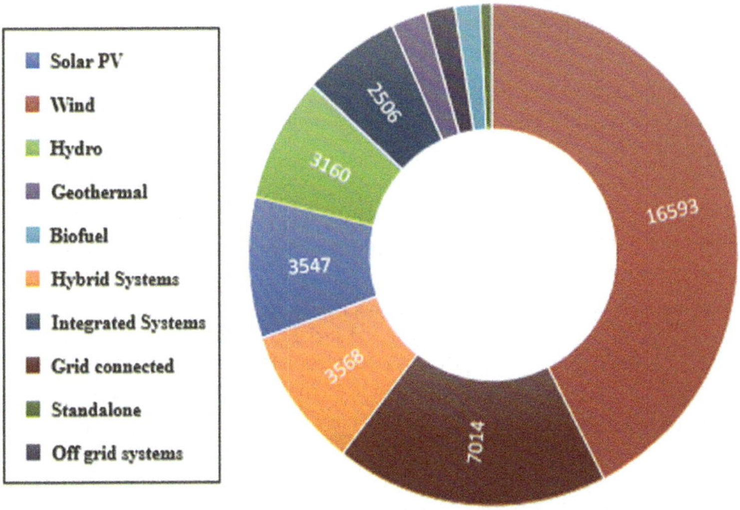

Fig. 14 Share of various domains of renewable power in optimization applications

Table 10 Contribution of various domains of renewable power in optimization applications

Area of optimization	Number of publications
Solar PV	3547
Wind	16,593
Hydro	3160
Geothermal	896
Biofuel	660
Hybrid systems	3568
Integrated systems	2506
Grid connected	7014
Standalone	316
Off-grid systems	751

3.2 *Legacy of Nature-Inspired Algorithms—PSO and GA*

Nature-inspired algorithms either swarm based or evolutionary based are in a way or other linked to PSO and GA. These two paradigms have set the standards and continued to reign over the domain of optimization. The application of PSO and GA in the domain of energy is numerous with several variants of these paradigms being developed to maximize their efficiency for a particular application. PSO is swarm inspired, while GA is an evolutionary algorithm. However, both of these algorithms have been in one way or the other are inspired by the forces of the nature. PSO draws its inspiration form the swarm behaviour of birds/fish referred to as particles, while GA implements the mutation and crossover of the genetic material (DNA) to

Table 11 Brief description of PSO and GA

Details	PSO	GA
Year of development	1995	First introduced in 1960, developed in 1989
Developers	James Kennedy and Russell Eberhart	First introduced by John Holland in 1960. Developed by David E. Goldberg
Inspiration	Social behaviour of bird/fish	Charles Darwin's theory of evolution
Tuning parameters	Inertia weight (w), cognitive parameter ($c1$) and social parameter ($c2$), population size (Np), number of iterations (T)	Population size (Np), Number of iterations (T), crossover probability, mutation probability
Number of fitness evaluations	$Np \times T$	$Np \times T$ *(Approximately)*

optimize the solution. A brief description of GA and PSO is provided in Tables 11. The domain of renewable energy has seen a surge in the applications of GA and PSO to various sub-domains. PSO has over 12,000 documents published in the renewable energy domain with all its variants considered in the Scopus database. GA has over 25,000 documents relating to all the sub-domains in the energy domain. The year-wise growth of PSO and GA to the energy domain is represented in Figs. 15 and 16 (data retrieved from SCOPUS® database), and a comparison of the two paradigms in the renewable energy domain and its deployment to the optimization of various modes of renewable power generation is provided in Figs. 17 and 18, respectively. The literature survey regarding the notable applications of PSO, GA and their hybrid variants to the domain of renewable energy is presented in Table 12. The publications related to the applications of the hybridized/combinatorial variants of PSO-GA over the last two decades are provided in Fig. 19. The publication works concerning the deployment of some of the other popular nature-inspired meta-heuristic optimization algorithms other than PSO and GA is shown in Fig. 20.

Fig. 15 Application of GA to power systems

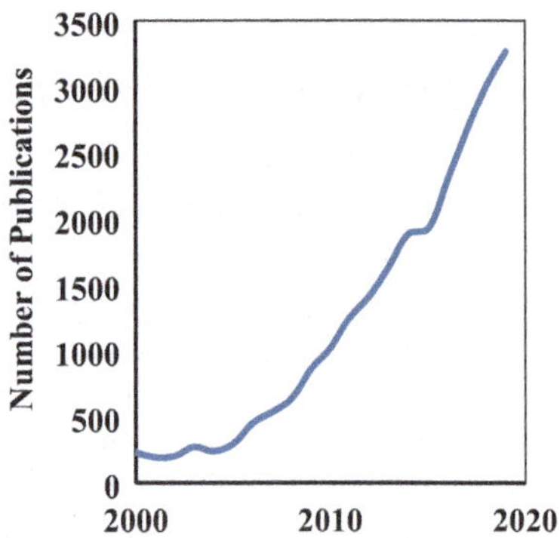

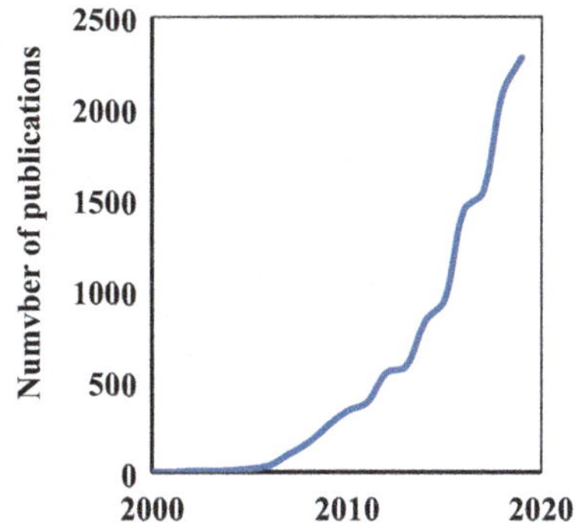

Fig. 16 Application of PSO to power systems

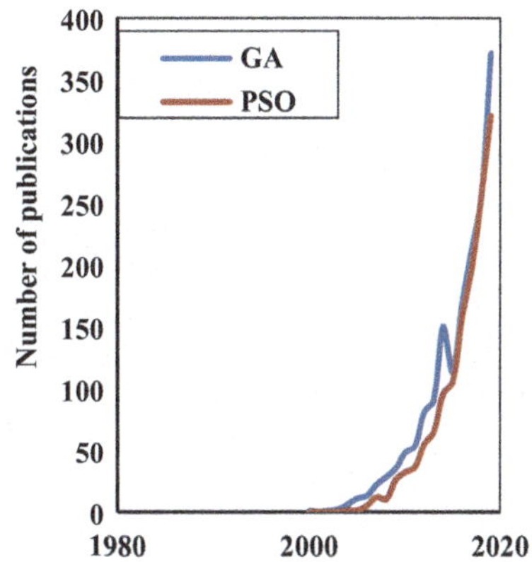

Fig. 17 PSO versus GA in renewable systems

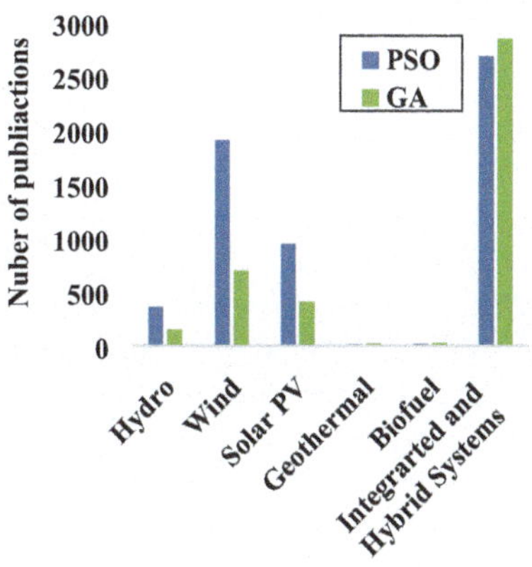

Fig. 18 PSO versus GA for various domains

Table 12 Literature survey of the applications of PSO and GA along with their hybrid variants (data retrieved from SCOPUS® database)

S. No.	Algorithm deployed	Year of publication	Algorithm description	Problem description
Hydro power optimization				
01.	HPSO-GA	2013	Hybrid PSO-GA combines the strengths of both PSO and GA to attain a good balance between the natural selection and information exchange for an effective exploration of the search space	An optimal model to efficiently utilize the water resources of the Tao river basin in China with multi-reservoir model to minimize water shortages and maximize hydro power generation
02.	Refined PSO	2004	The refined PSO organizes particles into several clusters and the idea of cataclysm is introduced to improve the quality of exploration and convergence speed	A short-term hydro-thermal power scheduling with optimal load distribution is achieved
03.	PSO	2017	A comparative analysis between GA, PSO and DE (differential evolution) is performed	Hydro-generation scheduling is optimized considering the nonlinearities associated with the water discharge and net hydraulic head
04.	PSO	2012	PSO algorithm with dynamic range on the velocities of the particles on each dimension to enhance global exploration is proposed	The optimization of hydro power plant design considering various cost factors towards the maximization of the benefit cost ratio for the economic viability of the plant is proposed

(continued)

Table 12 (continued)

S. No.	Algorithm deployed	Year of publication	Algorithm description	Problem description
05.	PSO	2014	A solving algorithm based on PSO to solve the mixed integer nonlinear programming (MINLP) model	The optimal reactive power compensation for a small hydro power plant with different operating conditions is investigated for an IEEE 33 system
06.	Improved PSO	2015	An improved PSO with linearly decrement of the inertial weight coefficient to avoid local entrapment and slow convergence is developed	The maximization of daily hydro power production in coordination with the outflows from the reservoir and optimal scheduling period is developed
07.	GA-NLP	2019	A hybrid GA and nonlinear programming (NLP) model is developed to obtain the initial feasible solution for NLP which is further developed to achieve the global optimum	The Nagarjuna Sagar reservoir in India is considered with the intent to maximize the hydro power generation and irrigation potential while considering various constraints and restrictions
08.	Refined GA	2007	Refined GA is developed to tune the rule sets and the membership functions of the fuzzy control through approximate optimization method	The area control error formulated for load frequency control is minimized through optimal gain setting of the PI controller

(continued)

3.3 *Role of Other Nature Inspired Algorithms in Renewable Energy Optimization*

Besides PSO and GA, there are a lot of other NIA's whose contribution to the optimization of renewable power is enormous. Each NIA is unique with different foraging

Table 12 (continued)

S. No.	Algorithm deployed	Year of publication	Algorithm description	Problem description
09.	Binary coded GA	1999	Binary coded GA with the crossover and mutation operations defined for the strings of binary variables decoded to decimal system is deployed	The economic load dispatch with optimal unit commitment of hydro-thermal power system comprising of various operational constraints is investigated
10.	Real coded GA	2013	Real coded GA with mutation, crossover and selection operation for decimal numeric system with convergence check system is deployed	The optimal operation of hydro stations to maximize the profit in the electricity markets based on an hourly basis is investigated with various case studies in Portugal
Wind power optimization				
01.	PSO	2010	The PSO algorithm with the weighting factor adjusted automatically between 0.4 and 0.9 is formulated	The optimal setting of rotor speed and tip-speed ratio to maximize the annual power extracted from the wind for fixed-speed operation and variable-speed operation is investigated

(continued)

techniques with their implementation to the diversification and intensification to explore and exploit the search space. The growth of NIA's in the renewable sector has sky rocketed in the recent time with the last decade having seen a surge in the number of publications. The promising nature for higher-dimensional problems coupled with the flexibility of improving the algorithm for reliable and optimal performance, while handling both equality and inequality constraints with a low computational requirement has led to the researchers and scientist to prioritized for NIA's compared to other computational techniques.

Table 12 (continued)

S. No.	Algorithm deployed	Year of publication	Algorithm description	Problem description
02.	Chaos PSO	2015	Chaos-based PSO to initialize the random population and assign the inertia weight of the velocity component is discussed	The economic dispatch for a standard IEEE 30-bus system with 6-generators and 41-transmission lines to minimize the total operating costs is considered
03.	PSO	2012	An efficient PSO to handle constraints with two new strategies namely, first-come-first-removed and worst-first removal is designed	The multi-objective optimization of maximization of energy output while minimizing the cost of the turbines and wind farm area is investigated
04.	PSO	2016	The basic PSO algorithm without any modifications or changes is utilized	The optimal sliding mode control of squirrel cage induction generator of the wind turbine is achieved through the minimization of the mean square error (MSE) of the desired and actual speed is implemented
05.	ANFIS-GA	2018	A hybrid GA and adaptive neuro-fuzzy inference system (ANFIS) is proposed with GA implemented to identify the optimal parameters for the ANFIS system	The performance of a variable-speed wind energy conversion system (WECS) through the design of an optimal controller is proposed

(continued)

Table 12 (continued)

S. No.	Algorithm deployed	Year of publication	Algorithm description	Problem description
06.	PSO-GA	2019	A hybrid parallel PSO-GA with both the algorithms synchronized to each other working parallel to attain the global optimal solution	The minimization of the total cost of a wind energy storage system based on the energy flow management between the battery and supercapacitor is proposed
07.	GA-Wavelet framework	2013	A wavelet-GA-MLP (multilayer perceptron) based on the classical time series analysis is formulated to improve the efficiency and quality of solutions	Wind speed prediction is investigated through a combination of wavelet theories and evolutionary algorithm for accurate and forecasting performance
08.	El-PSO	2015	A combination of Elman neural networks and PSO is designed to improve the prediction accuracy	An accurate prediction of wind power output of wind farm considering the intermittencies associated with the wind speed
Wind power optimization				
01.	PSO	2010	The PSO algorithm with the weighting factor adjusted automatically between 0.4 and 0.9 is formulated	The optimal setting of rotor speed and tip-speed ratio to maximize the annual power extracted from the wind for fixed-speed operation and variable-speed operation is investigated

(continued)

Table 12 (continued)

S. No.	Algorithm deployed	Year of publication	Algorithm description	Problem description
02.	Chaos PSO	2015	Chaos-based PSO to initialize the random population and assign the inertia weight of the velocity component is discussed	The economic dispatch for a standard IEEE 30-bus system with 6-generators and 41-transmission lines to minimize the total operating costs is considered
03.	PSO	2012	An efficient PSO to handle constraints with two new strategies namely, first-come-first-removed and worst-first removal is designed	The multi-objective optimization of maximization of energy output while minimizing the cost of the turbines and wind farm area is investigated
04.	PSO	2016	The basic PSO algorithm without any modifications or changes is utilized	The optimal sliding mode control of squirrel cage induction generator of the wind turbine is achieved through the minimization of the mean square error (MSE) of the desired and actual speed is implemented
05.	ANFIS-GA	2018	A hybrid GA and adaptive neuro-fuzzy inference system (ANFIS) is proposed with GA implemented to identify the optimal parameters for the ANFIS system	The performance of a variable-speed wind energy conversion system (WECS) through the design of an optimal controller is proposed

(continued)

Table 12 (continued)

S. No.	Algorithm deployed	Year of publication	Algorithm description	Problem description
06.	PSO-GA	2019	A hybrid parallel PSO-GA with both the algorithms synchronized to each other working parallel to attain the global optimal solution	The minimization of the total cost of a wind energy storage system based on the energy flow management between the battery and supercapacitor is proposed
07.	GA-Wavelet framework	2013	A wavelet-GA-MLP (multilayer perceptron) based on the classical time series analysis is formulated to improve the efficiency and quality of solutions	Wind speed prediction is investigated through a combination of wavelet theories and evolutionary algorithm for accurate and forecasting performance
08.	El-PSO	2015	A combination of Elman neural networks and PSO is designed to improve the prediction accuracy	An accurate prediction of wind power output of wind farm considering the intermittencies associated with the wind speed
Solar PV power optimization				
01.	PSO and GA	2019	A comparative study with the training of artificial neural networks (ANN) using PSO and GA	The maximization of energy output from PV panels under various shading conditions with prediction through ANN is implemented

(continued)

Table 12 (continued)

S. No.	Algorithm deployed	Year of publication	Algorithm description	Problem description
02.	PSO	2019	Standard PSO without any modifications	The minimization of grid congestion for the high-demand areas through the manipulation of surplus PV power
03.	IPSO	2019	PSO is coupled with support vector regression (SVR) with PSO optimizing the weights and biases of SVR is proposed	The tracking of the maximum power point (MPPT) with respect to temperature, irradiance, wind speed and relative humidity is implemented
04.	PSO	2009	Standard PSO with no modifications	The solar cell parameters form the current–voltage (IV) characteristics extracted with good accuracy
05.	GA	2010	Real coded GA with selection, reproduction/pairing and mutation operations is utilized	The identification of various electrical parameters of PV module used to extract the maximum power working points is implemented
06.	GA and pattern search (PS)	2016	GA and PS are paired to search for the optimal combination of the solution with in a short iterative count	The optimization of the PV plant location with respect to the seasonal and diurnal cycles for load profiling and generation dispatch is implemented

(continued)

Table 12 (continued)

S. No.	Algorithm deployed	Year of publication	Algorithm description	Problem description
07.	GA	2014	GA is used as an optimization tool without any modifications	The MPPT under partial shading conditions for the total cross tied (TCT) connected modules in a PV array is considered for optimization
08.	HOGA	2005	Hybrid optimization by GA (HOGA) where the best solutions (descendants) have the highest probability of reproducing is presented	The cost minimization of a PV-diesel plant with respect to various cost constraints and storage limitations is addressed
Geothermal power optimization				
01.	PSO	2020	Standard PSO with no modifications	The thermodynamic performance analysis and optimization of geothermal power station based on organic Rankine cycles is implemented
02.	PSO	2010	Standard PSO with no modifications	The optimal configuration of elements of a geothermal system to enhance the power extraction is proposed
03.	MO-PSO	2015	Multi-objective PSO (MO-PSO) to optimize multiple objective function while handling various constraints is utilized	The optimal utilization of a superheater with a tradeoff between specific work output and specific heat exchanger area is realized

(continued)

Table 12 (continued)

S. No.	Algorithm deployed	Year of publication	Algorithm description	Problem description
04.	GA	2014	Real coded GA with tournament selection strategy is proposed	The optimization of the simulation-based study of a double-flash geothermal unit is carried out
05.	GA	2006	GA following the strategy of evaluate-selection-crossover-mutation is chosen	The optimization of low-enthalpy geothermal unit with cost minimization under non-uniform water temperature distribution is studied
Biofuel power optimization				
01.	GA	2008	Penalty function-based genetic algorithm to handle various inequality constraints is deployed	The optimal circuit parameter estimation of a biohydrogen power generating system with respect to the varying operating temperature and current density is implemented
02.	Multi-objective NSGA	2019	Multi-objective non-dominant sorting genetic algorithm II with pareto-optimal solution is utilized	The optimal gas exhaust recirculation system correlating to the engine performance of a biodiesel system is proposed
Integrated and hybrid renewable systems				

(continued)

Second to PSO and GA, the mantle of renewable energy optimization is dominated by the artificial bee colony (ABC) algorithm followed by the whale optimization algorithm (WOA) and grey wolf optimization algorithm (GWO). To further improve the performance of these algorithms and attain a perfect balance between the conflicting exploration and exploitation, several hybrid combinations with other meta-heuristics, improved version through Levy flights and chaotic maps, opposition-based learning

Table 12 (continued)

S. No.	Algorithm deployed	Year of publication	Algorithm description	Problem description
01.	PSO	2014	Standard PSO with parameter settings based on multiple monitoring conditions is used	The optimal design of a hybrid renewable energy system (HRES) minimizing the total cost of the system including various generators and storage devices is implemented
02.	PSO-GWO	2018	PSO is hybridized with grey wolf optimizer (GWO) with PSO for initial population updation followed by an additional population updation through GWO is implemented	The optimization of different components of a grid-connected hybrid renewable energy system PV and with turbines backed up by a diesel generating system is proposed
03.	PSO	2019	Standard PSO with no modifications	The optimal component selection and their contribution to the overall cost and power reliability for an off-grid-integrated renewable system is implemented
04.	PSO	2015	Inertia-weighted PSO with the tuning parameters set through empirical analysis	The optimal tuning of PI controller for the control of the grid-connected inverter for a hybrid PV-wind system is carried out

(continued)

Table 12 (continued)

S. No.	Algorithm deployed	Year of publication	Algorithm description	Problem description
05.	GA	2016	Real coded GA with no modifications is used	The optimization total net present cost (TNPC) of an integrated system based on the irradiance, wind sweep and state of charge (SOC) of batteries is implemented
06.	GA	2016	Real coded GA with no modifications is used	The minimization of the cost generation costs and cost of energy (COE) with the optimal sizing of integrated systems is proposed
07.	GA	2020	Real coded GA with variation operator is utilized	The optimal configuration of a hybrid AC/DC system with solid-state transformers with minimum operating loss is developed
08.	GA/BPSO	2017	Real coded GA and binary PSO are implemented for a comparative analysis	The optimal configuration to minimize the home electric bill with the integration of renewable systems for home energy management is proposed

techniques have surfaced over the past two decades. A comparative description of the popular NIA's for renewable power optimization is presented in Tables 13 and 14.

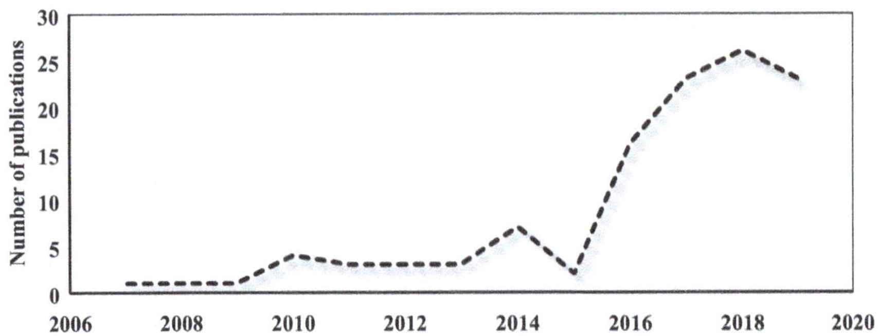

Fig. 19 Hybrid/combinatorial PSO-GA algorithms applied to renewable power optimization

Fig. 20 Contribution of other NIA's to renewable power optimization

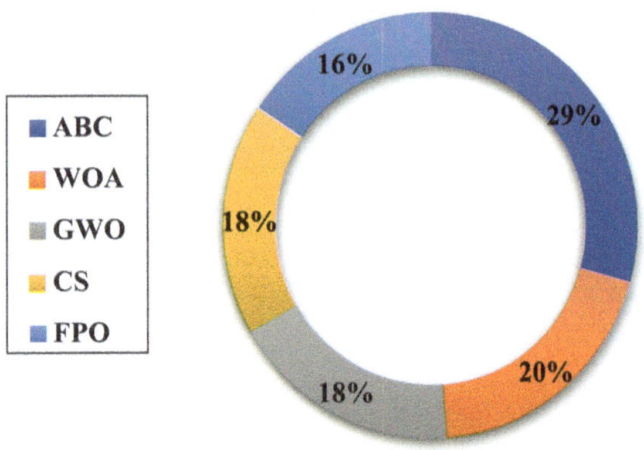

4 Role of Other Meta-Heuristics and Optimization Techniques in Renewable Power Sector

Apart from the NIAs, the contribution of other meta-heuristics and numerical methods has been significant and the undisputed crown goes to differential evolution (DE) followed by teaching learning-based optimization (TLBO). A detailed survey of the application of various meta-heuristics and optimization techniques in the last two decades with the development of the optimization technique and the problem applied to is provided in Tables 15 and 16.

The data from Table 15 makes it obvious that DE and TLBO have maintained their stance as the leaders for renewable power optimization. Several hybrid and improved variants of these paradigms have been proposed over time to balance the conflicting case of exploration versus exploitation. The key reason for the success of DE is its popularity and familiarity within the researchers and the tuning parameters play a key role. On the other hand, TLBO is a relatively simple and effective paradigm with no tuning required with a promising reliability. Besides DE and TLBO, HS and

Table 13 The top ten ranked NIAs based on their usage data in the renewable power sector

Rank	Algorithm	Year of publication	Authors	Inspiration	Phases	Tuning parameters	Number of fitness evaluations	Description
1	Artificial bee colony (ABC)	2007	Dervis Karaboga and Bahriye Basturk	Intelligent behaviour of honey bee swarm	i. Employed bee phase ii. Onlooker bee phase iii. Scout bee phase	Population Size (*Np*), Number of iterations (*T*)	$Np \times T$ *(approx.)*	The colony of artificial bees with food sources, employed forager bees and unemployed forager bees with a trail counter to efficiently explore the search space is developed [36]
2	Whale optimization algorithm (WOA)	2016	Seyedali Mirjalili and Andrew Lewis	Bubble-net hunting strategy of hump-back whales	i. Encircling prey ii. Bubble-net attacking method iii. Search for prey	Population Size (*Np*), Number of iterations (*T*)	$Np \times T$	WOA is based on the social behaviour of hump-back whales as they hunt for they prey collectively through a unique technique [37]

(continued)

Table 13 (continued)

Rank	Algorithm	Year of publication	Authors	Inspiration	Phases	Tuning parameters	Number of fitness evaluations	Description
3	Grey wolf optimizer (GWO)	2014	Seyedali Mirjalili, Seyed Mohammad Mirjalili and Andrew Lewis	Leadership hierarchy and hunting mechanism of grey wolves	i. Social hierarchy ii. Encircling prey iii. Hunting iv. Attacking prey v. Search for prey	Population Size (*Np*), Number of iterations (*T*),	$Np \times T$	The collective hunting mechanism of grey wolves (canis lupus) classified as alpha, beta, delta and omega as they hunt for their prey is modelled into an optimization algorithm [38]
4	Cuckoo search (CS)	2009	Xin-She Yang and Suash Deb	Brood parasitic behaviour of cuckoo and Levy flight behaviour of birds and fruit flies	i. Random generation through Levy flights ii. Choosing nest iii. Forwarding the best solutions	Population Size (*Np*), Number of iterations (*T*)	$Np \times T$	The breeding behaviour of cuckoos along with the levy flight-based behaviour of birds and fruit flies with the choice of best solutions for the next iterations is implemented [39]

(continued)

Table 13 (continued)

Rank	Algorithm	Year of publication	Authors	Inspiration	Phases	Tuning parameters	Number of fitness evaluations	Description
5	Flower pollination algorithm (FPA)	2012	Xin-She Yang	Pollination process of flowers	i. Global pollination through Levy flights ii. Local pollination iii. Reproduction	Population Size (*Np*), Number of iterations (*T*)	$Np \times T$	The self and cross pollination of flowers is modelled for global and local search optimization [40]
6	Salp-swarm optimization (SSO)	2017	Seyedali Mirjalili, Amir H. Gandomi, Seyedeh Zahra Mirjalili, Shahrzad Saremi, Hossam Faris and Seyed Mohammad Mirjalili	Navigating and foraging behaviour of salps in oceans	i. Moving salp chains ii. Swarm behaviour	Population Size (*Np*), Number of iterations (*T*)	$Np \times T$	The swarming behaviour of salps through salp chain and their position updation through Newton's law of motion is implemented [41]

(continued)

Table 13 (continued)

Rank	Algorithm	Year of publication	Authors	Inspiration	Phases	Tuning parameters	Number of fitness evaluations	Description
7	Moth flame optimization (MFO)	2015	Seyedali Mirjalili	Navigation technique of moths through transverse orientation	i. Initialization ii. Transverse motion iii. Updation	Population Size (Np), Number of iterations (T)	$Np \times T$	The special navigation of moths at night towards the light sources through transverse orientation is modelled as an optimization algorithm [42]
8	Grasshopper optimization algorithm (GOA)	2017	Shahrzad Saremi, Seyedali Mirjalili and Andrew Lewis	The behaviour of grasshopper swarms	i. Swarming ii. Interaction of grasshoppers iii. Convergence	Population Size (Np), Number of iterations (T)	$Np \times T$	The swarming nature of adult and nymph grasshoppers is modelled to explore and exploit the search space [43]
9	Dragonfly algorithm (DA)	2015	Seyedali Mirjalili	The static and dynamic swarming behaviour of dragonflies.	i. Separation ii. Alignment iii. Cohesion iv. Attraction to food v. Distraction from enemy	Population Size (Np), Number of iterations (T)	$Np \times T$	The sub-swarming and combined swarming nature of dragonflies towards the exploration of food sources is modelled as optimization technique [44]

(continued)

Table 13 (continued)

Rank	Algorithm	Year of publication	Authors	Inspiration	Phases	Tuning parameters	Number of fitness evaluations	Description
10	Crow search algorithm (CSA)	2016	Alireza Askarzadeh	Intelligent behaviour of crows	i. Flocking ii. Memorizing iii. Thievery iv. Protection	Population Size (*Np*), Number of iterations (*T*)	$Np \times T$	CSA is based on the concept of crows hiding their surplus food and retrieving it [45]

Table 14 Literature survey of the applications of various NIAs to different domains of the renewable power sector (data retrieved from SCOPUS® database)

S. No.	Algorithm used	Year	Description of the algorithm	Description of the problem
Hydro power optimization				
01.	ANN-ABC	2014	Artificial bee colony (ABC) is used to train the feed-forward artificial neural network (ANN) to obtain the optimal synaptic weights and biases	The estimation of annual hydro-electric power generation of Turkey with respect to the energy consumption, gross power demand, population of the country and annual average temperature is done
02.	H-GWO	2019	Hybrid grey wolf optimizer (H-GWO) for both continuous and discrete multi-objective optimization with a synchronized continuous and discrete variable updation system is proposed	The optimal multi-objective scheduling of a hydro-thermal power system in order to minimize the emission and dispatch of load satisfying various constraints on storage and power generation is addressed
03.	M-CSA	2020	Modified crow search algorithm (M-CSA) with adaptive chaotic awareness probability for a tradeoff between intensification and diversification is developed	The optimal system operation for a PV-diesel system with pumped hydro storage system through the minimization of operating and fuel costs through penalty constraint handling technique is developed
04.	BFOA	2013	Bacterial foraging optimization algorithm (BFOA) inspired by the human foraging behaviour of E. coli bacteria is utilized	The economic dispatch of a hydro power station through the minimization of operational costs of the generators with respect to reservoir, standby power and storage constraints is carried out
05.	ALO	2016	Ant lion optimization (ALO) based on the hunting movement of the ant lions with an adaptive exploration and exploitation system is adopted for optimization	The optimal combined hydro-thermal-wind scheduling with multi-reservoir cascaded hydro plants optimization to minimize the operational costs, emissions and power loss is implemented

(continued)

Table 14 (continued)

S. No.	Algorithm used	Year	Description of the algorithm	Description of the problem
06.	FPA	2019	Flower pollination algorithm (FPA) based on the evolutionary tactic that the pollination results in generation of fittest flowering plants with Levy flight to generate population is adopted	The short-term optimal scheduling of hydro-thermal power plant through the minimization of the running costs with constraints laid down on prohibited operating zones is implemented
07.	WOA	2019	Whale optimization algorithm (WOA) inspired by the unique "bubble-net foraging" technique of hump-back whales through an adaptive exploration and intensification system is used	The optimal constrained emission dispatch for the hydro-thermal system with wind power integration considering the intermittencies associated with the wind power with a focus on environmental aspects is carried out
08.	Levy-MFO	2019	Moth flame optimization (MFO) based on the moth flame attraction concept with the addition of Levy flights to enhance the global exploratory characteristics of MFO is developed	The load frequency control of an interconnected hydro-thermal system through the optimal parameter tuning of the fractional order PID controller with respect to the constraints on the speed governor and generation limits is addressed
09.	NS-WOA	2019	A modified version of multi-objective non-sorting whale optimization algorithm (NS-WOA) based on the non-sorting genetic algorithm (NSGA-II) is developed to generate an optimal set of non-dominant solutions	The long-term multi-objective optimization of hydro-PV-wind system with hydropower compensation to maximize the annual power production with reduced power fluctuations is implemented
10.	WOA	2017	Standard whale optimization algorithm without any modifications is used	The fixed-head short-term hydro-thermal-PV power scheduling problem with the multi-reservoir cascaded hydro plants system is considered for optimization while dealing with the uncertainties with the PV power
Wind power optimization				

(continued)

Table 14 (continued)

S. No.	Algorithm used	Year	Description of the algorithm	Description of the problem
01.	Elitist ABC	2011	Elitist ABC (EABC) with a slower rate of search in order to improve the exploration of the search space and to avoid premature convergence is proposed	The combined economic load dispatch and economic emission dispatch is optimized by the minimization of the operating costs and emissions of a wind-thermal power system
02.	DA	2016	Dragonfly algorithm (DA) based on the feeding and migration behaviour of the dragonflies with Levy flights is adopted	The economic load dispatch of a wind-thermal unit considering the valve points effects for a modified IEEE-30 bus system with six thermal generators and one wind farm is addressed
03.	H-FPA	2015	Hybrid-flower pollination algorithm (H-FPA) is a hybrid of FPA and differential evolution (DE) intended to optimize the fuzzy selection index is proposed	The multi-objective dynamic economic load dispatch for a wind integrated system through emission and cost minimization while tackling uncertainties of the load and wind speed is presented
04.	SSA	2019	Salp-swarm algorithm (SSA) based on the chaining behaviour of salp swarm with a position updating system based on the Newton's law of motion is adopted for optimization	The optimization of an airfoil-based savonius wind turbine to maximize the power coefficient based on the class transfer function is tackled
05.	RTO	2019	Root tree optimization (RTO) algorithm based on the social behaviours of roots of a tree in search of underground water is modelled and utilized for optimization	The maximum power point tracking (MPPT) of a doubly-fed induction generator of a wind turbine and the minimization of harmonic currents with active and reactive power control is presented as an optimization problem
06.	E-WOA	2020	Enhanced whale optimization algorithm (E-WOA) with modification in the exploration and exploitation system of the original algorithm to improve the quality of solution is presented	The MPPT of a variable-speed wind generator is improved through the design of an optimal Takagi-Sugeno fuzzy logic controller (FLC)

(continued)

Table 14 (continued)

S. No.	Algorithm used	Year	Description of the algorithm	Description of the problem
07.	NSCS	2017	The parallel exploratory capabilities of cuckoo search algorithm (CSA) and the non-dominant sorting technique of NSGA-II is combined to develop a non-dominant cuckoo search (NSCS) algorithm.	The design of an optimal controller to improve the frequency regulation aspects of wind-thermal power unit through a multi-objective optimized tuning of various controller parameters is addressed.
08.	MO-GWO	2020	Multi-objective grey wolf optimizer (MO-GWO) to optimize the kernel-based nonlinear extension of the ARPS model is used	The optimal prediction accuracy of wind farm power with higher prediction accuracy and stability with the historical data of wind farms in Belgium taken up for the case study
09.	WOA	2020	Standard whale optimization algorithm without any modifications is used	The optimization of a Sugeno FLC for the faut ride-through implementation for a grid-integrated variable-speed wind generator for the balanced and unbalanced load grid fault conditions
Solar PV power optimization				
01.	CAO	2019	Coyote optimization algorithm (COA) inspired by the survivability techniques of the coyotes (*Canis latrans* species) through social norms adopted by them	The optimal parameter extraction of the single-diode model and two-diode model through the minimization of root mean square error (RMSE) is implemented
02.	GWO-CS	2020	A hybrid grey wolf optimizer and cuckoo search algorithm (GWO-CS) through opposition-based learning strategy with an ability to balance exploration and exploitation is presented	The extraction of the parameters of a PV module by minimizing the difference between the measured parameters and simulated parameters is presented
03.	GO-FL	2020	Grasshopper optimized fuzzy logic (GO-FL) system with the grasshopper optimization algorithm tuning the membership functions of the FLC is presented	The MPPT of solar cells considering the stochasticity of the load, uncertainties associated with the irradiance and ambient temperature is addressed

(continued)

Table 14 (continued)

S. No.	Algorithm used	Year	Description of the algorithm	Description of the problem
04.	SSO	2020	Salp-swarm optimization (SSO) without any modifications is considered	The optimal energy harvesting through MPPT from PV modules considering the partial shading conditions and non-uniform weather conditions is carried out
05.	H-ABC PSO	2020	The local convergence issue of PSO has been addressed through a hybrid artificial bee colony particle swarm optimization (H-ABC PSO) while incorporating the advantages of both the algorithms	The capacity optimization of a solar-fuel cell-based energy system under grid-connected conditions with consideration on component sizing and total net present costs is implemented
06.	IO-WOA	2018	An improved opposition-based whale optimization algorithm (WOA) is proposed with the opposition-based learning strategy to enhance the exploration capability of the existing WOA	The optimal solar cell parameter estimation is done for single-diode model, two-diode model and three-diode model for industrial applications
07.	DE-WOA	2018	A hybrid differential evolution-based WOA is developed combining the exploratory capabilities of DE with the exploitation of WOA to avoid premature convergence is presented	The solar cell parameter extraction with a practical implementation to a PV power station in Guizhou power grid in China is presented
08.	RL-WOA	2020	Refraction learning-based WOA with a logistic model to balance diversity and convergence inspired by the refraction of the light is developed for solving high-dimensional optimization problems	The parameter extraction of PV model for the single-diode model is implemented
09.	WDO	2019	Wind driven optimization (WDO) based on the air movement in the earth's atmosphere based on the temperature imbalance modelled after Newton's second law is utilized	The MPPT of a PV system under non-uniform solar irradiance (partial shading conditions) with various test cases is carried out
Geothermal and biofuel power optimization				

(continued)

Table 14 (continued)

S. No.	Algorithm used	Year	Description of the algorithm	Description of the problem
01.	ABC	2016	Standard ABC algorithm without any modifications is deployed	The optimization and thermodynamic analysis of a Kalina cycle for a geothermal unit to obtain the optimal thermal and exergy coefficients for the Husavik geothermal plant in Iran is conducted
02.	ABC	2018	A comparative analysis between ABC and the physics-inspired GSA (gravitational search algorithm) is performed	The maximization of exergy efficiency of geothermal power plant with the exergy balance condition for the vaporizers, preheaters, turbines, condensers and pumps is considered
03.	ABC	2017	Basic ABC algorithm with a threshold limiting control parameter is deployed	The thermodynamic optimization of the organic Rankine cycle system to increase the system efficiency and minimize the power losses under various cases is studied
04.	ABC	2016	Standard ABC algorithm is deployed	The optimization of biodiesel percentage in fuel mixture with limitation on emissions is implemented
Integrated and hybrid renewable power optimization				
01.	ABC	2012	Standard ABC algorithm without any modifications is deployed	The dynamic economic dispatch (DED) problem with the optimization of power generation through renewable units while dealing with load demand dispatch is tackled
02.	ABC	2012	Standard ABC algorithm is deployed	The optimization of the performance of a smart grid with renewables through the minimization of fuel costs to achieve high controllability and reliability in the load flow is implemented
03.	GWO	2017	Grey wolf optimizer (GWO) inspired by the hierarchical leadership and hunting mechanism of grey wolves is implemented	The optimization of a hybrid renewable PV-diesel-battery system with the application to Djanet city in Algeria considering the minimization of total cost of the hybrid system is carried out

(continued)

Table 14 (continued)

S. No.	Algorithm used	Year	Description of the algorithm	Description of the problem
04.	WOA	2018	Standard WOA without any special modifications is implemented	The optimal renewable resource placement for a distribution network with the power loss index for an IEEE 15, 33, 69, 85 and 118-bus test systems with focus on voltage profile improvement and enhancing reliability is implemented
05.	IGWO	2020	An improved grey wolf optimization (IGWO) algorithm with the improvements to the local and global exploration system while improving the rate of convergence is utilized	The optimal placement and the sizing of electric power storage systems for microgrid with the uncertainties associated with the renewable energy sources and the defect occurrences in the grid-connected system is tackled
06.	IDHO	2020	An improved deer hunting optimization (IDHO) algorithm inspired by the hunting techniques while escaping predators is adopted for the optimization process	The optimal designs of solar chimney and fuel cell improvements for the surplus power storage for both summer and winter conditions is presented
07.	HNMCS	2018	A hybrid Nelder–Mead cuckoo search (HNMCS) algorithm with the Nelder–Mead algorithm solving the nonlinear functions while the cuckoo search aids in local search is developed	The optimization of renewable power generation in AC/DC microgrid system by maximizing the power output of the renewable energy distributed generators (REDG) is tackled

Jaya algorithms hold major positions with a good degree of optimization for both single- and multi-objective optimizations compared to the numerical optimization problems. The recent trend of modelling the renewable systems as a mixed integer linear program (MILP) and mixed integer nonlinear program (MINLP) has also seen good success, but the usage of advanced optimization solvers is required. While much of the research in renewables is focused towards the cost and energy management, the incorporation of distribution system optimization to renewable power has seen a considerable growth with the optimization algorithms producing good results compared other schemes and methods. The environmental concern towards reducing emissions through the maximization of the utilization of renewables over thermal units is also integrated into optimization. Apart from optimization and constraint handling, there is a growth of optimization algorithms towards the controller tuning and training of prediction systems for accurate and reliable performance.

Table 15 Literature survey of the applications of various meta-heuristics and other optimization techniques to different domains of the renewable power sector (data retrieved from SCOPUS® database)

S. No.	Algorithm used	Year	Description of the algorithm/solving mechanism	Description of the problem
Hydro power optimization				
01.	DE	2014	Differential evolution (DE), an evolutionary algorithm based on the natural selection with mutation, crossover, evaluation and selection phases is chosen for optimization	The optimal hydro power dispatch along with the storage through reservoir operation of an eight-cascade hydro power plant in Slovenia is tackled and compared with literature
02.	Implicit stochastic optimization	2018	Implicit stochastic optimization (Monte Carlo optimization) is a dynamic programming (DP) technique to derive the rules for optimization based on the uncertainties through a regression equation	The optimal operating rules for a large-scale hydro-PV unit to address the intermittencies related to reservoir inflow and PV power based on China's Longyangxia power system is studied
03.	QOTLBO	2013	Quasi-oppositional teaching learning-based optimization (QOTLBO) to with oppositional learning for a better approximation to improve the performance and convergence speed of TLBO is proposed	The optimal short-term hydro-thermal scheduling considering the cascading nature of reservoirs, water transport delay and scheduling errors is addressed through a robust optimization algorithm
04.	Enhanced DE	2008	An enhanced differential evolution algorithm through the introduction of chaos theory with a self-adaptive parameter tuning to improve the convergence characteristics of the standard DE is proposed	The daily optimal hydro power generating scheduling with four interconnected cascaded hydro power plants satisfying various constraints on operation and storage is addressed

(continued)

A generalized optimization model proving the description of the various objective functions, constraints, decision variables and inputs to optimization is shown in Table 17.

Table 15 (continued)

S. No.	Algorithm used	Year	Description of the algorithm/solving mechanism	Description of the problem
05.	Stochastic optimization	2019	Stochastic optimization based on global sensitivity analysis (GSA) is implemented as a mixed integer linear programming problem solved using continuous probability of variables	The optimized adaption strategies for a hydropower system with respect to the climate change laws for planning, operation and management of hydro units is implemented
06.	Integer programming	2018	Integer programming is optimization/feasibility technique requiring all the decision variables to be integers. It is also referred to as integer linear programming (ILP)	The optimization of micro-hydro power units for generic river profiles through Pelton wheel system considering the power, flow and gap constraints is formulated and optimized
07.	Three-layer nested approach	2018	A three-layer nested approach with direct search algorithm, cuckoo search algorithm and dynamic programming to optimize each subsequent layer of optimization is presented	The daily generation optimization of a hydro-PV unit with problems considerations on power outputs, hydro unit allocation and load dispatch as a three-tier system is implemented
08.	Adapted DE	2014	A modified DE known as adapted DE with dynamic population generation and with self-adapting crossover and differential factors to improve the global search is implemented	The optimization of hydro energy storage units by minimization of water quantity to generate power for a 24-hour cycle while satisfying the load demand is carried out
09.	Nonlinear and logical optimization	2019	A combination of logical and nonlinear optimization with decision variables optimized based on the previous data and classic optimization is presented	The optimal operation of a hydro power unit without conflicting the increased capacities of solar and wind installations to maximize the revenue subject to various constraints is presented

(continued)

Table 15 (continued)

S. No.	Algorithm used	Year	Description of the algorithm/solving mechanism	Description of the problem
10.	Probability interval optimization (PIO) model	2017	A probability interval optimization (PIO) model is one which optimizes the profits and risk associated with the uncertainties modelled as distribution function through stochastic optimization	The short-term hydro-wind-thermal scheduling to maximize the profit and minimize the risks associated with the uncertainty of wind power is tackled
11.	Stochastic scenario-tree simulation optimization	2016	This optimization procedure requires the previous data to generate a sequence of problems referred to as scenarios and extracts the optimal scenario based on the constraints and decision variables	The long-term hydro-thermal scheduling based on the operating conditions, market prices and socio-economic surplus considering interconnected power markets is studied
Wind power optimization				
01.	Evolution strategy algorithm and data mining algorithm	2010	Differential evolution coupled with data mining algorithm to extract the dynamic models based on classification, regression tree model and support vector machine regression is utilized	The optimization of wind turbine energy and power factor of a 1.5 MW wind turbine through the control of active and reactive power to achieve unity power factor is done
02.	Numerical optimization algorithm	2012	A numerical approach known as line search numerical algorithm, a derivative of steepest direction method where in the algorithm searches for the optimal solution based on the direction derived by steepness of position with respect to the objective function	The MPPT for wind energy conversion system (WECS) based on the wind profile through optimal torque technique is implemented

(continued)

5 Conclusion

This book chapter presents a comprehensive survey of various NIA's, meta-heuristic algorithms and other optimization algorithms and techniques applied to the optimization of generation, management, operation, distribution, scheduling of various

Table 15 (continued)

S. No.	Algorithm used	Year	Description of the algorithm/solving mechanism	Description of the problem
03.	IS-DE	2016	An improved self-adaptive differential evolution with an elitism operator to guide the direction of mutation towards feasible solutions with a self-learning strategy to vary the scaling factor and crossover rate is proposed	The reactive power optimization of a distribution network with the integration of wind power for an IEEE 33-node system considering the reactive power limits of the wind turbine is implemented
04.	APO	2018	Artificial physical optimization (APO) algorithm is a swarm-based stochastic algorithm based on the natural physical forces in nature formulated based on the artificial physics (AP) framework	The optimal power flow in a wind energy system while considering the issues of power quality is tackled for an IEEE -30,118 and 300 bus systems
05.	Hybrid probabilistic optimization algorithm	2020	A hybrid probabilistic optimization algorithm combining the properties of better diversity and elitism based on multi-objective PSO and NSGA-III is proposed	The optimal allocation of energy storage systems for correlated wind farms considering the high variability and unpredictability of power generation for an IEEE-30 and 57 bus system is implemented
06.	CA	2018	Cultural algorithm (CA) inspired by the normative, domain-specific, situational, temporal and spatial knowledge aspects in a belief space (search space) is utilized	The parameter optimization of a doubly-fed induction generator (DFIG) to improve the efficiency and the dynamic response of the system is carried out
07.	Lazy greedy algorithm	2015	A lazy greedy optimization algorithm benefitting from the sub-modular property wherein the knowledge of all the test criteria is not required to search for the optimal solution is used	The optimization of the power output of a wind turbine based on its positioning on a complex terrain encompassing virtual particle wake flow models based on the wind dynamics for an accurate evaluation is performed

(continued)

Table 15 (continued)

S. No.	Algorithm used	Year	Description of the algorithm/solving mechanism	Description of the problem
08.	G-TLBO	2015	A hybrid genetic-teaching learning-based optimization (G-TLBO) algorithm to decrease the run time of the overall algorithm with a superior convergence is proposed	The optimization of power flow for a 19 bus, 7336 MW Turkish wind-thermal system under different loading conditions while reducing the fuel costs is performed
09.	Hybrid CS-HS-SA algorithm	2019	A hybrid chaotic search (CS), harmony search (HS) and simulated annealing (SA) is implemented to improve the exploration, exploitation and the rate of convergence	The optimal sizing of a standalone PV-wind system with hydrogen energy storage based on the weather data is implemented and tested for the city of Khorasan, Iran
10.	EL-EHO	2019	A hybrid ensemble learning-elephant herding optimization (EL-EHO) algorithm wherein the least square support machine optimized by the EHO is utilized	The optimization of an ultra-short-term wind power forecasting considering the fluctuations of the wind power for an American area is proposed
11.	LBBO-DE	2016	A hybrid linearized biogeography-based optimization (LBBO) and DE algorithm to enhance the evolutionary characteristics of DE to generate better optimal solutions is proposed	The maximization of wind power through the optimization of the sliding mode controller while keeping in check the power loss of the system is implemented
12.	Cooperative game theory	2018	A collaborative approach known as *Game theory* where every member (players) jointly contributes to the optimality to reap maximum benefits without conflicting with other is opted	The optimization, control and coordination of a wind power system with hybrid energy storage system (HESS) subject to the state of charge of batteries (SOC) is implemented

(continued)

Table 15 (continued)

S. No.	Algorithm used	Year	Description of the algorithm/solving mechanism	Description of the problem
13.	IDE	2018	An improved DE is proposed through a novel strategy known as double best mutation operator (DBMO) to speed up the convergence by accelerating mutation	The congestion management with the inclusion of wind turbines in power systems through the optimal location of wind turbines and rescheduling is analysed
14.	BCDE	2012	Bi-population chaotic DE with sub-population approach through rough and meticulous selection with chaos to update the population is proposed for the optimization	The optimal dynamic economic dispatch for wind-thermal power system based on the probability distribution function (PDF) of the wind power output is considered
Solar PV power optimization				
01.	Monte Carlo algorithm	2018	Monte Carlo algorithm is based on the technique of random sampling to obtain optimality for problems with probabilistic interpretation. These techniques are accompanied by other optimization algorithms like PSO to enhance the quality of the solution	The optimization of solar tower systems in solar thermal plants through the optimal design for reduced capital investment is done
02.	EA-PSO	2013	An evolutionary algorithm-aided PSO (EA-PSO) combining the exploration and evolutionary capabilities into one hybrid algorithm for neural network training is proposed	The solar radiation prediction accuracy is improved using recurrent neural networks based on the historical data and the developed system is deployed for solar radiation monitoring
03.	Jaya algorithm	2020	Jaya algorithm (*Jaya* signifying victory) is a stochastic optimization algorithm which proceeds only with the best solutions while avoiding any worst solution	The reactive power dispatch problem through the incorporation of solar power for an IEEE-14 and 30 bus system for an interconnected power system is proposed

(continued)

Table 15 (continued)

S. No.	Algorithm used	Year	Description of the algorithm/solving mechanism	Description of the problem
04.	BBO	2013	Biography-based optimization (BBO) algorithm is inspired by the demographic disparity classified by habitat suitability index (HSI) and survivability index variables (SIV) that contribute to the flourishing of a species	The optimal power management and economic analysis for a solar-wind power system with hybrid energy storage through cost minimization and maximization of the energy availability is performed
05.	GA-TLBO	2020	A genetic-algorithm-assisted TLBO (GA-TLBO) for training the artificial neuro-fuzzy inference systems (ANFIS) is proposed	The optimal design parameter modelling of solar power tower system for stations in India, Bangladesh, Pakistan and Afghanistan is implemented to maximize annual power production and minimize the cost of energy
06.	HIS and SAA	2020	An improved harmony search (IHS) algorithm through the introduction of special weighting factors to improve the solution quality and simulated annealing algorithm (SAA) based on the annealing of metals is adopted	The optimal sizing of hybrid solar unit with battery storage to generate the required power demanded while minimizing the total life cycle costs (TLCC) is implemented
07.	TLBO	2020	Teaching learning-based optimization (TLBO) algorithm inspired by the teacher and student (learner) based on the knowledge transfer in a classroom is used	The techno-economic analysis of a PV system with batteries integrated to a grid to minimize the total net present cost (TNPC) of the system and cost of energy (COE) is dealt with

(continued)

renewable power sources for both standalone, hybrid and grid-integrated systems. The various objective functions and constraints pertaining to various domains in renewable power sector are studied and the different algorithms opted to optimize them are classified. The following aspects have been extensively covered in this comprehensive review.

Table 15 (continued)

S. No.	Algorithm used	Year	Description of the algorithm/solving mechanism	Description of the problem
08.	Improved TLBO	2014	An improved TLBO with elitism parameter to reduce the number of fitness evaluations while maintain a good convergence rate is proposed	The optimal parameter identification of a proton exchange membrane (PEM) fuel cells and solar cells under different operating conditions is done
09.	TLBO	2014	Standard TLBO algorithm with the "teacher" and "learner" phases is implemented	The extraction of solar cell parameters based on the single illuminated current–voltage characteristics for various fabrication material used like silicon, plastic and de-sensitized solar cells is performed
10.	NSCDE	2019	Non-dominated sorting culture differential evolution (NSCDE) algorithm based on NSGA-II and cultural algorithm (CA) to ensure optimal set of solutions for multi-objective optimization is developed	The optimal operation of PV-wind-hydro system to maximize the net power output and the ten-day joint output while dealing with intermittencies associated with each system is presented
11.	L-SHADE	2019	Linear population-success history-based differential evolution (L-SHADE) with dynamically adaptive parameter tuning based on the success history of previous generation is adopted	The datasheet information-based solar cell parameter estimation for single-diode and double-diode models through the minimization of root mean square error (RMSE) is performed
12.	Prediction algorithm-based DE	2011	A prediction-aided differential evolution algorithm to help minimize the error between the measured data and predicted data is considered	The optimal prediction of solar angles for PV panels through the solar radiation data for dual-axis sun tracking system by error minimization to improve the efficiency of the solar cells is presented

(continued)

Table 15 (continued)

S. No.	Algorithm used	Year	Description of the algorithm/solving mechanism	Description of the problem
13.	TL-ABC	2018	A hybrid paradigm based on TLBO and ABC to balance the exploration and exploitation of the search process is proposed	The PV panel parameter estimation for single-diode and double-diode models to improve the efficiency of the solar cell is done
14..	MO-DP	2018	A multi-objective dynamic programming optimization technique based on NSGA-II is with a non-dominant approach such that the optimality of one objective doesn't conflict or dominate the other	The performance management of a solar thermal unit with suitable charge management and thermal energy storage system to help improve the reliability and profitability of the system is proposed
Integrated and hybrid renewable power optimization				
01.	DEa-ER	2019	An efficient differential evolution (DEa-ER) algorithm incorporating a novel mutation operation with archive strategy especially tailored for constraint handling is proposed	The optimal active and reactive power dispatch integrating renewable generators for an IEEE-57 bus system with various stochastic scenarios is tackled
02.	TLBO	2016	Standard TLBO algorithm with the "teacher" and "learner" phases is implemented	The automatic generation control of a multi-area power system comprising of various energy sources through the optimal tuning of controller parameters for various energy sources is considered
03.	Hybrid TLBO and pattern search (hTLBO-PS)	2020	A two-phase hybrid algorithm to explore the multi-modal search-scape with TLBO improving the global search and pattern search (PS) enhancing the local search is implemented	The automatic generation control of hybrid power system deregulated environment considering the nonlinearities and storage limitations associated with each area is considered

(continued)

Table 15 (continued)

S. No.	Algorithm used	Year	Description of the algorithm/solving mechanism	Description of the problem
04.	MOEA/D	2018	Multi-objective evolutionary algorithm based on decomposition (MOEA/D) is a technique of splitting a multi-objective problem into a series of sub-problems and simultaneously individually optimizing them	The optimal economic-environmental power dispatch of a wind-PV-hydro power system with its integration into the grid system for a standard IEEE-30 bus system is implemented
05.	QOTLBO	2014	Multi-objective quasi-oppositional TLBO (QOTLBO) based on opposition-based learning resulting in faster convergence is considered for the optimization process	The optimal placement of distributed generators with renewables to optimize power loss, voltage stability and voltage deviations for an IEEE-33,69 and 118 bus radial distribution network is considered
06.	CFCEP	2019	A chaotic fast convergence evolutionary programming (CFCEP) based on DE to handle multiple objective with a decent converge with self-adapting parameter tuning through chaos theory is proposed	The multi-region dynamic economic dispatch for a PV-wind-hydro-thermal system with pumped hydro power storage through multi-cascaded reservoir operation is done
07.	MDE	2019	A modified differential algorithm (MDE) through the incorporation of the sigmoid function into the mutation function of DE is proposed for prevention of local entrapment is used	An optimal peak shaving strategy for wind-PV-hydro-based hybrid power system to improve the safety and stability is analysed

(continued)

(a) The current scenario of renewable power systems and their development in contrast to the conventional sources of power generation is studied with the data from International Energy Agency (IEA).
(b) The share of renewable power generation based on each domain and contribution of various countries to renewable power usage is outlined.
(c) The estimation in the growth of renewable power generation for the next two decades based on the flagship report "World Energy Outlook 2019" is covered.

Table 15 (continued)

S. No.	Algorithm used	Year	Description of the algorithm/solving mechanism	Description of the problem
08.	SCA	2020	Sine cosine algorithm (SCA) with the sine function to enhance the exploration while the cosine function enhances the exploitation with a balance between the two functions is opted for optimization	The optimal power flow-based hydro-thermal-wind scheduling through the minimization of the fuel costs and emissions for a standard 9 bus system is studied

(d) The challenges in the renewable power industry with respect to hydro, wind, solar PV, geothermal, biofuel, integrated and hybrid renewables system are studied.
(e) The various objective functions and constrains for each of the renewable domain is tabulated, and the detailed mathematical modelling is carried out.
(f) The role of various NIA's to the optimization of the renewable power domain along with an analysis of various work contributing to the renewable sector in the literature is studied for the last two decades, and the important trends and challenges are brought out.
(g) The legacy of PSO and GA along with their hybrid/improved counterparts for various problems in the literature is highlighted.
(h) A comprehensive and detailed survey report comprising of the contributions of various NIAs to different problems relating to every domain of the renewables is prepared.
(i) The ranking of the NIAs based on their application and usage data over the last decades is performed, and their description is provided.
(j) The contribution of other optimization algorithm and techniques alongside their ranking based on their usage is provided through the literature survey.
(k) A generalized optimization model comprising of all the different objectives, constraints, input data and decision variables based on the data from the entire literature survey is presented.

Table 16 Top five ranked meta-heuristics and optimization techniques based on their usage data in the renewable power sector

Rank	Algorithm	Year of publication	Authors	Inspiration	Phases	Tuning parameters	Number of fitness evaluations	Description
1	Differential evolution (DE)	1997	Rainer Storn and Kenneth Price	Natural selection by Charles Darwin	i. Initialization ii. Mutation iii. Crossover iv. Selection	Population Size (*Np*), Number of iterations (*T*), Crossover rate, Mutation rate, Scaling factor	$Np \times T$ *(approx.)*	The evolution of species through natural selection serves as a basis to evolve the solution through a series of genetic operators to achieve optimality
2	Teaching learning-based optimization (TLBO)	2011	R. V. Rao, V. J. Savsani and D. P. Vakharia	Influence of teacher on learners	i. Teacher phase ii. Student phase	Population Size (*Np*), Number of iterations (*T*)	$2(Np \times T)$	A population-based optimization technique based on the interaction of teachers and students in a classroom is modelled for optimization of real-world problems [46]
3	Harmony search (HS)	2001	Zong Woo Geem, Joong Hoon Kim and G. V. Loganathan	The improvisation of music players	i. Initialization ii. Improvisation iii. Inclusion/Exclusion iv. Stopping	Population Size (*Np*), Number of iterations (*T*), Memory consideration rate, Pitch rate, Number of improvisations	$Np \times T$	The improvisation of musical performances to create aesthetic music based on various instrumental combinations serves as the base of optimization
4	Jaya algorithm	2016	R. Venkata Rao	The movement towards the best solution avoiding the worst solution	i. Identification ii. Modification iii. Termination	Population Size (*Np*), Number of iterations (*T*)	$Np \times T$	An algorithm-specific-parameter-less optimization based on the solution movement towards the best solution avoiding the worst solutions is proposed [47]

(continued)

Table 16 (continued)

Rank	Algorithm	Year of publication	Authors	Inspiration	Phases	Tuning parameters	Number of fitness evaluations	Description
5	Monte Carlo optimization	Mid-1960s	Henry P. McKean Jr.	Random sampling to obtain numerical results	i. Pseudo random number generation ii. Sampling iii. Validity	Number of random number generations, Number of samples	–	This is a numerical optimization based on repeated random sampling and statistical analysis based on test, run and re-run methods

Table 17 Generalized optimization model of the renewable power system including all domains

Inputs to the optimization algorithm		
The number of renewable units Number of renewable generators, turbines, PV panels The area of the power generating unit Expected operation time Desired power generation Environmental state Demographic/weather profile Market pricings Location data Intermittency data		
Decision variables		
Power generated through renewable sources Operation period/dispatch period Speed and toque of the wind turbines Tuning parameters to enhance controller optimization Speed governor control Water inflow rate/outflow rate (Discharge rates) State of charge (SOC) of batteries Number of batteries The mode of operation and control Placement of generators/turbines Bus selection Profit/loss		
Objective		
Minimize	Maximize	Optimize
Total net present cost (TNPC) of the system Integral square error (ISE) Total squared deviation (TSD) Annual total cost (ATC) Cost of energy (COE) Emission rate Control error Total system cost Power loss Root mean square error (RMSE) Mean square error (MSE) Total life cycle costs	Annual energy production Lifetime of the equipment Net present value (NPV) Specific work output Thermal efficiency System efficiency	Controller tuning parameters Wind farm layout Maximum power point tracking (MPPT) Economic load dispatch Active and reactive power dispatch Training of FLC controllers Configuration of renewables Power reliability Training performance of neural networks
Constraints		

(continued)

Table 17 (continued)

Water storage/water level
Power balance equation
Power output restrictions
Power generation limits
Generation unit ramp rate
Discharge rate
Water balance equation
Reservoir storage
Battery storage
Tie-line power
Prohibited zones of operation
Loss of energy
Loss of load
Number of equipment including turbines, generators, PV panels, inverters, controllers, biomass digestors, batteries, hydrogen tanks, reservoirs etc.
Feasible zones
Boundary conditions
Power output oscillations
Time of operation
Total online/offline time
Ageing/depreciation factors
Power loss
Exergy balance equation
Load uncertainty
Emission limits
Thermal limits
Design constraints
Altitude difference
Mean wind power density
Apparent power flow
Power ratings
Pricing and cost constraints

References

1. Liu, J., Li, D., Wu, Y., Liu, D.: Lion swarm optimization algorithm for comparative study with application to optimal dispatch of cascade hydropower stations. Appl. Soft Comput. J. **87**, 105974 (2020). https://doi.org/10.1016/j.asoc.2019.105974
2. Sahani, A.K., Raj, U., Shankar, R., Mandal, R.K.: Firefly optimization based control strategies for combined load frequency control and automatic voltage regulation for two-area interconnected power system. Int. J. Electr. Eng. Inf. **11**, 746–757 (2019). https://doi.org/10.15676/ijeei.2019.11.4.8
3. Mohammadi, M., Farzin, S., Mousavi, S.F., Karami, H.: Investigation of a new hybrid optimization algorithm performance in the optimal operation of multi-reservoir benchmark systems. Water Resour. Manag. **33**, 4767–4782 (2019). https://doi.org/10.1007/s11269-019-02393-7
4. Bozorg Haddad, O., Afshar, A., Mariño, M.A.: Honey-bee mating optimization (HBMO) algorithm in deriving optimal operation rules for reservoirs. J. Hydroinf. **10**, 257–264 (2008). https://doi.org/10.2166/hydro.2008.018
5. Haddad, O.B., Mariño, M.A.: Dynamic penalty function as a strategy in solving water resources combinatorial optimization problems with honey-bee optimization (HBMO) algorithm. J. Hydroinf. **9**, 233–250 (2007). https://doi.org/10.2166/hydro.2007.025
6. Basu, M.: Artificial bee colony optimization for short-term hydrothermal scheduling. J. Inst. Eng. Ser. B **95**, 319–328 (2014). https://doi.org/10.1007/s40031-014-0119-7
7. Kar, S, Das, D.P., Sanyal, S.K.: Enhanced artificial bee colony optimization for fixed head hydrothermal power system. In: Proceedings 2019 international conference applied machine learning ICAML, pp. 213–221 (2019). https://doi.org/10.1109/ICAML48257.2019.00047
8. Dubey, H.M., Pandit, M., Panigrahi, B.K.: Ant lion optimization for short-term wind integrated hydrothermal power generation scheduling. Int. J. Electr. Power Energy Syst. **83**, 158–174 (2016). https://doi.org/10.1016/j.ijepes.2016.03.057
9. Soroudi, A.: Robust optimization based self scheduling of hydro-thermal Genco in smart grids. Energy **61**, 262–271 (2013). https://doi.org/10.1016/j.energy.2013.09.014
10. Amer, M., Miloudi, A., Lakdja, F.: Optimal DTC control strategy of DFIG using variable gain PI and hysteresis controllers adjusted by PSO algorithm. Period Polytech. Electr. Eng. Comput. Sci. **64**, 74–86 (2020). https://doi.org/10.3311/PPee.14237
11. Qais, M., Hasanien, H.M., Alghuwainem, S.: Salp swarm algorithm-based TS-FLCs for MPPT and fault ride-through capability enhancement of wind generators. ISA Trans. 1–14 (2020). https://doi.org/10.1016/j.isatra.2020.01.018
12. John Hetzer, D.C.Y.K.B.: Wind power wind power. Econ. Dispatch Model Inc. Wind Power **23**, 56 (2012)
13. Jiang, R., Member, S., Wang, J., et al.: Power and pumped storage hydro. IEEE Trans. Power Syst. **27**, 800–810 (2012). https://doi.org/10.1109/TPWRS.2011.2169817
14. Hadidian Moghaddam, M.J., Kalam, A., Nowdeh, S.A., et al.: Optimal sizing and energy management of stand-alone hybrid photovoltaic/wind system based on hydrogen storage considering LOEE and LOLE reliability indices using flower pollination algorithm. Renew. Energy **135**, 1412–1434 (2019). https://doi.org/10.1016/j.renene.2018.09.078
15. Grady, S.A., Hussaini, M.Y., Abdullah, M.M.: Placement of wind turbines using genetic algorithms. Renew. Energy **30**, 259–270 (2005). https://doi.org/10.1016/j.renene.2004.05.007
16. Feng, J., Shen, W.Z.: Solving the wind farm layout optimization problem using random search algorithm. Renew. Energy **78**, 182–192 (2015). https://doi.org/10.1016/j.renene.2015.01.005
17. Ishaque, K., Salam, Z., Amjad, M., Mekhilef, S.: An improved particle swarm optimization (PSO)-based MPPT for PV with reduced steady-state oscillation. IEEE Trans. Power Electron. **27**, 3627–3638 (2012). https://doi.org/10.1109/TPEL.2012.2185713
18. Dufo-López, R., Bernal-Agustín, J.L.: Design and control strategies of PV-diesel systems using genetic algorithms. Sol. Energy **79**, 33–46 (2005). https://doi.org/10.1016/j.solener.2004.10.004

19. Arabali, A., Ghofrani, M., Etezadi-Amoli, M., et al.: Genetic-algorithm-based optimization approach for energy management. IEEE Trans. Power Deliv. **28**, 162–170 (2013). https://doi.org/10.1109/TPWRD.2012.2219598
20. Merei, G., Berger, C., Sauer, D.U.: Optimization of an off-grid hybrid PV-Wind-Diesel system with different battery technologies using genetic algorithm. Sol. Energy **97**, 460–473 (2013). https://doi.org/10.1016/j.solener.2013.08.016
21. Alam, D.F., Yousri, D.A., Eteiba, M.B.: Flower pollination algorithm based solar PV parameter estimation. Energy Convers. Manag. **101**, 410–422 (2015). https://doi.org/10.1016/j.enconman.2015.05.074
22. Eseye, A.T., Zhang, J., Zheng, D.: Short-term photovoltaic solar power forecasting using a hybrid Wavelet-PSO-SVM model based on SCADA and meteorological information. Renew. Energy **118**, 357–367 (2018). https://doi.org/10.1016/j.renene.2017.11.011
23. Clarke, J., McLay, L., McLeskey, J.T.: Comparison of genetic algorithm to particle swarm for constrained simulation-based optimization of a geothermal power plant. Adv. Eng. Inf. **28**, 81–90 (2014). https://doi.org/10.1016/j.aei.2013.12.003
24. Saffari, H., Sadeghi, S., Khoshzat, M., Mehregan, P.: Thermodynamic analysis and optimization of a geothermal Kalina cycle system using artificial bee colony algorithm. Renew. Energy **89**, 154–167 (2016). https://doi.org/10.1016/j.renene.2015.11.087
25. Özkaraca, O., Keçebaş, A.: Performance analysis and optimization for maximum exergy efficiency of a geothermal power plant using gravitational search algorithm. Energy Convers. Manag. **185**, 155–168 (2019). https://doi.org/10.1016/j.enconman.2019.01.100
26. Bemani, A., Xiong, Q., Baghban, A., et al.: Modeling of cetane number of biodiesel from fatty acid methyl ester (FAME) information using GA-, PSO-, and HGAPSO- LSSVM models. Renew. Energy **150**, 924–934 (2020). https://doi.org/10.1016/j.renene.2019.12.086
27. Amirante, R., Tamburrano, P.: Novel, cost-effective configurations of combined power plants for small-scale cogeneration from biomass: feasibility study and performance optimization. Energy Convers. Manag. **97**, 111–120 (2015). https://doi.org/10.1016/j.enconman.2015.03.047
28. Gholami, K., Dehnavi, E.: A modified particle swarm optimization algorithm for scheduling renewable generation in a micro-grid under load uncertainty. Appl. Soft. Comput. J. **78**, 496–514 (2019). https://doi.org/10.1016/j.asoc.2019.02.042
29. Zheng, Y., Jenkins, B.M., Kornbluth, K., Træholt, C.: Optimization under uncertainty of a biomass-integrated renewable energy microgrid with energy storage. Renew. Energy **123**, 204–217 (2018). https://doi.org/10.1016/j.renene.2018.01.120
30. Ghasemi, A., Enayatzare, M.: Optimal energy management of a renewable-based isolated microgrid with pumped-storage unit and demand response. Renew. Energy **123**, 460–474 (2018). https://doi.org/10.1016/j.renene.2018.02.072
31. Yang, H.T., Wu, Y.S., Liao, J.T.: Economic dispatch and frequency-regulation reserve capacity integrated optimization for high-penetration renewable smart grids. IEEE PES Innov. Smart Grid Technol. Asia, ISGT 2980–2985 (2019). https://doi.org/10.1109/ISGT-Asia.2019.8881726
32. Rajanna, S., Saini, R.P.: Development of optimal integrated renewable energy model with battery storage for a remote Indian area. Energy **111**, 803–817 (2016). https://doi.org/10.1016/j.energy.2016.06.005
33. Suresh, V., Muralidhar, M., Kiranmayi, R.: Modelling and optimization of an off-grid hybrid renewable energy system for electrification in a rural areas. Energy Rep. **6**, 594–604 (2020). https://doi.org/10.1016/j.egyr.2020.01.013
34. Nyeche, E.N., Diemuodeke, E.O.: Modelling and optimisation of a hybrid PV-wind turbine-pumped hydro storage energy system for mini-grid application in coastline communities. J. Clean. Prod. **250**, 119578 (2020). https://doi.org/10.1016/j.jclepro.2019.119578
35. Bansal, A.K., Kumar, R., Gupta, R.A.: Economic analysis and power management of a Small autonomous hybrid power system (SAHPS) using biogeography based optimization (BBO) algorithm. IEEE Trans. Smart Grid **4**, 638–648 (2013). https://doi.org/10.1109/TSG.2012.2236112

36. Karaboga, D., Basturk, B.: A powerful and efficient algorithm for numerical function optimization: artificial bee colony (ABC) algorithm. J. Glob. Optim. **39**, 459–471 (2007). https://doi.org/10.1007/s10898-007-9149-x
37. Mirjalili, S., Lewis, A.: The whale optimization algorithm. Adv. Eng. Softw. **95**, 51–67 (2016). https://doi.org/10.1016/j.advengsoft.2016.01.008
38. Mirjalili, S., Mirjalili, S.M., Lewis, A.: Grey wolf optimizer. Adv. Eng. Softw. **69**, 46–61 (2014). https://doi.org/10.1016/j.advengsoft.2013.12.007
39. Yang, X.S., Deb, S.: Cuckoo search via Lévy flights. In: Proceedings of 2009 world congress national biology inspired computing NABIC, pp. 210–214 (2009). https://doi.org/10.1109/NABIC.2009.5393690
40. Yang, X.S., Karamanoglu, M., He, X.: Flower pollination algorithm: a novel approach for multiobjective optimization. Eng. Optim. **46**, 1222–1237 (2014). https://doi.org/10.1080/0305215X.2013.832237
41. Mirjalili, S., Gandomi, A.H., Mirjalili, S.Z., et al.: Salp swarm algorithm: a bio-inspired optimizer for engineering design problems. Adv. Eng. Softw. **114**, 163–191 (2017). https://doi.org/10.1016/j.advengsoft.2017.07.002
42. Mirjalili, S.: Moth-flame optimization algorithm: a novel nature-inspired heuristic paradigm. Knowled. Based Syst. **89**, 228–249 (2015). https://doi.org/10.1016/j.knosys.2015.07.006
43. Saremi, S., Mirjalili, S., Lewis, A.: Grasshopper optimisation algorithm: theory and application. Adv. Eng. Softw. **105**, 30–47 (2017). https://doi.org/10.1016/j.advengsoft.2017.01.004
44. Mirjalili, S.: Dragonfly algorithm: a new meta-heuristic optimization technique for solving single-objective, discrete, and multi-objective problems. Neural Comput. Appl. **27**, 1053–1073 (2016). https://doi.org/10.1007/s00521-015-1920-1
45. Askarzadeh, A.: A novel metaheuristic method for solving constrained engineering optimization problems: crow search algorithm. Comput. Struct. **169**, 1–12 (2016). https://doi.org/10.1016/j.compstruc.2016.03.001
46. Rao, R.V., Savsani, V.J., Vakharia, D.P.: Teaching-learning-based optimization: a novel method for constrained mechanical design optimization problems. CAD Comput. Aided Des. **43**, 303–315 (2011). https://doi.org/10.1016/j.cad.2010.12.015
47. Venkata Rao, R.: Jaya: a simple and new optimization algorithm for solving constrained and unconstrained optimization problems. Int. J. Ind. Eng. Comput. **7**, 19–34 (2016). https://doi.org/10.5267/j.ijiec.2015.8.004

Learning Automata and Soft Computing Techniques Based Maximum Power Point Tracking for Solar PV Systems

S. Sheik Mohammed, D. Devaraj, and T. P. Imthias Ahamed

1 Introduction

Electricity has become the essential need for day-to-day life. Generation of electricity has been through a long way from the time Edison's DC electricity power station in Pearl Street, New York, began its operation in 1882 [1]. The non-renewable sources such as wood, coal, fossils fuels and gas are used as the fuel for power generation over centuries. Nuclear power generating stations are also under operation in many countries. Using such resources as fuel for power generation threatens the natural living conditions and the environment. It is reported in Emissions Gap Report (EGR) 2019 that the greenhouse gas emission (GHG) has raised at the rate of 1.3% per year between 2009 and 2018. Emissions have reached a new record of 55.3 GtCO2e in 2018. Two critical solutions suggested in EGR 2019 for emission reduction are electrification of transportation and expansion of renewable energy-based electricity generation [2]. Solar PV-based power generation is a promising alternative energy solution and has been widely accepted as a primary source of energy by many countries in the last two decades. By the end of 2018, the global installation capacity has reached 512 GW and with an addition of 141 GW; the estimated capacity of solar PV power plant for the year 2019 was 653 GW. Solar PV system offers several advantages such as easy installation and less maintenance. Technological advancements

S. Sheik Mohammed (✉) · T. P. Imthias Ahamed
TKM College of Engineering, Kollam, India
e-mail: ssheikmd@yahoo.co.in

T. P. Imthias Ahamed
e-mail: imthaisa@gmail.com

D. Devaraj
Kalasalingam Academy of Research and Education, Srivilliputhur, Tamil Nadu, India
e-mail: deva230@yahoo.com

B. Vinoth Kumar et al. (eds.), *Intelligent Paradigms for Smart Grid and Renewable Energy Systems*, Algorithms for Intelligent Systems,
https://doi.org/10.1007/978-981-15-9968-2_7

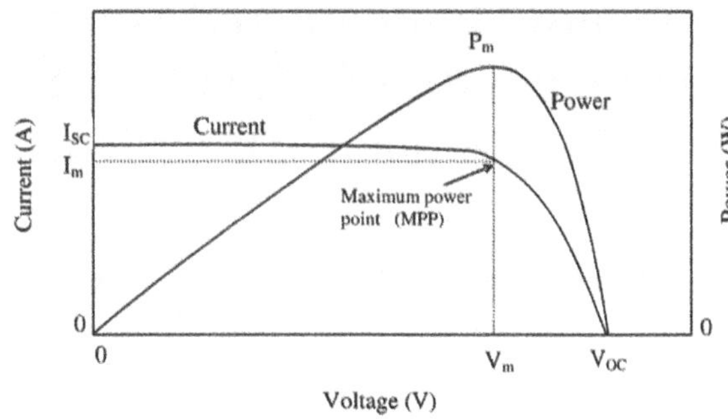

Fig. 1 Characteristic curves of SPV cell

in PV cell technology and inverter technologies had brought a significant reduction in the system cost during the last decade.

The characteristic curves of SPV module are depicted in Fig. 1. At any instant, the efficiency and output power of SPV cell are maximum at a point which is called as maximum power point (MPP). MPP is denoted as P_m in the figure. MPP is dependent on the input conditions of SPV module. To be more specific, photovoltaic current, i.e., the current generated by the SPV module and irradiation, is directly proportional to each other. On the other hand, voltage across the module decreases with rise in temperature. Hence, the MPP varies whenever the input condition changes. Therefore, in order to improve the generation efficiency, the SPV system should be always operated at the MPP under all input conditions.

In solar PV systems, maximum power point tracker (MPPT) is incorporated to operate them in optimal conditions for maximum power harvesting under all input conditions. MPPTs are electronic controllers aided by an algorithm for controlling the operation of power electronic converter connected to the PV arrays in a solar PV system.

This chapter discusses the implementation of MPPT algorithm for stand-alone SPV system. In Sect. 2, components of solar PV system are explained. In Sect. 3, a detailed review on maximum power point techniques is presented. The basic concept of learning automata and finding optimal solution using learning automata are outlined in Sect. 4. Optimization is explained with the help of *n-arm* bandit problem. In Sect. 5, development of hybrid MPPT algorithm using P&O MPPT and LA is explained. A comprehensive comparison between the P&O MPPT, VSS P&O MPPT and the hybrid MPPT is presented in Sect. 6. Section 7 discusses the development of soft computing technique-based MPPT controllers. Fuzzy MPPT, GA-optimized fuzzy MPPT and ANFIS MPPT are the main topics of the study in this section. Simulation and analysis of the soft computing techniques-based MPPT are also discussed. The chapter is concluded in Sect. 8.

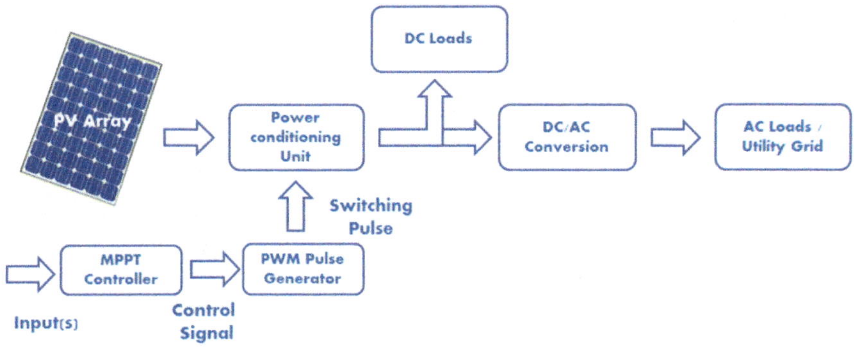

Fig. 2 Functional block diagram of solar PV system

2 Components of Solar PV System

Operation of solar PV system and its basic types are briefly explained in this section. The functional block diagram of SPV system is depicted in Fig. 2.

Solar PV array produces DC electricity. The output of SPV array is processed in the power conditioning unit (DC–DC converter). A DC–DC boost converter is selected as power condition unit (PCU) in this work. The MPPT produce appropriate control signals depends on the input signals received. The input signals to the MPPT can be physical parameters such as irradiation, temperature or the electrical parameters like current, voltage output and power of the SPV array. Based on the control signal produced by the MPPT controller, the PWM pulse generator produces the switching pulse for the converter. The input(s) are continuously monitored by the MPPT controller. With respect to the changes in input(s), the operating point of the converter is adjusted to harvest maximum power at all input conditions. The converter output can be directly fed to the DC loads with voltage regulation. To feed into the AC loads/utility services, the DC power should be converted into AC. A stand-alone SPV system has battery storage system and does not supply electricity to the grid. The on-grid PV system works with grid supply as reference. It supplies power to the grid when generation is excessive and power is taken from the grid during deficit generation. The on-grid PV system does not have energy storage.

3 Maximum Power Point Tracking Techniques

In Fig. 3a, *P–V* characteristics of SPV module for different input conditions are depicted. MPPs are marked as $A1$, $A2$, $B1$ and $B2$. For the input conditions $C1$, i.e., for 25 °C and 1000 W/m^2 input, the maximum power point is $A1$. Compared to $C1$, in $C2$ the irradiation is decreased. With reduction in irradiation, the power output is reduced and the MPP is now $A2$. For $C3$, the maximum power point is $B1$. At $C4$,

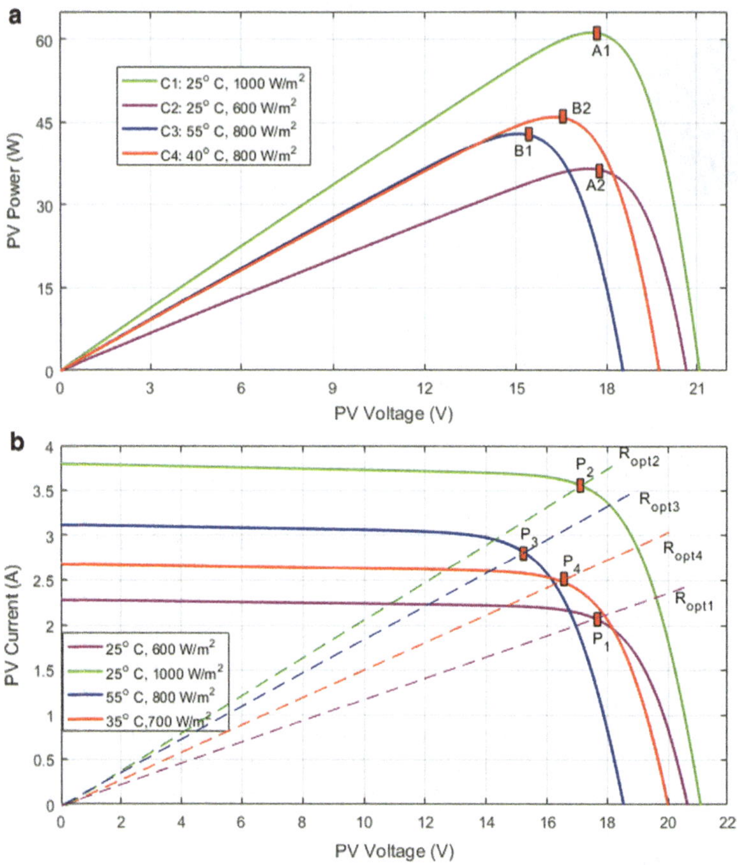

Fig. 3 **a** *P–V* characteristics of SPV module for different input conditions. **b** *I–V* characteristics of SPV module for different input conditions

with increase in temperature from *C*3, power generated by PV has reduced and the MPP shifted from *B*1 to *B*2. From the above discussion, it can be understood that the output power of SPV module is dependent on the input conditions and hence the MPP too.

According to maximum power transfer theorem, maximum power will be transferred from the source to load only when the load and source impedance are equal. To analyze the effect of optimal resistance, *I–V* characteristics of SPV module for different input conditions depicted in Fig. 3b are considered. R_{opt} is the optimal load resistance with respect to the input conditions of the SPV module. For the given input conditions (25 °C, 600 W/m^2), maximum power point is P_1 and the system will deliver maximum power when optimal resistance is R_{opt1}. For every change in input conditions, MPP of SPV module shifts as P_2,P_3 and P_4as shown in Fig. 3b. Therefore, R_{opt} also needs to be changed, respectively, as R_{opt2},$R_{opt3}$and R_{opt4} for every change in input conditions for maximum power extraction. It is practically impossible to change the load resistance for every change in the input conditions. Hence, MPPT controllers are integrated in SPV systems to control the operation of system for maximum power extraction at all input conditions.

Numerous MPPT techniques are presented in literatures [3–5]. The proposed MPPT techniques are mainly classified as (a) online MPPT techniques, (b) offline MPPT techniques and (c) hybrid MPPT techniques. A detailed review on each category is discussed in the following subsections.

3.1 Online MPPT Techniques

The online MPPT techniques determine the position of the MPP by sensing instantaneous voltage and current of SPV module/array. P&O MPPT technique is the most popular online MPPT technique. Incremental conductance (InC) technique, hill climbing MPPT, open-circuit voltage (OCV), extremum seeking control (ESC) and short-circuit current (SCC) are some of the other online MPPT techniques [3–10]. In P&O method, the operating point is varied in steps after every perturbation. The perturbation will be decided based on the observation of $\frac{dP}{dV}$ in each step. P&O method is a very popular and most widely accepted technique among the conventional online MPPT techniques [6, 7]. However, conventional P&O MPPT suffers from slow tracking since the step size is fixed (usually 0.01), and it produces oscillation at MPP. Another major drawback of this method is that under rapidly varying environmental conditions, the tracking is improper and it deviates from MPP. Variants of conventional P&O technique are proposed in literatures to improve the tracking performance [11, 12]. The adaptive and VSS P&O MPPT technique work with multiple step size values. These techniques help to improve the tracking speed. But still oscillations exist at MPP. Moreover, tracking becomes improper especially under frequently changing input conditions.

To track MPP using OCV technique, the V_{oc} of PV array should be measured in a regular interval. Similarly, $I_{sc}$of PV array should be measured in a regular interval for SCC technique. To do so, PV array must be disconnected from the system. This leads to power loss, and hence, these techniques are inefficient. Other MPPT techniques like InC and ESC have implementation complexity.

3.2 Offline MPPT Techniques

The offline MPPT techniques are knowledge-based or training-based algorithm. In most cases, a thorough knowledge about the dynamics of system is essential for the algorithm development. The offline MPPT techniques use either the pre-assigned electrical/physical parameters or the mathematical functions obtained from empirical data for the estimation of MPP. The conventional offline MPPT technique is the look-up table method [13]. The most common MPPT algorithms based on soft computing techniques are fuzzy logic control (FLC)-based MPPT. MPPT techniques based on computational intelligence-based optimization techniques like particle swarm optimization (PSO), genetic algorithm (GA) and artificial neural network (ANN) are some other offline MPPT techniques [3–5, 14–16]. In LUT method, the input and output are pre-assigned by the user based on the data collected through experiments conducted on the system. So, this technique does not require any training or computational algorithm for its implementation. But, this technique provides accurate results only for the pre-assigned conditions. Hence, the method is not popular. The soft computing technique-based MPPTs provide better performance. They have good tracking speed with better accuracy and are capable to outperform under fast changing weather conditions also. However, such performance can be attained upon performing trainings with large number of data and with better hardware. This leads to implementation complexity and high cost. But still the soft computing techniques are popular due to their faster and accurate performance.

3.3 Hybrid MPPT Techniques

Hybrid MPPT algorithms are the combination of more than one MPPT technique. In most cases, hybrid MPPT techniques are developed by combining an online technique with an offline technique [17]. In [18], P&O and PSO technique-based hybrid MPPT is proposed. This algorithm is effective and accurate, but its implementation is not easy. The hybrid algorithm proposed in [19] combines OCV and P&O MPPT technique. Initially, MPP is tracked from V_{oc} measurement, and then P&O fine-tunes to reach the MPP. Another hybrid MPPT proposed in [20] is the combination of SCC technique with P&O MPPT. The works presented in [19] and in [20, 21] are very similar. In [19], short-circuit current method is adopted, and in [20, 21] open-circuit voltage method is used along with P&O MPPT method.

Most of the hybrid MPPT methods proposed in literatures combine an offline algorithm with P&O MPPT. Hybridization improves the tracking speed and reduces the oscillation also. In most cases, the hybrid algorithms work in two stages. The first stage helps to reach near the MPP quickly, and the second stage fine-tunes to attain MPP with reduced oscillation. A hybrid MPPT using learning automata and P&O MPPT is presented in this chapter.

4 Learning Automata

Learning automaton is an optimization tool suitable for finding optimal solutions for a single-stage decision-making problem (SSDMP) in a random environment. The concept of learning automata (LA) was introduced by Tsetlin [22]. A detailed survey

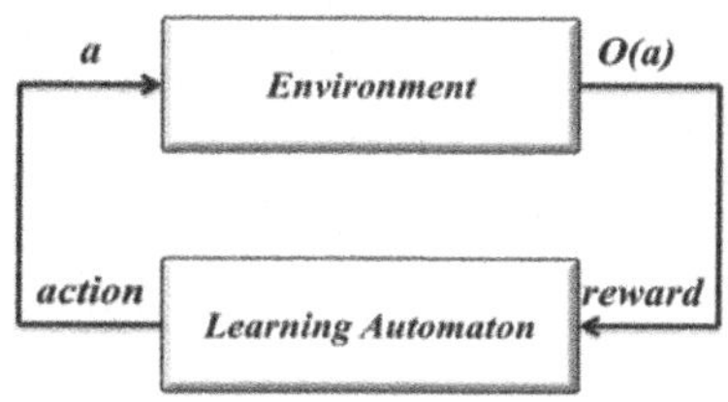

Fig. 4 Block diagram of learning automaton

on LA-based researches and different models of LA presented in various literatures are reported in [23, 24]. LA models have been adopted mostly in communication systems to take adaptive decisions. In [25], LA are used for conflict avoidance in star networks. Few other applications where LA are adopted are dynamic channel allocation [26] and for ATM switches [27]. Another key area where LA have been used for decision making and to find optimal solutions is electrical engineering problem such as optimal load scheduling [28], economic dispatch of the power network [20] and automatic generation control [29].

4.1 Learning Automaton

The block diagram of learning automaton is illustrated in Fig. 4. LA receive the response ($O(a)$) from the environment for sending an action (a). For every action sent by LA, environment will respond and this response can be either reward or penalty. LA are assigned with a set of actions (A). The objective of LA is to find the optimal action from "A" which provides the best reward. So, LA will repeatedly collect responses by sending the actions randomly. The response given by the environment may also be random. From the random rewards received through repeated random actions, LA will learn and identify the optimal action which provides the best reward.

To analyze the optimization of best solution for a given problem and the need for a machine learning tool like learning automata for finding optimal solutions, n-arm bandit problem is taken as an example.

4.2 n-arm Bandit Problem

Figure 5 shows a typical n-arm slot machine. It is a gambling game machine, and it has n number of arms. In this game, whenever an arm is played it gives a reward. The arm is denoted as "a" where $a = 1, 2, \ldots n$. Each arm in the machine is assigned with a set of rewards (X_a). Therefore, playing the same arm again will not give the same reward; i.e., reward will be random for playing the same arm again and again. The goal of the player is to find the arm which gives the highest reward.

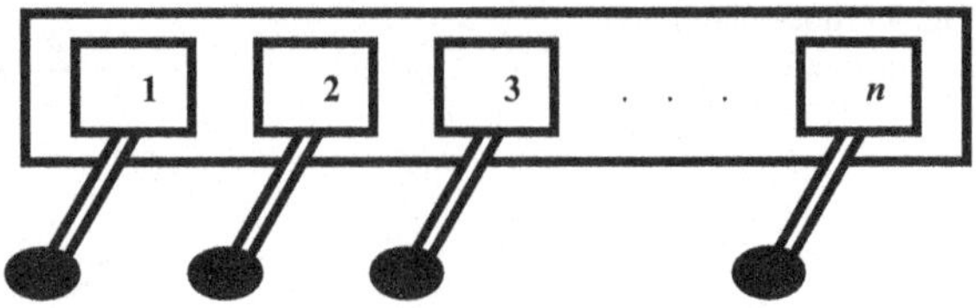

Fig. 5 *n-arm* slot machine

Let us assume any one of the arms can be played once by paying an amount of \$1. When the player plays an "*a*", a random reward of $O(a)$ will be given for playing arm "*a*". The set of rewards assigned for arm "2" are [0.75\$, 0.8\$, 0.85\$, 0.9\$, 0.95\$, 1.0\$, 1.05\$, 1.1\$, 1.15\$]. The probability density function (pdf) of the arm is denoted as P_a. The uniformly distributed *pdf* of arm "2" can be expressed as

$$P\{O(1) = M_2(1)\} = P\{O(1) = M_2(2)\} = \ldots = P\{O(1) = M_2(m)\} = \frac{1}{m}$$

where m is the number of rewards for each arm. Therefore,

$$P_2 = \left[\frac{1}{9}, \frac{1}{9}, \frac{1}{9}, \frac{1}{9}, \frac{1}{9}, \frac{1}{9}, \frac{1}{9}, \frac{1}{9}, \frac{1}{9}\right]$$

The expected mean value of rewards for the second arm of the slot machine is $q(2) = \$0.95$. The random rewards for each arm are $Y_{1a}, Y_{2a}, \ldots Y_{ma}$, and the expected mean value of arm a is $q(a)$. Therefore, to find the best arm, i.e., to find the arm that provides the highest reward among the arms in the slot machine, every arm should be played for several numbers of times. This is a tedious process and, practically, a non-feasible solution.

To understand the complexity, let us consider that arm "*a*" is played for "*N*" times. Then, the estimated mean of arm "*a*" can be

$$q^N(a) = \frac{\sum_{i=1}^{N} O^i(a)}{N} \tag{1}$$

Using the above equation, the estimated mean $q^N(a)$ of all the arms can be found. Then, from the observed estimated mean values of all the arms, the arm with maximum estimated mean can be identified. The arm with maximum estimated mean is the best arm, and it is denoted as "greedy arm". Greedy arm is expressed as

$$q^N(a_g) = \max_{a \in A} q^N(a) \tag{2}$$

The argument corresponding to the maximum estimated mean "argmax" is given by

$$q^N(a_g) = \max_{a \in A} q^N(a) \Rightarrow a_g = \arg\max_{a \in A} q^N(a) \tag{3}$$

This method is suitable only for the problems, which are having less number of actions. For example, consider the machine has 5 arms (a) and the number of trials (N) for each arm is 10, and then the total number of plays is $x = a \times N$ which is $x = 50$. But, the number of plays will increase exponentially as the number of trails and number of arms increase. Consider $a = 50$ and $N = 500$; then, x will be 25,000. Therefore, a player should play 25,000 trials in order to find the arm which gives best reward.

4.3 Optimization Using Learning Automata

LA approach the same problem in a different way. In LA, the arms will not be played for "ax N" times. Because here the goal is not to find the mean of all the arms, rather it is only to find the best arm. Thus, instead of playing all the arms for "x" times and finding their estimated mean, LA approach the problem differently. Rather than playing all the arms, LA play the arms which are having more possibility for being the best. LA use *pursuit algorithm* (PA) [29] as an algorithm for finding the optimal action. PA is a very simple algorithm, and its convergence toward the optimal action is rapid. At each iteration, PA allows to "pursue" the best solution or best action corresponds to the largest value of the estimates of reward probabilities.

5 Hybrid MPPT Algorithm Using Perturb and Observe Algorithm and Learning Automata

As discussed earlier, the MPP will vary depending on the environmental conditions. However, for any given input conditions there is only one MPP and therefore the duty corresponding to the input conditions to achieve the MPP will also be unique. Due to this reason, MPPT can be treated as a single-stage decision-making problem. This has led to the idea of studying the adaptability of LA for MPPT development. The development of hybrid MPPT algorithm using P&O and LA is explained in this section followed by the brief overview of P&O and VSS P&O MPPTs. P&O and LA-based hybrid MPPT is modeled and built in MATLAB. Simulation of solar PV system for different input conditions with P&O, VSS P&O and the proposed hybrid MPPT is carried out, and a comprehensive comparison is made.

5.1 P&O MPPT Technique

P–V characteristic curve with different operating points is illustrated in Fig. 6. It can be noted that $\frac{\Delta P_{\text{pv}}}{\Delta V_{\text{pv}}}$ is always positive at left of MPP and the ratio is always negative at right of MPP. At MPP, the ratio is equal to zero (Fig. 7).

P&O MPPT observes $\frac{\Delta P_{\text{pv}}}{\Delta V_{\text{pv}}}$ after every perturbation and decides the next perturbation based on the value from the previous perturbation. The operating points toward the MPP are indicated with green arrow, and the operating points away from MPP are indicated by red arrow. Consider the operating point has shifted from "A" to "B" upon applying a perturbation. It means, for the applied perturbation, the operating point of SPV has moved toward the MPP. Therefore, perturbation can be proceeded in the same direction. Instead, if the new operating point becomes "BB" for the applied perturbation, then the MPPT is proceeding in the opposite direction of MPP. In this case, the perturbation should be reversed to reach MPP.

The operation at right side of MPP also should be realized in the same way. The system is at point "C", and for the applied perturbation the operating point has moved to "D" and then the applied perturbation is correct and can be proceeded further in the same direction to attain MPP. Otherwise, direction of perturbation must be reversed to attain MPP.

Figure 6 illustrates the P&O MPPT flowchart. The algorithm continuously perturbs and observes until it reaches the MPP. The duty cycle step size is a fixed value for P&O MPPT. Hence, tracking performance is low. Secondly, since the algorithm continuously perturbs, the condition for MPP $\left(\frac{\Delta P_{\text{pv}}}{\Delta V_{\text{pv}}} = 0\right)$ cannot be achieved. Therefore, the duty cycle oscillates around the MPP.

The tracking performance of P&O MPPT is explained with an example here. The irradiation of a particular SPV system is changed after every 20 ms as shown in Fig. 8a. The expected (actual) output from an efficient MPPT and the MPP tracking of P&O MPPT for the given inputs (irradiation) are depicted in Fig. 8b

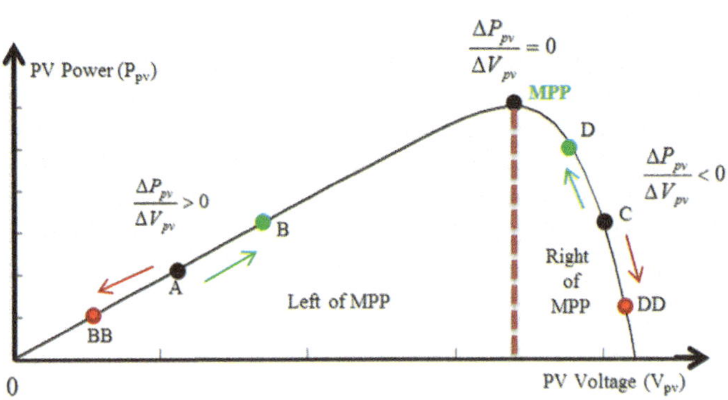

Fig. 6 Operating points of MPPT

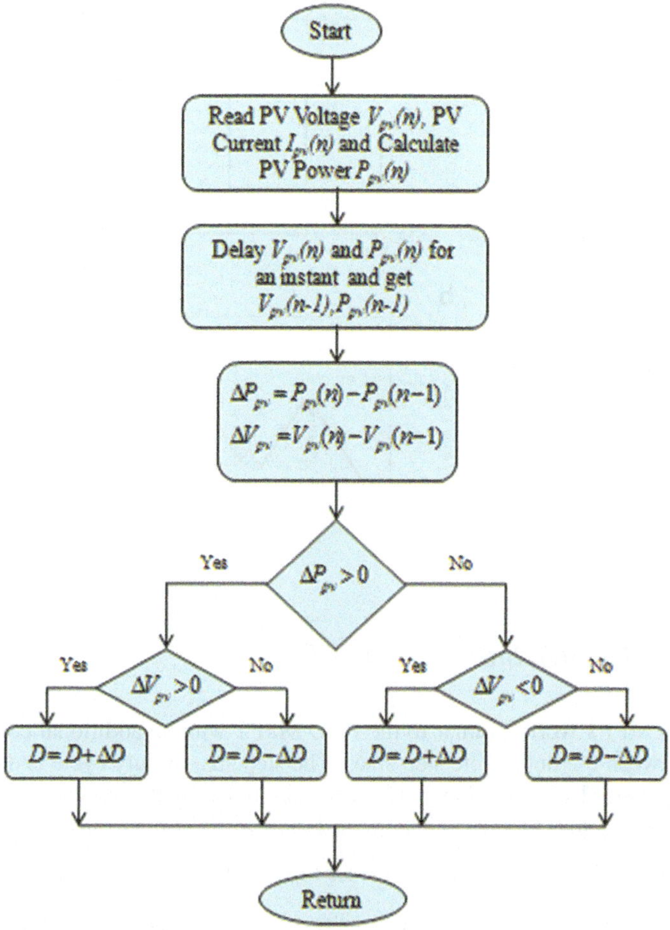

Fig. 7 Flowchart of P&O MPPT

From the figure, it can be realized that the tracking performance of P&O MPPT is slow due to fixed step size value. Moreover, the P&O MPPT failed to reach the MPP under fast changing input conditions. Further, for the cases where it reached the MPP, the response is poor and it oscillates around MPP. Variants of P&O MPPT such as VSS and adaptive step size P&O are introduced to overcome the above-said issues.

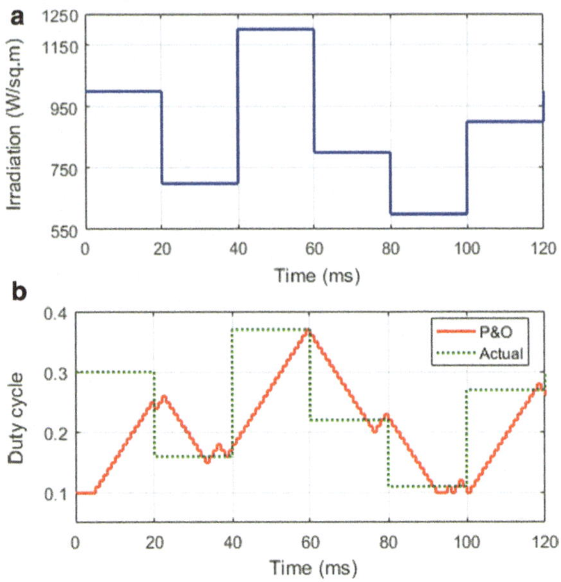

Fig. 8 **a** Change of irradiation. **b** MPP tracking of P&O MPPT

5.2 *Variable Step Size P&O MPPT*

VSS P&O MPPT works similar to the P&O MPPT with an additional condition to switch between the duty cycle step sizes. The step size of duty cycle is decided by the error value. The error value and step size are tuned by the user in trial and error basis. When a sudden change occurs in the inputs, $\frac{\Delta P_{\text{pv}}}{\Delta V_{\text{pv}}}$ will become large. This will satisfy the condition, $\frac{\Delta P_{\text{pv}}}{\Delta V_{\text{pv}}} >$ error. Under this condition, MPPT applies large step size value. As the operating point approaches close to the MPP, the ratio $\left(\frac{\Delta P_{\text{pv}}}{\Delta V_{\text{pv}}}\right)$ will get reduced and become less than the error value. Then, the MPP switch to small step size value. The tracking speed of the MPPT is improved by this way. The flowchart of VSS P&O MPPT is shown in Fig. 9. Though the VSS P&O has improvement in performance when compared with P&O, its performance is not satisfactory under frequently changing input conditions.

A two-stage hybrid MPPT using P&O and LA is proposed to overcome the above-mentioned issues. The proposed LA-based hybrid algorithm has improved tracking performance under rapidly changing input conditions, and the oscillation is also reduced.

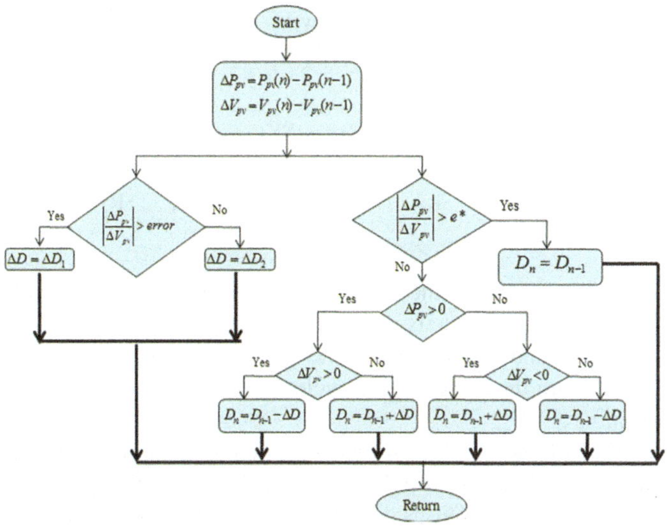

Fig. 9 VSS P&O MPPT flowchart

5.3 *Optimization of MPPT Using Learning Automata*

During the learning process, LA send an action and receives a response from the environment. In the case of SPV system, environment is SPV array; action is duty cycle, and the response received is output power. In Fig. 10, LA-based learning model of the SPV system is depicted.

For every input condition, there is an optimal duty cycle at which maximum power can be obtained from the system. The function of learning model is to find the optimal action for maximum power output at different operating conditions. PV

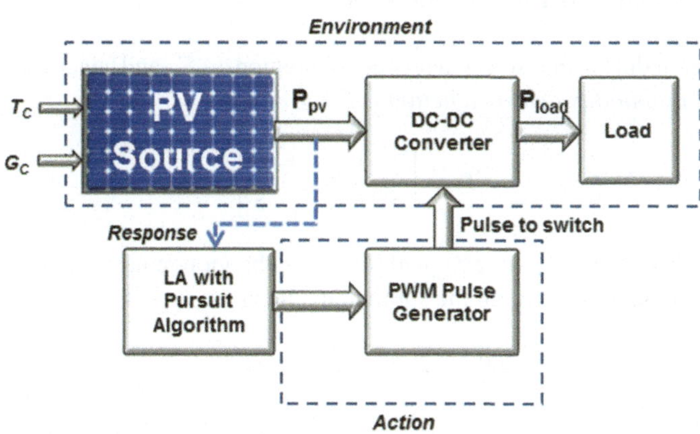

Fig. 10 LA model of SPV system for learning

Table 1 Subregions for learning the optimal duty cycle

G (W/m^2)→ T (°C)↓	500	600	700	800	900	1000	1100	1200	1300
20	SR_1	SR_2	SR_3	SR_4	SR_5	SR_6	SR_7	SR_8	SR_9
30	SR_{10}	SR_{11}	SR_{12}	SR_{13}	SR_{14}	SR_{15}	SR_{16}	SR_{17}	SR_{18}
40	SR_{19}	SR_{20}	SR_{21}	SR_{22}	SR_{23}	SR_{24}	SR_{25}	SR_{26}	SR_{27}

has wide operating region. So, the operating region of PV array is divided into 27 subregions and set of duty cycle is assigned to each subregion. The subregions are presented in Table 1, and a set of duty cycles is assigned for each region as given in Table 2. LA will learn the optimal duty cycle from the given set of duty cycles for each subregion.

The optimal action is found from the *set of actions* by LA through an iterative learning process. LA adjust the *probability density function* of the actions in the action set based on the *mean value.* This is done by the *pursuit algorithm.* The components of LA and optimization of MPPT using LA are explained here.

5.3.1 Action Set (*A*)

Action set contains the number of actions. In the case of solar PV system, action is the duty cycle. The number of duty cycles selected for each subregion is 5. Therefore, the action set for the subregions can be expressed as

$$A_i = \{a_{i1}, a_{i2} \ldots a_{i5}\} \quad \text{for} \quad i = 1, 2, \ldots, 27$$

5.3.2 Probability Density Function (*pdf*)

The number of duty cycles selected for each subregion is “5”, and hence the uniformly distributed *probability density function* is

$$P = \left[\frac{1}{5}, \frac{1}{5}, \frac{1}{5}, \frac{1}{5}, \frac{1}{5}\right]$$

The number of duty cycles assigned for each subregion remains same. Therefore, *P* also will remain same for all the subregions. It can be expressed as

$$P_1 = P_2 = \ldots = P_i = P$$

Table 2 Action set for subregions

Subregion	Set of duty cycles				
1	0.1	0.11	0.12	0.13	0.14
2	0.12	0.13	0.14	0.15	0.16
3	0.18	0.19	0.2	0.21	0.22
4	0.2	0.21	0.22	0.23	0.24
5	0.25	0.26	0.27	0.28	0.29
6	0.3	0.31	0.32	0.33	0.34
7	0.34	0.35	0.36	0.37	0.38
8	0.37	0.38	0.39	0.4	0.41
9	0.37	0.38	0.39	0.4	0.41
10	0.1	0.11	0.12	0.13	0.14
11	0.13	0.14	0.15	0.16	0.17
12	0.23	0.24	0.25	0.26	0.27
13	0.24	0.25	0.26	0.27	0.28
14	0.28	0.29	0.3	0.31	0.32
15	0.29	0.3	0.31	0.32	0.33
16	0.34	0.35	0.36	0.37	0.38
17	0.36	0.37	0.38	0.39	0.4
18	0.39	0.4	0.41	0.42	0.43
19	0.1	0.11	0.12	0.13	0.14
20	0.16	0.17	0.18	0.19	0.2
21	0.21	0.22	0.23	0.24	0.25
22	0.26	0.27	0.28	0.29	0.3
23	0.31	0.32	0.33	0.34	0.35
24	0.33	0.34	0.35	0.36	0.37
25	0.37	0.38	0.39	0.4	0.41
26	0.4	0.41	0.42	0.43	0.44
27	0.41	0.42	0.43	0.44	0.45

5.3.3 Mean Value (Q)

The mean value $q(a)$ is initially assigned as zero for all the arms. Therefore, initially the expected mean for all the actions is zero.

$$q_i = [0\,0\,0\,0\,0] \quad \text{for} \quad i = 1, 2, \ldots, 27$$

The algorithm starts the optimization process from subregion 1 (Z_1). In every iteration, upon sending a duty cycle (a), the output power is obtained as reward $O(a)$ for $Z_{1.}$ Based on the obtained reward and its position, "q" value corresponding to the action Z_1 will be updated using the recursive equation given below [30]

$$q^{N+1}(a) = q^{N}(a) + \alpha_N\left[O(a) - q^{N}(a)\right] \tag{4}$$

Once $q(a)$ is updated, based on the present estimates of $q^N(a)$ the algorithm will determine the best action (a_g) for Z_1 using Eq. (2). The algorithm updates the probability of all the actions in every iteration. That is, the algorithm will increase the probability for the selection of best duty cycle and the probability of other duty cycles in the action set will be decreased in every iteration. As the iterations increase, it converges toward the optimal action and the iterations continue until the probability of one of the actions becomes $0.99 \approx 1$. In every iteration, the probability of the actions is updated using Eqs. (5) and (6), respectively.

$$p^{N+1}(a_g) = p^{N}(a_g) + \beta(1 - p^{N}(a_g)) \tag{5}$$

$$p^{N+1}(a) = p^{N}(a) - \beta p^{N}(a) \quad \forall a \in A \neq a_g \tag{6}$$

where β is the probability learning rate ($0 < \beta < 1$) [31]. From the obtained results, best duty cycle for the selected subregion is found using the following equation.

$$P_{\max} = \max_{a \in A}[P(a)] \tag{7}$$

This way, the algorithm ensures selection of best action among the possible actions. The algorithm stops the optimization process when $P_{\max} = \max_{a \in A}[P(a)] > 0.999$. Therefore, the algorithm terminates naturally once the condition is met.

Once the process is completed for the selected subregion, the algorithm starts over from the beginning to find the optimal action for next subregion say Z_2, and the same process is repeated until the optimization of duty cycle for all the 27 subregions is completed.

5.3.4 Algorithm

The flowchart for learning the optimal action using pursuit algorithm is shown in Fig. 11.

To analyze how PA converges toward the optimal solution through iteration process, subregion Z_3 is selected as an example. At Z_3, the inputs to the SPV module are $T_3 = 20$ °C and $G_3 = 700$ W/m^2. The duty cycle values assigned for the selected subregion are [0.18 0.19 0.2 0.21 0.22]. At the initial stage, the *pdf* of all actions, i.e., for all the duty cycle, is same (0.2) as shown in Fig. 12a.

After every iteration, the *pdf* of greedy action is increased and other actions will be decreased. At the initial stage, the algorithm selects the best action randomly based on initial assumptions and the rewards are received. Figure 12b shows the *pdf*

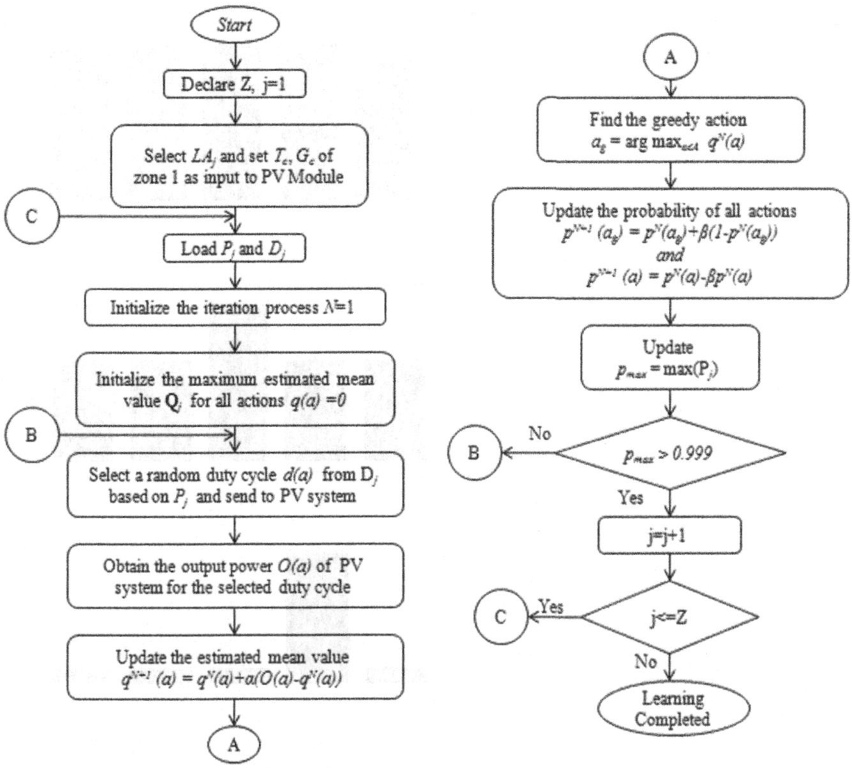

Fig. 11 Flowchart of duty cycle optimization for P&O and LA-based hybrid MPPT

of actions after few number of iterations. It can be seen that *pdf* of 3rd action is high comparing to others. It means that the algorithm identified $D = 0.2$ as the best duty cycle at this stage. But, it has not converged. Hence, the algorithm will continue the iteration process until the condition $P > 0.999$ is satisfied for any one of the duty cycles in the given set of duty cycles.

The probability density function of the action set for Z_3 at intermediate stage is shown in Fig. 12c. The algorithm has changed its optimal selection, and the optimal duty cycle is now $D = 0.19$. At this stage, also the condition for optimal action ($P > 0.999$) is not satisfied and hence, the process will continue. In Fig. 12, optimized *pdf* upon completion of learning is depicted. The algorithm has converged to $D = 0.19$ by satisfying the condition $P > 0.999$. Thus, the best action for subregion Z_3 is 0.19. PA identifies the optimal duty cycle for all the regions in the same way. The optimal duty cycle for all the 27 subregions is highlighted in Table 2.

Now, the LA is trained for all subregions and ready to use in the system to be controlled. So, the trained LA is integrated to the P&O MPPT. The block diagram of proposed P&O and LA-based hybrid MPPT technique is illustrated in Fig. 13.

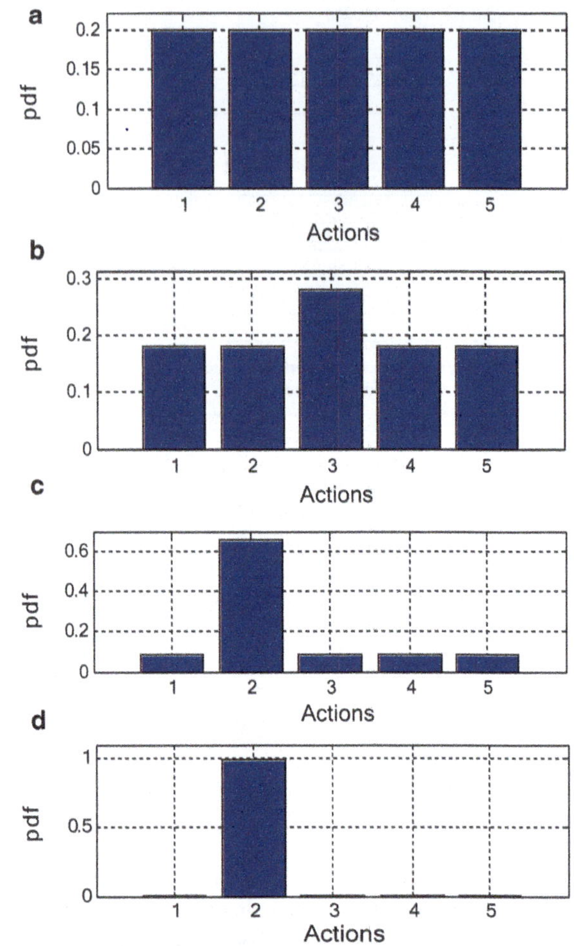

Fig. 12 **a** *pdf* at the beginning of learning. **b** *pdf* after few iterations of learning. **c***pdf* of action set at intermediate stage of learning. **d** Optimized *pdf* of subregion Z_3

In LA, the input values of each subregion have a range. For example, the values of subregion Z_1 are $G = 500$ W/m^2 and $T = 20$ °C. LA read the input values and identify the subregion to which the values belong to. Once the region is identified, then it will switch to it and fetch the duty cycle of the subregion.

In LA for Z_1, the range of input is assigned as $G_1 = 450 - 550$ W/m^2 and $T_1 = 15 - 25$ °C. Thus, for any values of input between these ranges, LA will provide the duty cycle of subregion Z_1 as output. Then, the SPV system will start operating at this duty cycle. Then, P&O MPPT observes and perturbs the duty cycle in steps until MPP is reached. Initially, LA will send the duty cycle. Then, P&O will follow and fine-tune to attain the maximum power point. This way the tracking performance is improved. In conventional P&O MPPT, the step size of duty cycle is set as $dD = 0.01$. But, in the hybrid MPPT the step size is set as $dD = 0.005$ to reduce the oscillation at MPP.

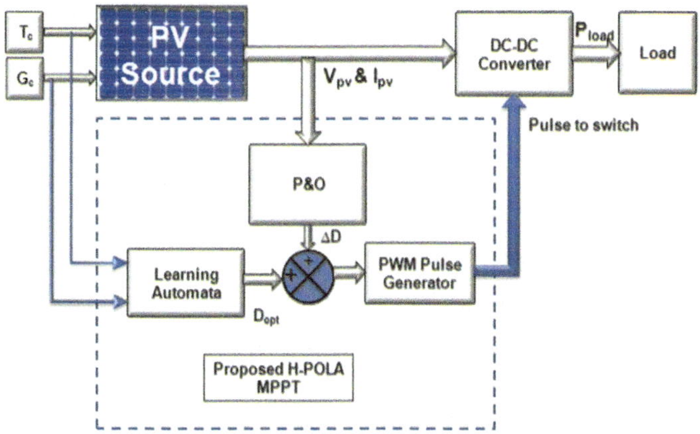

Fig. 13 Conceptual block diagram of P&O and LA-based hybrid MPPT

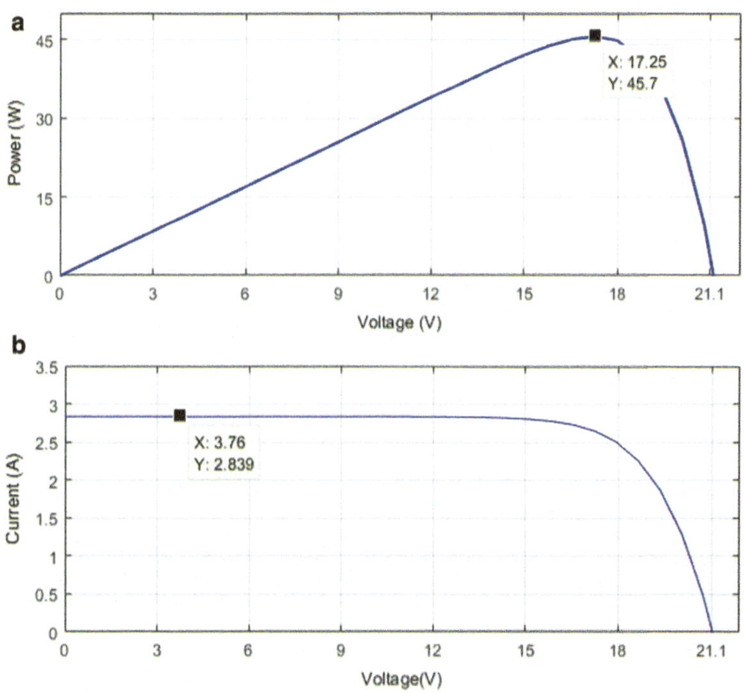

Fig. 14 **a** P–V curve of MSX 60 PV at $G = 750\ \text{W/m}^2$ and $T = 20\ °\text{C}$. **b** I–V curve MSX 60 PV at $G = 750\ \text{W/m}^2$ and $T = 20\ °\text{C}$

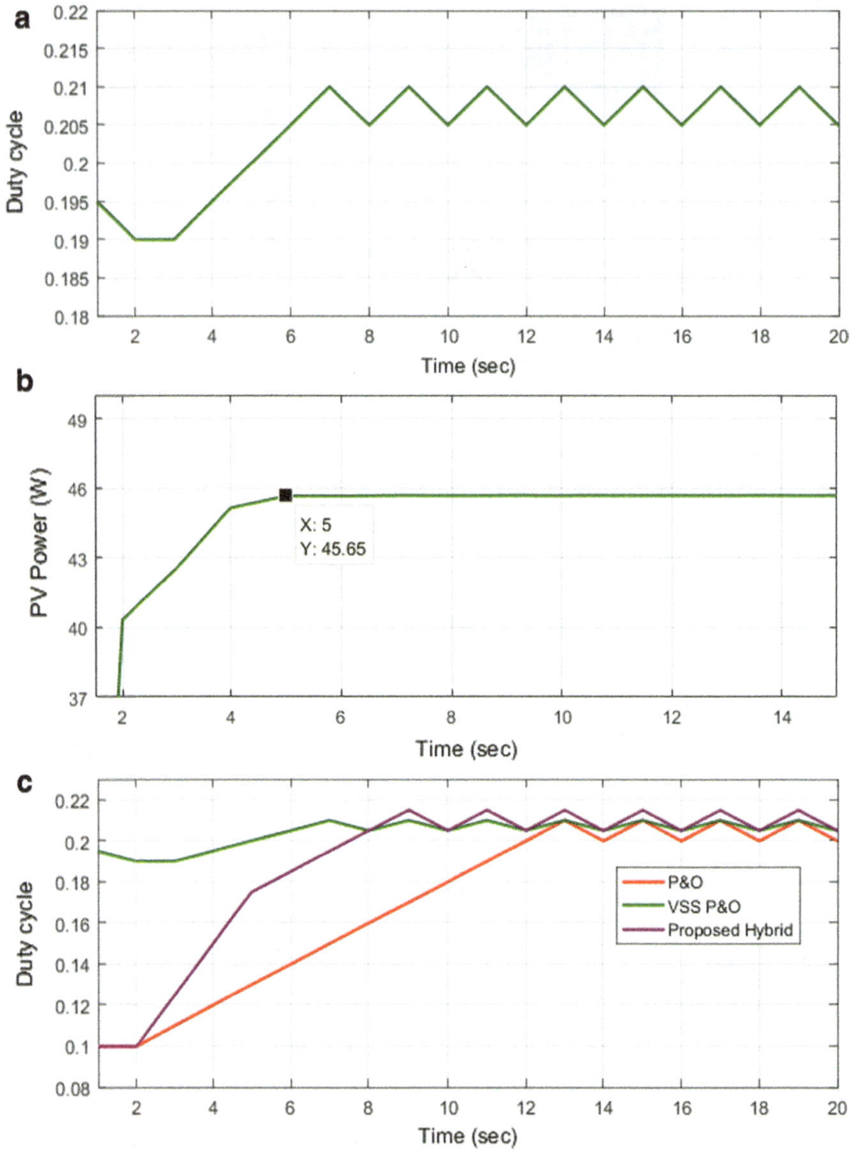

Fig. 15 **a** Duty cycle tracking at $G = 750\ \text{W/m}^2$ and $T = 20\ °\text{C}$. **b** Output power at $G = 750\ \text{W/m}^2$ and $T = 20\ °\text{C}$. **c** Duty cycle tracking of different MPPT techniques at $G = 750\ \text{W/m}^2$ and $T = 20\ °\text{C}$

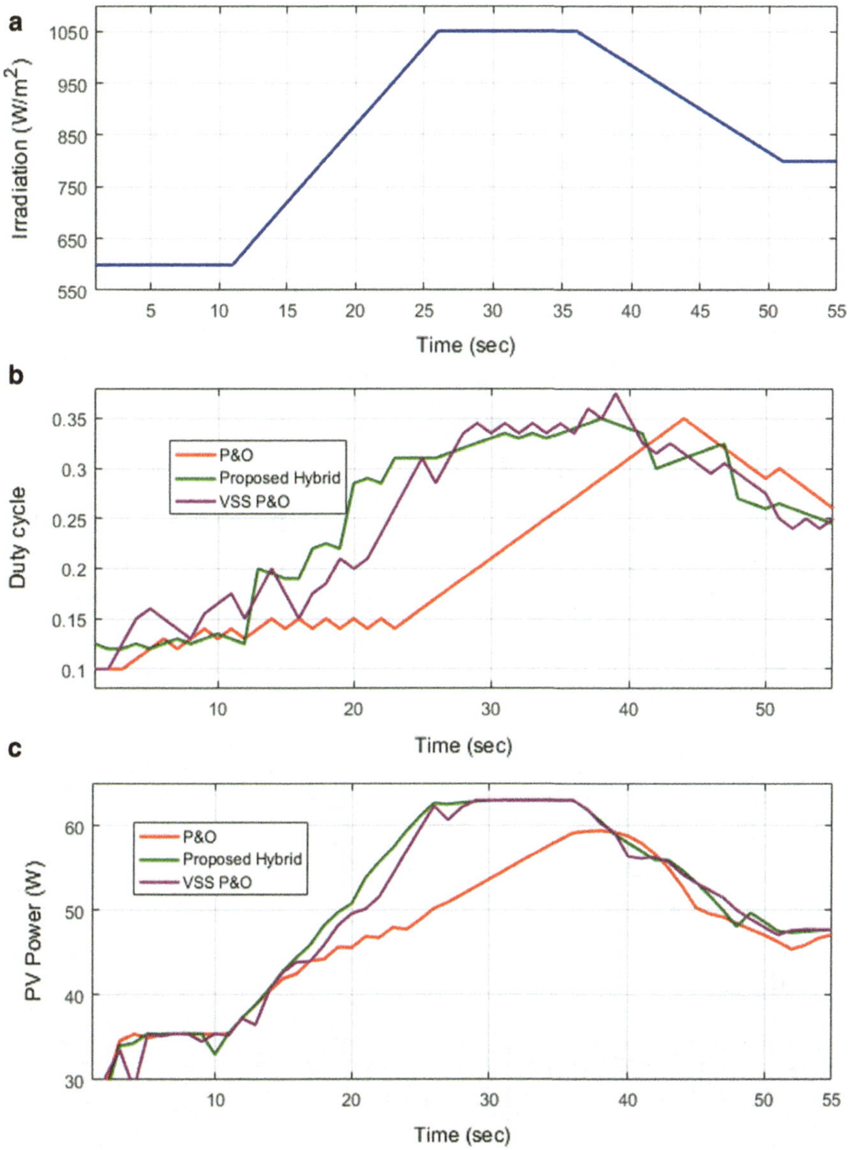

Fig. 16 **a** Slow changing irradiation pattern. **b** MPP tracking for slow changing irradiation. **c** PV output for slow changing irradiation

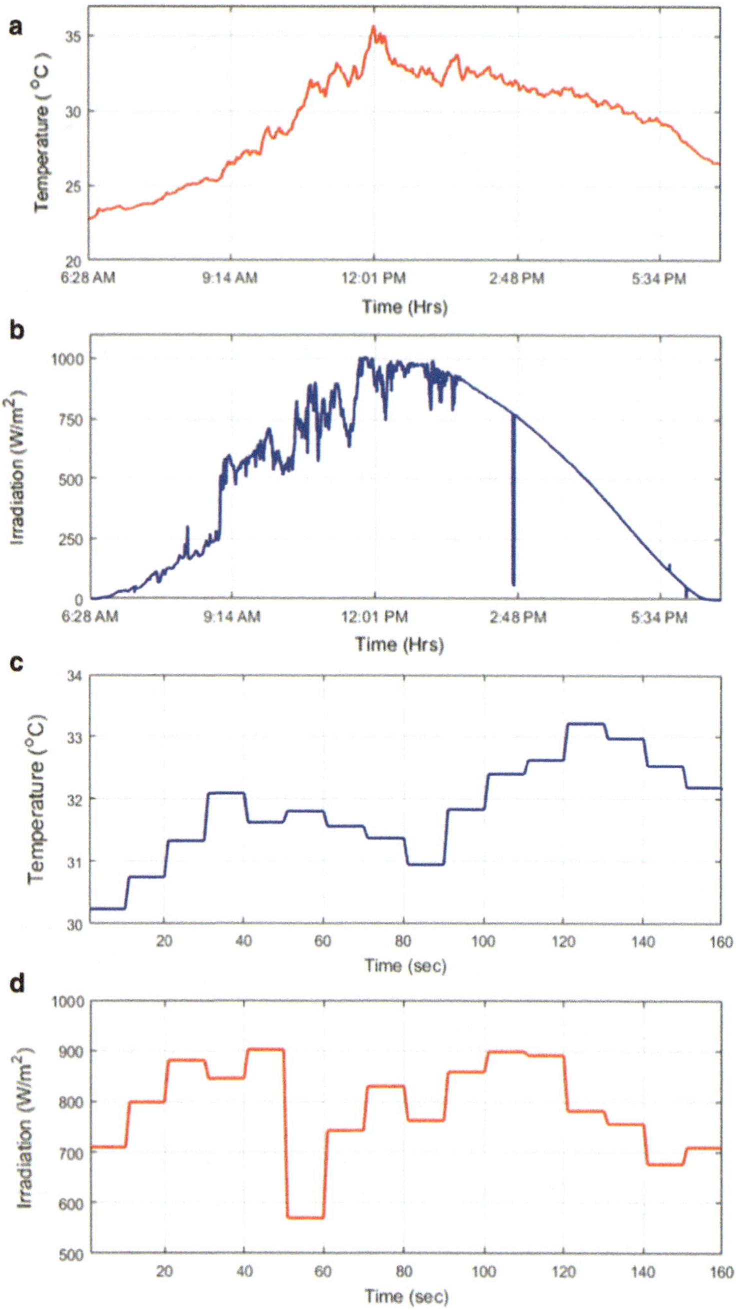

Fig. 17 **a** Real-time temperature. **b** Real-time irradiation. **c** Samples of temperature. **d** Samples of irradiation. **e** Tracking performance under frequently changing inputs. **f** Output power of PV under frequently changing inputs

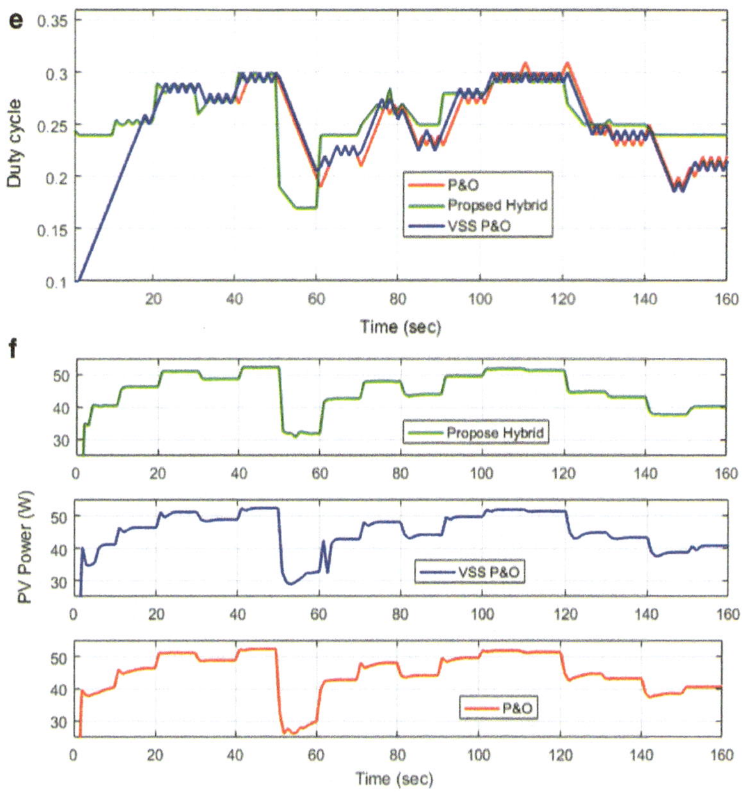

Fig. 17 (continued)

6 Performance Analysis of MPPT Algorithms

A 60 W SPV module (MSX60 PV) is used as PV source in the system. The standard test condition (STC) of PV module is 25 °C and 1000 W/m^2. For the selected PV module, the open-circuit voltage is 21.1 V and it has a short-circuit current of 3.8 A at STC. The maximum voltage (V_{max}@MPP), maximum current (I_{max}@MPP) and maximum output power (P_{max}@MPP) of the module at STC are 17.5 V, 3.5 A and 60 W, respectively. The DC–DC boost converter is designed for the input and output voltage of $V_{in} = 17$ V and $V_{out} = 24$ V, respectively. The power rating of the converter is 60 W, and the switching frequency is 10 kHz.

The performance analysis of proposed MPPT algorithm is carried out by performing exhaustive simulation under different input conditions. For the selected SPV, characteristic curves are obtained at $G = 750$ W/m^2 and $T = 20$ °C. P–V curve is depicted in Fig. 14a and I–V curve is depicted in Fig. 14b. The maximum power output is 45.7 W for the given input conditions.

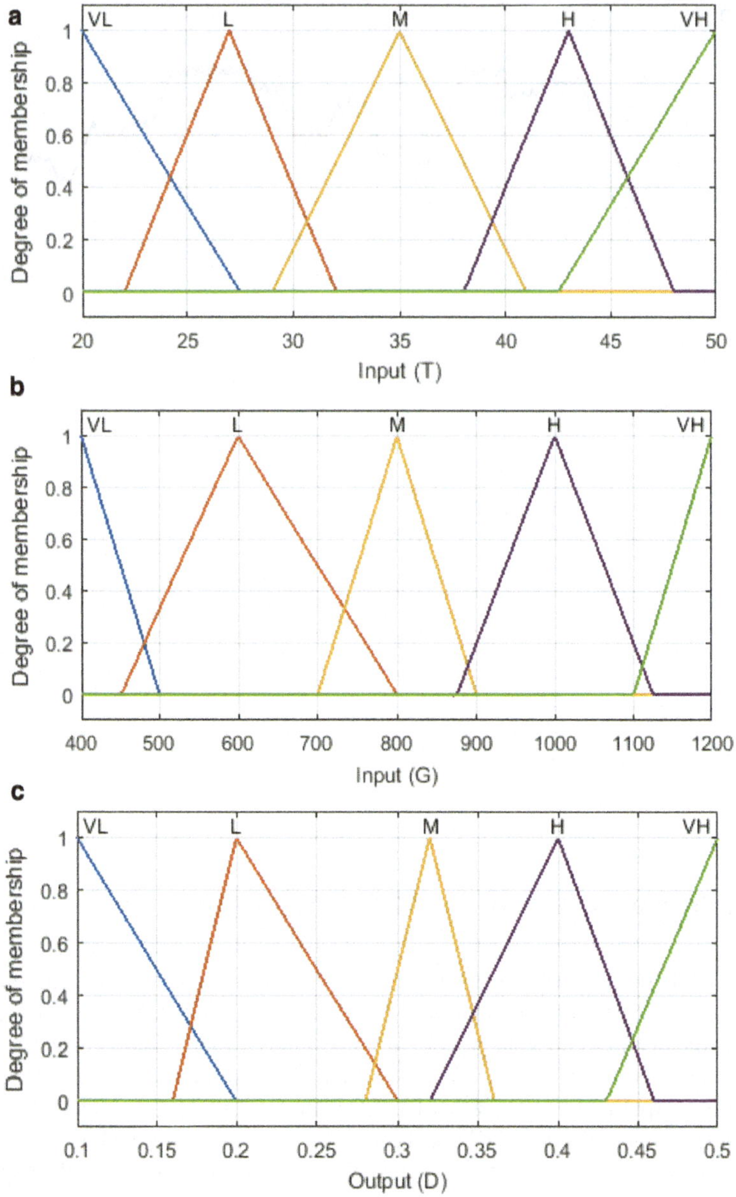

Fig. 18 **a** Membership functions of *T*. **b** Membership functions of *G*. **c** Membership functions of *D*

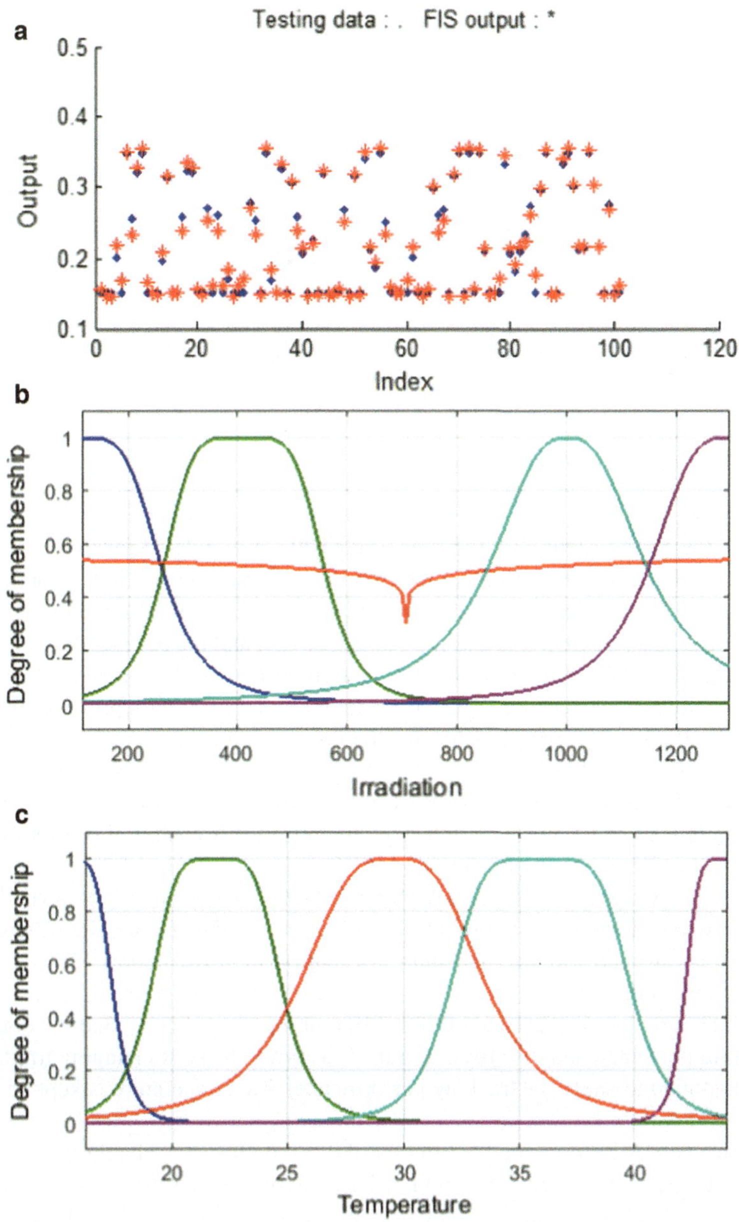

Fig. 19 **a** Testing data and FIS output. **b** MFs of input (*G*) for ANFIS. **c** MFs of input (*T*) for ANFIS

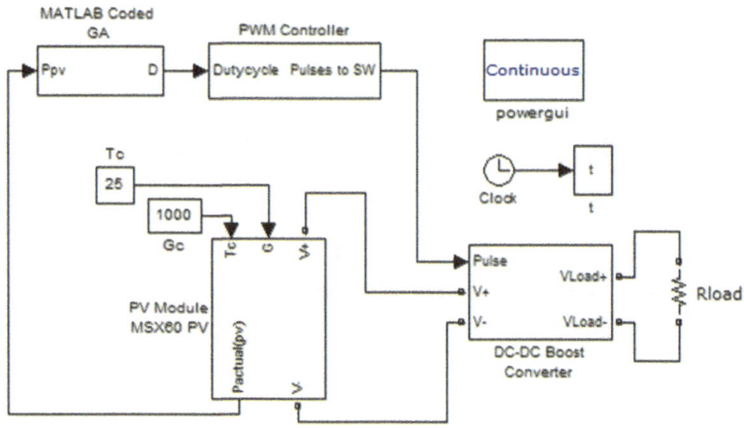

Fig. 20 MATLAB-coded GA model for training

Now, the SPV system with the proposed MPPT is simulated for the same input conditions. The algorithm initially identifies the region of input(s) in LA and fetches the optimal duty cycle. The irradiation and temperature values coincide with the subregion Z_3 ($G = 700$ W/m^2 and $T = 20$ °C) of LA, and the optimal duty cycle for Z_3 is 0.19. Thus, it forces the system to operate at $D = 0.19$ instantly. Then, duty cycle is increased in steps by P&O MPPT to attain the maximum power point. The duty cycle tracking of hybrid MPPT for the input conditions $G = 750$ W/m^2 and $T = 20$ °C is shown in Fig. 15a.

In Fig. 15b, the power output of PV is shown. SPV system with proposed MPPT has an output of 45.65 W. The efficiency obtained at this condition is 99.89%, and this confirms the accuracy of proposed MPPT.

The MPP tracking of P&O and VSS P&O MPPT algorithm are compared with the proposed technique. From Fig. 15c, it can be observed that other algorithms have slower tracking performance and more oscillation than the hybrid MPPT for the given input conditions.

Further, simulation is carried out for different input conditions, and the obtained results are presented and discussed. Figure 16a shows the slow changing irradiation pattern applied to study the tracking performance. The temperature is kept 25 °C in this case.

In Fig. 16b, MPP tracking of different algorithms for slow changing irradiation is presented. The proposed P&O and LA-based hybrid MPPT effectively senses the changes and track accordingly. The tracking response is faster and accurate also.

In Fig. 16c, power output of SPV system for slow charging irradiation condition with different MPPTs is illustrated. Throughout the operation, the power output of the SPV system is comparatively higher with the proposed MPPT. A study on frequently varying input conditions is carried out using the real-time data. Temperature and solar irradiation on a particular day with frequent variations are depicted in Fig. 17a, b, respectively.

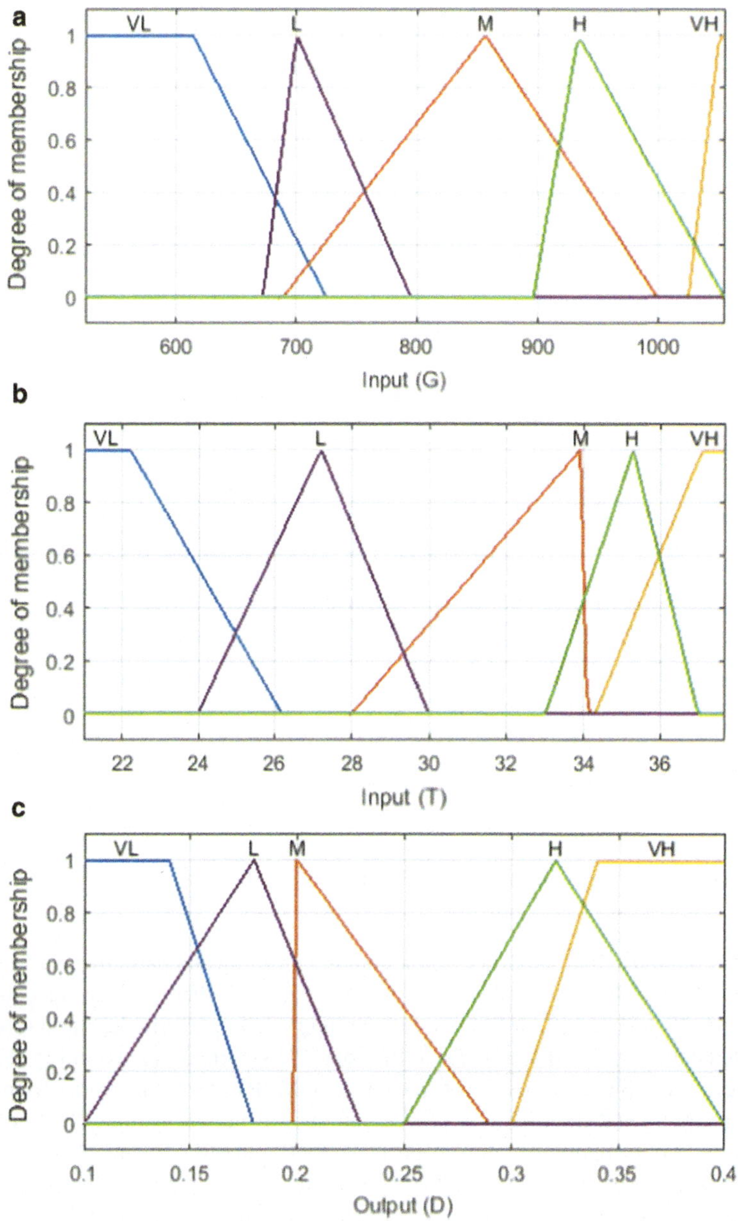

Fig. 21 **a** Membership functions of *G* of GA fuzzy MPPT. **b** Membership functions of *T* of GA fuzzy MPPT. **c** Membership functions of *D* of GA fuzzy MPPT

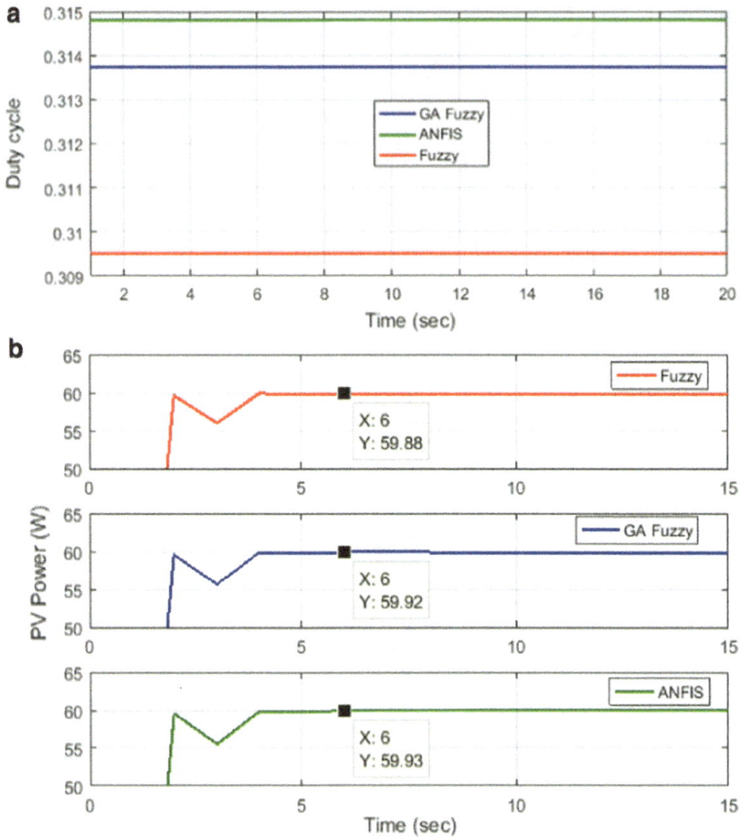

Fig. 22 **a** Duty cycle output of fuzzy, ANFIS and GA fuzzy MPPTs at STC. **b** PV power output

Variation in irradiation is very frequent in the morning and particularly from 9:40 AM to 12:00 PM. Six samples of temperature and irradiation data are collected between 9:40 AM and 12:00 PM in a regular interval. To perform the simulation, the collected data is given as step input to the solar PV system as shown in Fig. 17c, d.

Figure 17e shows the tracking performance of MPPT algorithms under frequently varying inputs, and the obtained power outputs are shown in Fig. 17f. For every change in input conditions, the proposed MPPT quickly responded and MPP is reached in all cases. The other MPPTs selected for the study have managed to reach MPP in few cases. But, the tracking performance is comparatively poor and oscillates around MPP. This confirms the effectiveness, tracking speed and accuracy of the proposed algorithm under frequently changing input conditions.

From the above discussions, it can be inferred that the overall tracking performance is improved by the proposed P&O and learning automata-based hybrid MPPT, and it has good tracking accuracy and is suitable for frequently varying input conditions also. The MPPT controllers considered for comparison in this study work well

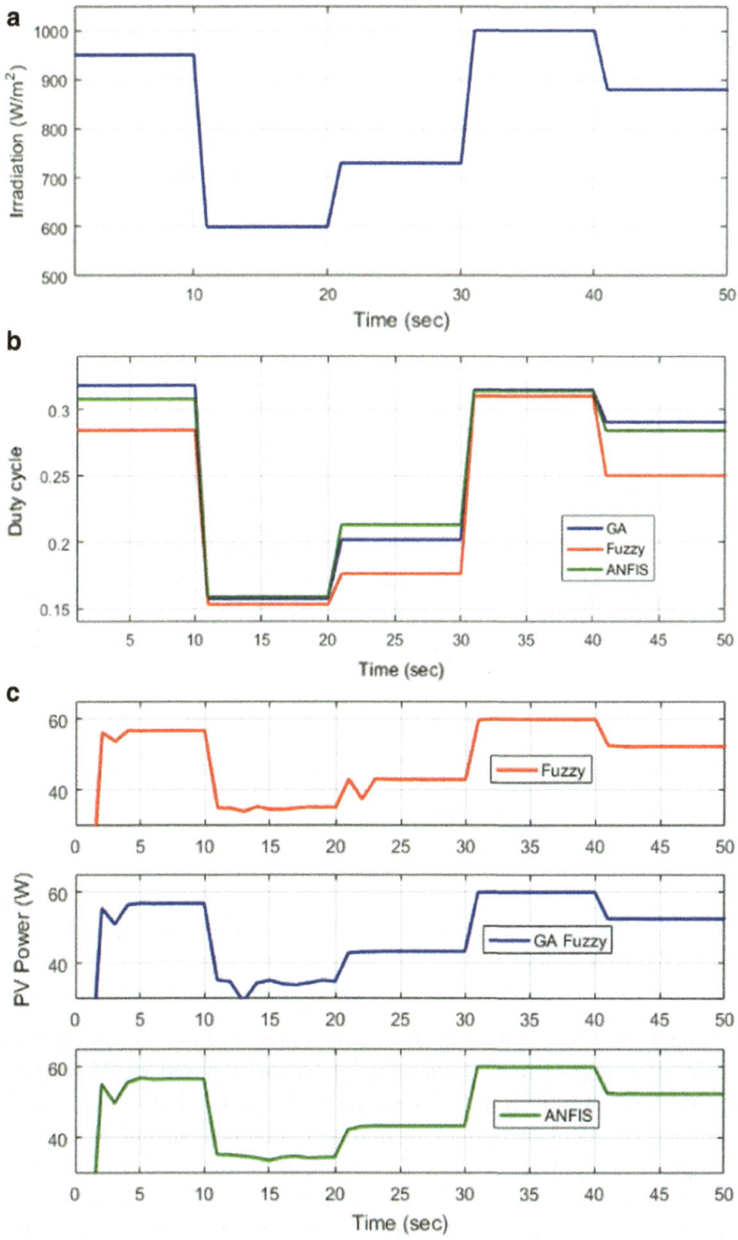

Fig. 23 **a** Change of irradiation. **b** MPP tracking under change of irradiation. **c** Output power under change of irradiation

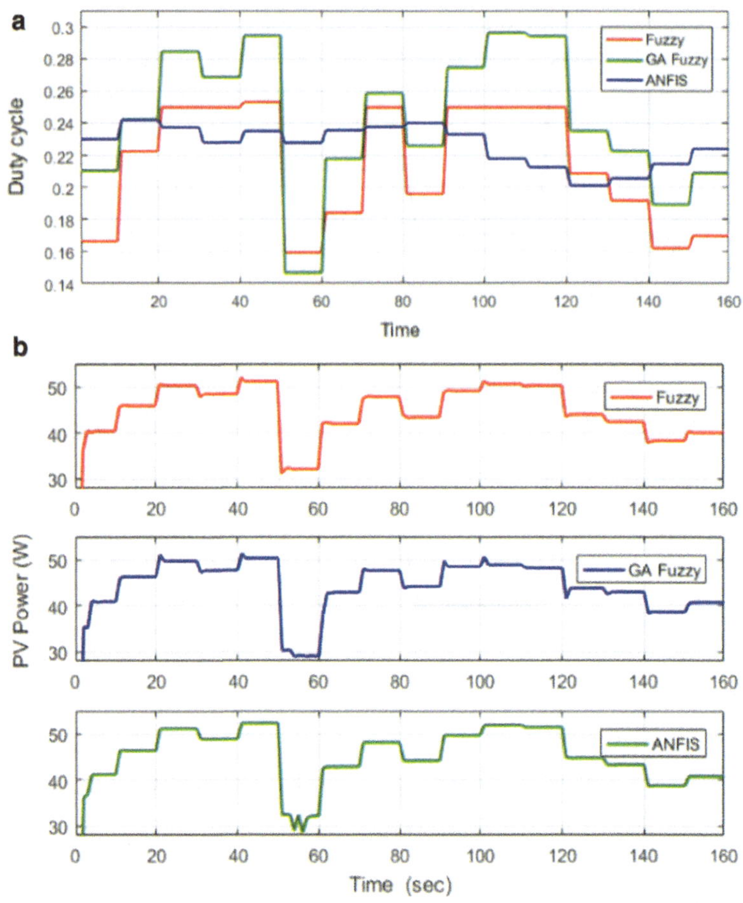

Fig. 24 **a** Tracking performance of fuzzy MPPTs under frequently changing inputs. **b** Output power with fuzzy MPPTs under frequently changing inputs

under normal conditions, and the average output power of the PV system with these MPPT controllers is close to the values obtained from proposed MPPT-based system. However, the key advantages of the proposed P&O and LA-based hybrid system are the tracking speed, accuracy of tracking under frequently changing inputs and reduction of ripple in the power. In Table 3, the output power ripple for different input conditions is given. Since the step size of P&O and VSS P&O is same at MPP, only VSS P&O is taken into account for comparison. The ripple generated by MPPT in the output power has significant reduction with the proposed hybrid MPPT-based system.

Table 3 Comparison of output power ripple

Inputs	T (°C)	25	32	25	27
	G (W/m^2)	750	1000	900	1000
VSS P&O		9.29	12.32	10.09	12.16
Hybrid MPPT		8.87	11.88	9.67	11.6
Improvement (%)		4.52	3.57	4.16	4.61

7 Soft Computing MPPT Techniques

Due to nonlinear characteristics, the solar PV system is treated as a nonlinear system. The soft computing techniques such as fuzzy and neural network are suitable for the control of nonlinear system, and hence they are used for MPPT developments. Numerous soft computing techniques-based MPPTs are proposed in literatures. The soft computing MPPT techniques are categorized as offline MPPT techniques. These MPPTs are optimized either by the user knowledge about the dynamics of the system or through the training process using the available data. The soft computing techniques are preferred for their performance, accuracy and efficiency. Fuzzy-based MPPT is the most popular soft computing technique-based MPPT. Development and implementation of fuzzy MPPT is comparatively easier than the other soft computing techniques. However, fuzzy-based MPPT may not offer better performance since the membership functions and rule base are optimized by the user based on his experience and knowledge about the system.

FLC is generally optimized by genetic algorithm (GA) or adaptive neuro-fuzzy inference system (ANFIS). In this section development of fuzzy MPPT, ANFIS-optimized fuzzy MPPT and GA-optimized fuzzy MPPT are discussed. Development of soft computing technique-based MPPTs considered for discussion in this chapter have been presented in many literatures. Hence, only the MPPT modeling using fuzzy MPPT, GA-optimized fuzzy and ANFIS-optimized fuzzy MPPT is discussed in this chapter. Simulation study of solar PV system with selected soft computing technique-based MPPT algorithms is carried out, and the results obtained under several conditions are presented and discussed.

7.1 Fuzzy MPPT for SPV System

The main components of FLC are input and output membership function (MF), inference system and rule base. The inputs are irradiation (G) and temperature (T), and the output of the FLC is duty cycle (D). The inputs and output have five membership functions each, and 25 rules are derived.

Mamdani method is considered as fuzzy inference system, and a max–min composition technique is adopted for inference. For defuzzification, centroid method/center of gravity method is selected. The membership functions of input "T" are illustrated in Fig. 18a. Figure 18b shows the membership functions of input "G" and output of the fuzzy MPPT are depicted in Fig. 18c.

7.2 Adaptive Neuro-fuzzy Inference System (ANFIS)-Based MPPT

ANFIS is a combination of fuzzy logic controller and neural network. The membership functions and rule base of the FLC are developed through a training process in ANFIS. This training is carried out by the in-built NN of ANFIS. Similar to the fuzzy MPPT, ANFIS-based MPPT with two inputs and one output is developed. The data required for training the MPPT model is collected by conducting simulation on the SPV system with conventional MPPT for wide range of input conditions.

The training details of ANFIS are as follows:

Number of epochs:	2000 epochs
Training error:	5%

Figure 19a shows the index between the testing data and FIS output. The ANFIS-trained input MFs are shown in Fig. 19b, c.

7.3 GA-Optimized Fuzzy MPPT

Genetic algorithm is a computational intelligence-based optimization technique. Biological evolution is the fundamental theory for genetic algorithm. GA finds the candidate solution which fits the most to the objective function of the problem through iteration process. In each iteration, the most fit individuals of the previous generation and mutated and new generation are formed. This process is repeated for many generations (iterations), and the optimal solution for the problem is identified.

The MPPT problem can be defined as maximization problem since the objective is harvesting of maximum power from SPV arrays. Therefore, the error between the power delivered to the load (P_{load}) and power generated by PV (P_{pv}) should be reduced. This can be expressed by

$$\min(\text{error}) = P_{\text{pv}} - P_{\text{load}} \tag{8}$$

GA training for optimization is carried out with a population size of 25 for 50 generations. The crossover and mutation probability are set as 0.6 and 0.06, respectively. MATLAB-coded GA model is integrated to solar PV system for training in order to find the optimal solutions for different inputs as shown in Fig. 20. The MATLAB-coded GA sends the duty cycle to the system and receive the output power. Through continuous population, GA optimizes the control parameters for every input conditions applied to the system. Upon completion of training, based on the obtained solutions GA generates the optimized membership functions for the inputs and output of FLC along with the optimal rule base.

GA-optimized fuzzy MFs for the inputs are depicted in Fig. 21a, b, and the output is given in Fig. 21c.

Performance of SPV system is tested by embedding the developed soft computing technique-based MPPT techniques in the system. Initially, the system is simulated at STC and the results are obtained. The duty cycle output of fuzzy, GA-optimized fuzzy and the ANFIS MPPT is presented in Fig. 22a, and the power output with the MPPTs is given in Fig. 22b.

As mentioned earlier, the actual duty cycle at STC is $D = 0.32$. The duty cycle generated by the optimized fuzzy MPPTs coincides very well with the desired value, and the output power is also very close to the expected value (60 W). The efficiency under this condition is 99.88% with optimized fuzzy MPPT techniques. The varying irradiation is applied to the solar PV system with constant temperature (25 °C). The solar irradiation level is varied for every 10 s as shown in Fig. 23a.

MPP tracking is illustrated in Fig. 23b, and the power output is illustrated in Fig. 23c. The optimized fuzzy system produces more accurate duty cycles in all conditions than the conventional fuzzy.

The system is further simulated for the frequently changing input conditions derived from the real-time data. The input conditions applied for the analysis in the previous section are considered here also. Over a period of 160 s, the irradiation and temperature are varied after every 10 s. The values at each step are the values extracted from the real-time data. Since the same input conditions are applied, irradiation pattern and temperature are not presented again. Tracking of MPP for the given input conditions with the selected MPPT techniques is shown in Fig. 24a. Figure 24b shows the PV output of the SPV system with selected MPPT techniques. The optimized fuzzy MPPT outperforms in all conditions than the fuzzy MPPT developed based on the expert knowledge.

A detailed analysis on stand-alone solar PV system with soft computing technique-based MPPT technique is presented in this section. To validate the tracking accuracy and efficiency of the LA-based hybrid MPPT over the soft computing technique-based MPPT, a comparison is made between LA-based hybrid MPPT and ANFIS MPPT.

The output obtained with ANFIS MPPT and LA-based hybrid MPPT for the frequently changing input conditions is presented in Fig. 25. The power output for few input conditions is listed in Table 4. From these results, it can be concluded that the proposed P&O and LA-based hybrid MPPT is capable of tracking accurately like the optimized fuzzy MPPT controllers also.

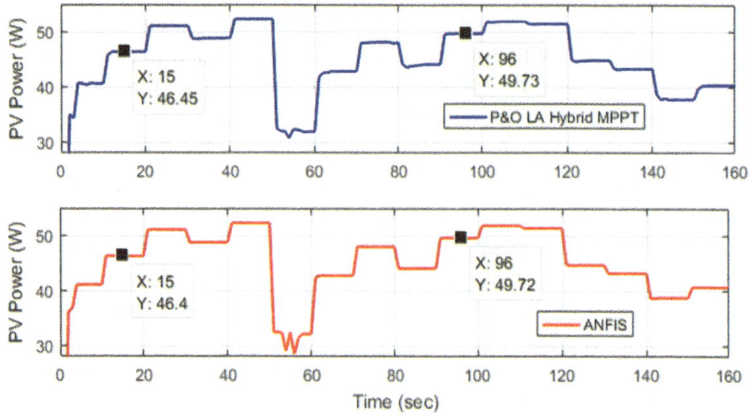

Fig. 25 Output power comparison

Table 4 Comparison of output power

Inputs						
	G (W/m^2)	800	903.4	830.08	891.8	756.9
	T (°C)	30.75	32.1	31.38	32.63	32.98
ANFIS (W)		46.4	52.41	48.12	51.95	43.33
P&O and LA hybrid (W)		46.45	52.41	48.16	51.95	43.32

8 Conclusions

Learning automata and soft computing technique-based MPPT for SPV systems are reviewed in this chapter. Optimization using learning automata and development of P&O and LA-based hybrid MPPT technique for stand-alone SPV system is explained in detail. Performance analysis of the proposed P&O and LA-based hybrid MPPT is carried out by conducting exhaustive simulation studies and by comparing the results with the conventional and variable step size P&O MPPT-based systems. Performance analysis of the soft computing-based MPPT techniques such as fuzzy MPPT, GA-optimized fuzzy MPPT and ANFIS-optimized fuzzy MPPT is also reviewed with the help of simulation studies. The P&O and LA-based hybrid MPPT is also compared with the ANFIS MPPT for performance validation, and the results are presented and discussed.

References

1. Sheik Mohammed, S., Ramasamy, K.: Solar power generation using SPS and wireless power transmission. In: International Conference on Energy and Environment, March 2009, Chandigarh, India (2009)

2. United Nations Environment Programme. Emissions Gap Report 2019. UNEP, Nairobi (2019)
3. Eltamaly, A., Abdelaziz, A.: Modern Maximum Power Point Tracking Techniques for Photovoltaic Energy Systems. Cham, Switzerland: Springer (2020)
4. Basha, C., Rani, C.: Different conventional and soft computing MPPT techniques for solar PV systems with high step-up boost converters: a comprehensive analysis. Energies **13**(2), 371 (2020)
5. Podder, A., Roy, N., Pota, H.: MPPT methods for solar PV systems: a critical review based on tracking nature. IET Renew. Power Gener. **13**(10), 1615–1632 (2019)
6. Yang, Y., Wen, H.: Adaptive perturb and observe maximum power point tracking with current predictive and decoupled power control for grid-connected photovoltaic inverters. J. Mod. Power Syst. Clean Energy **7**(2), 422–432 (2018)
7. Sheik Mohammed, S.: Multiple step size perturb and observe maximum power point tracking algorithm with zero oscillation for solar PV applications. In: 2018 IEEE International Conference on Current Trends towards Converging Technologies (IEEE ICCTCT 2018) (2018)
8. Lee, H., Yun, J.: Advanced MPPT algorithm for distributed photovoltaic systems. Energies **12**(18), 3576 (2019)
9. Baimel, D., Tapuchi, S., Levron, Y., Belikov, J.: Improved fractional open circuit voltage MPPT methods for PV systems. Electronics **8**(3), 321 (2019)
10. Ammar, H., Azar, A., Shalaby, R., Mahmoud, M.: Metaheuristic optimization of fractional order incremental conductance (FO-INC) maximum power point tracking (MPPT). Complexity **2019**, 1–13 (2019)
11. Ghassami, A.A., Sadeghzadeh, S.M., Soleimani, A.: A high performance maximum power point tracker for PV systems. Electr. Power Energy Syst. **53**, 237–43 (2013)
12. Belkaid, A., Colak, I., Kayisli, K.: Implementation of a modified P&O-MPPT algorithm adapted for varying solar radiation conditions. Electr. Eng. **99**(3), 839–846 (2016)
13. Bechouat, M., Sedraoui, M., Feraga, C., et al.: Modeling and fuzzy MPPT controller design for photovoltaic module equipped with a closed-loop cooling system. J. Electr. Mater. **48**, 5471–5480 (2019)
14. Chouay, Y., Ouassaid, M.: An experimental artificial neural network based MPP tracking for solar photovoltaic systems. In: Serrhini, M., Silva, C., Aljahdali, S. (eds.) Innovation in Information Systems and Technologies to Support Learning Research. EMENA-ISTL 2019. Learning and Analytics in Intelligent Systems, vol 7. Springer, Cham (2020)
15. Chaibi, Y., Allouhi, A., Salhi, M., et al.: Annual performance analysis of different maximum power point tracking techniques used in photovoltaic systems. Prot Control Mod Power Syst **4**(15) (2019)
16. Ibrahim, A., Aboelsaud, R., Obukhov, S.: Improved particle swarm optimization for global maximum power point tracking of partially shaded PV array. Electr. Eng. **101**(2), 443–455 (2019)
17. Ma, J., et al.: Improving power conversion efficiency via a hybrid MPPT approach for photovoltaic systems. Electron. Elect. Eng. **19**(7), 57–60 (2013)
18. Manickam, C., Raman, G., Raman, G., Ganesan, S., Nagamani, C.: A hybrid algorithm for tracking of GMPP based on P&O and PSO with reduced power oscillation in string inverters. IEEE Trans. Indust. Electron. **63**(10), 6097–6106 (2016)
19. Moradi, M.H., Reisi, A.R.: A hybrid maximum power point tracking method for photovoltaic systems. Solar Energy **85**(11), 2965–2976 (2012)
20. Sher, H., Murtaza, A., Noman, A., Addoweesh, K., Al-Haddad, K., Chiaberge, M.: A new sensor less hybrid MPPT algorithm based on fractional short-circuit current measurement and P&O MPPT. IEEE Trans. Sustain. Energy **6**(4), 1426–1434 (2015)
21. Sher, H.A., Addoweesh, K.E., Al-Haddad, K.: An efficient and cost-effective hybrid MPPT method for a photovoltaic flyback micro inverter. IEEE Trans. Sustain. Energy **9**(3), 1137–1144 (2018)
22. Tsetlin, M.L.: On the behavior of finite automata in random media. Autom. Remote Control **22**, 1210–1219 (1962)

23. Narendra, K.S., Thathachar, M.A.L.: Learning Automata: An Introduction. Prentice-Hall, Englewood Cliffs (1989)
24. Thathachar, M.A.L., Sastry, P.S.: Varieties of learning automata: an overview. IEEE Trans. Syst. Man Cybernet.—Part B: Cybernet. **32**(6) (2002)
25. Papadimitriou, G.I., Maritas, D.G.: Learning automata-based receiver conflict avoidance algorithms for WDM broadcast-and-select star networks. IEEE/ACM Trans. Network. **4**, 407–412 (1996)
26. Papadimitriou, G.I., Pomportsis, A.M.: On the use of stochastic estimator learning automata for dynamic channel allocation in broadcast networks. In: Proceedings of IEEE/CE (2000)
27. Obaidat, M.S., Papadimitriou, G.I., Pomportsis, A.S.: An efficient bus arbitration scheme for scalable shared-medium ATM switch. Comput. Commun. **24**, 790–797 (2001)
28. Thapa, R., Jiao, L., Oommen, B.J., Yazidi, A.: A learning automaton-based scheme for scheduling domestic shiftable loads in smart grids. IEEE Access **6**, 5348–5361 (2018)
29. Ali, S.Q., Imthias, A.T., Parambath, Nazar Malik, H.: Learning automata algorithms for load scheduling. Electr. Power Compon. Syst. **41**(3), 286–303 (2013)
30. Thathachar, M.A.L., Sastry, P.S.: Networks of Learning Automata: Techniques for Online Stochastic Optimization, p. 5. Kluwer Academic, Chap, Boston (2003)
31. Kushner, H.J., Yin, G.: Stochastic Approximation Algorithms and Applications, Applications of Mathematics. Springer, New York (1997)

Fuzzy Logic Controller Based Plug-In EV Battery Charger

N. Sujitha and S. Krithiga

1 Introduction

In the last decade, PHEVs and electric vehicles (EVs) have considerable attraction in transportation sector due to its effective way of reducing the usage of fossil fuels and environmental pollution issues like greenhouse gas emissions [1, 2]. Researchers and automobile industry authorities predicted that integration of vehicle to the power grid will be an indispensable part of electric vehicle in the future smart grids. In the V2G paradigm, EV battery can support the utility grid by delivering part of the stored energy during predefined schedule [3, 4]. In the near future, EV users will have major contribution in the energy market in order to have a payback with the help of V2G technology. In the next decade, sale of electric vehicles with V2G technology is expected to increase in millions globally [5]. Therefore, investigations of many researchers are focused in the development of battery charger topologies that allow the EVs integration to the smart grids in order to facilitate the V2G technology [6–8].

Battery chargers are mainly classified into two types, namely conductive and inductive chargers based on the mode of power transfer from utility grid to EV batteries. In the inductive method, power transfers in wireless mode without any electrical contact, whereas in the conductive method, grid power is transferred to the electric vehicle battery (EVB) through electrical contact. Although many researches are being focused on inductive chargers, majority of the electric vehicles are developed with conductive battery chargers as per the standard IEC 61851–1 [9, 10]. In

N. Sujitha · S. Krithiga (✉)
School of Electrical Engineering, Vellore Institute of Technology, Chennai, Tamil Nadu 600127, India
e-mail: s_krithiga@yahoo.com

N. Sujitha
e-mail: sujielanthirai@gmail.com

B. Vinoth Kumar et al. (eds.), *Intelligent Paradigms for Smart Grid and Renewable Energy Systems*, Algorithms for Intelligent Systems,
https://doi.org/10.1007/978-981-15-9968-2_8

order to facilitate V2G operation, there is a need for bidirectional power converter topologies.

Typically, two-stage configurations with inverter and isolated DC–DC converter are used in the bidirectional chargers [11–13]. The inverter is used for AC–DC power conversion, and the DC–DC converter is used for voltage matching. Due to simple circuitry and power reversal capability, dual active bridge (DAB) converter is used as isolated converter in the EV battery chargers [14–16]. DAB converters have the drawbacks like high circulating current, high turnoff loss and high reactive power under wide range of variation in voltage. Resonant converters with controlled frequency were introduced as bidirectional converter which has the advantage of low EMI and high efficiency [17, 18]. The switching pattern has to be changed based on the direction of flow of power in this converter which needs the complex control circuitry. IGBT-based AC–DC converter used in the EV chargers uses various pulse width modulation (PWM) techniques and also requires extra circuitry like phase-locked loop (PLL) and hysteresis current controller for grid synchronization [8]. However, multiple power stages eventually increase the weight and cost and reduce the efficiency of the charger.

Therefore, EV battery charger which uses single-stage bidirectional inverter is proposed in this chapter. Thyristor-based bidirectional line-commutated converter (BLCC) is utilized as bidirectional inverter in the proposed charger [19]. This bidirectional converter has the advantage of operating as rectifier and inverter based on the firing angle, α. BLCC does not require any extra circuitry for grid synchronization since it has inherent self-grid synchronizing capability.

The proposed charger requires a precise controller for managing the bidirectional flow of power between the utility grid and EVB depending on the mode of operation. AI controllers like fuzzy logic controller and genetic algorithm-based controller are more accurate and rapid in tracking the inputs and control the power flow efficiently [20, 21]. Hence, the proposed charger uses fuzzy logic controller for tuning the firing angle of BLCC to facilitate grid-to-vehicle (G2V) and V2G technology.

2 Description of the Proposed Charger

The proposed EVB charger consisting of single-phase utility grid, bidirectional line-commutated converter, fuzzy logic controller, EV battery and residential load is depicted in Fig. 1.

BLCC is composed of line-commutated converter and bidirectional configurator in order to facilitate bidirectional flow of power. The BLCC DC link voltage, E_{dc}, is provided by [22]

$$E_{dc} = \frac{2V_m}{\pi}\cos\alpha \tag{1}$$

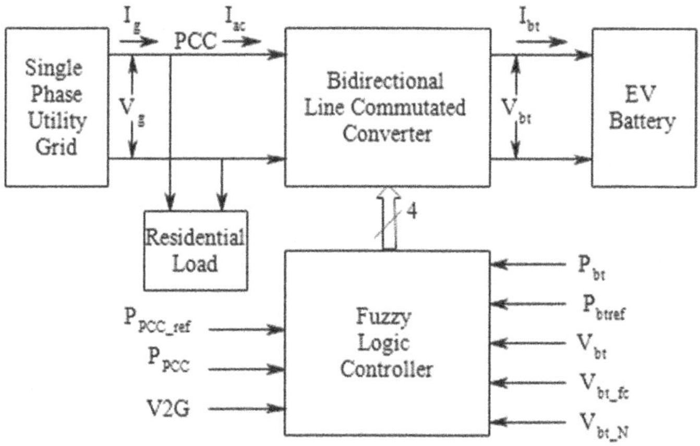

Fig. 1 Configuration of proposed charger

where V_{m} is the AC voltage peak value and α is the firing angle of BLCC. For continuous conduction of BLCC, the value of DC link inductance, L_{dc}, is calculated using the following equation [23].

$$L_{\mathrm{dc}} = 3.18 \frac{E_{\mathrm{dc}}}{I_{\mathrm{dclink}}} \tag{2}$$

where I_{dclink} is the dc link current. In the proposed charger, utility grid supplies power to the residential load and to charge the EVB through the BLCC in G2V mode and during predefined schedule, EV battery discharges to feed the residential load and utility grid through the BLCC in V2G mode. This is accomplished by appropriately tuning the firing angle of the SCR switches of BLCC by fuzzy logic controller. Different modes of operation of proposed charger are presented as follows.

2.1 Mode 1: (G2V)

In G2V mode, utility grid supplies power to the residential load in addition to charge the EVB. In G2V mode, BLCC works as rectifier with firing angle $\alpha < 90°$. Polarity of dc link voltage, E_{dc} is reversed which needs the polarity reversal to charge the EVB. Hence, the bidirectional configurator relays R_2 and R_3 are closed and relays R_1 and R_4 are opened to charge the EVB utilizing the grid power as shown in Fig. 2.

In this mode, the SCR switches T_2 and T_3 are triggered at a firing angle, $\alpha < 90°$, during positive half cycle (PHC) of grid supply and the current flow path between utility grid and EV battery is depicted in Fig. 2a, whereas during negative half cycle (NHC) of grid supply, SCR switches T_1 and T_4 are triggered and the corresponding flow of current path is depicted in Fig. 2b.

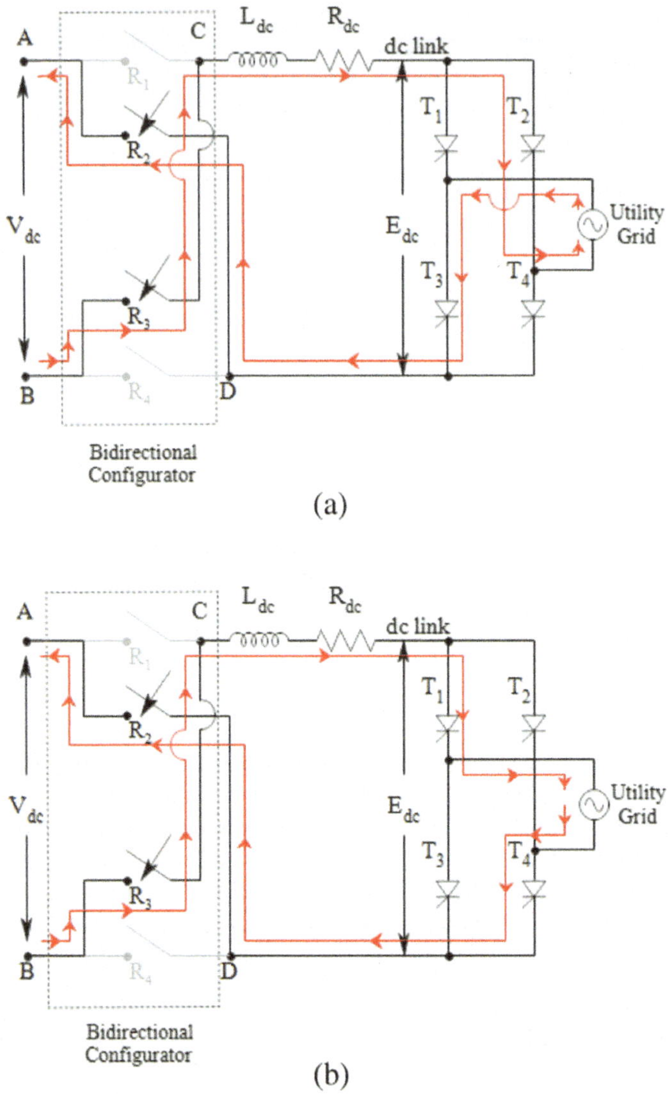

Fig. 2 Circuit diagram of BLCC in rectifier mode during **a** PHC and **b** NHC

2.2 *Mode 2: (V2G)*

In V2G mode, EV battery discharges to supply power to the utility grid and to the residential load during predefined scheduled hours. In this mode, BLCC operates as inverter with $\alpha > 90°$. The bidirectional configurator relays R_1 and R_4 are closed, and relays R_2 and R_3 are opened to ensure the power flows in reverse direction to feed the utility grid as shown in Fig. 3. In V2G mode, the SCR switches T_2 and T_3

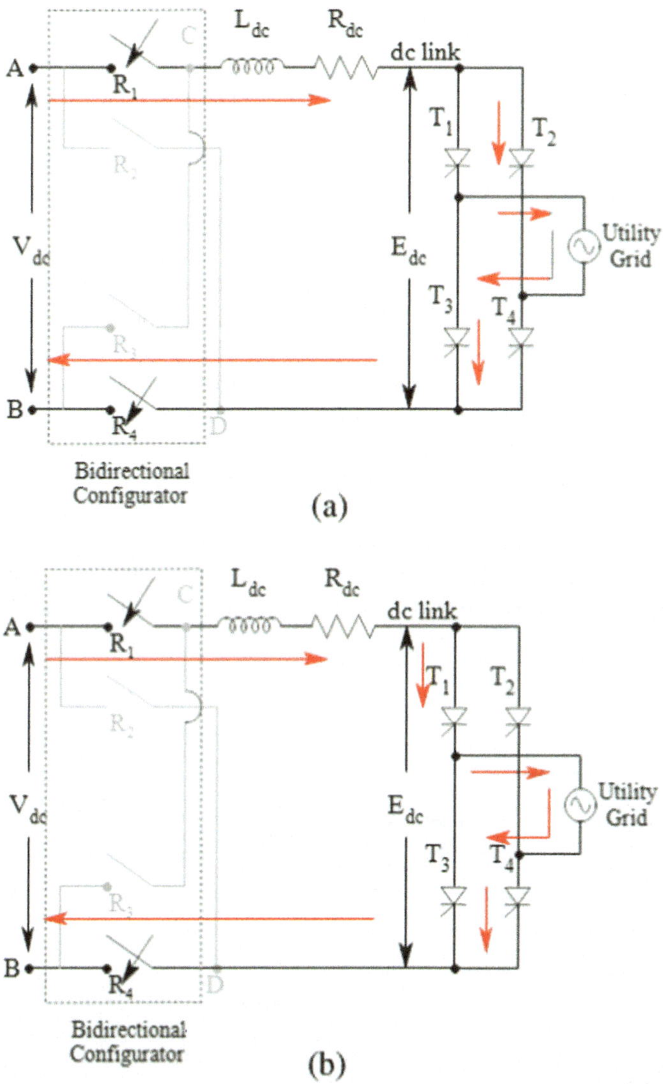

Fig. 3 Circuit diagram of BLCC in inverter mode during **a** PHC of grid voltage and **b** NFC of grid voltage

are triggered at a firing angle, $\alpha > 90°$, during PHC of grid voltage and the SCR switches T_1 and T_4 are triggered during NHC of grid voltage, and the current flow path between battery and utility grid is depicted in Fig. 3.

3 Working of the Controller

The main purpose of the FLC in the proposed charger is to tune the firing angle of the BLCC in order to facilitate bidirectional flow of power depending on the modes of operation. Initially, the proposed controller senses the clock and compares with predefined scheduled hours for V2G mode, and also compares the battery voltage, V_{bt}, with the predefined depth of discharge battery voltage, V_{bt_DOD}, and generates the signal, "V2G". Then, the controller senses the battery voltage, V_{bt}, compares with rated full charge voltage, V_{bt_fc}, and nominal voltage, V_{bt_N}, of EV battery and generates the control signals, $V_{g1} - V_{g4}$, for bidirectional configurator relays, $R_1 - R_4$, depending on the modes of operation as shown in Fig. 4.

During scheduled hours for V2G mode, BLCC works as inverter by closing the bidirectional configurator relays, R_1 and R_4, and opening relays, R_2 and R_3, in order to supply the EVB power to residential load and grid. If $V_{bt} < V_{bt\_DOD}$ or scheduled hours for V2G mode is over, the G2V mode is set. During G2V mode, BLCC works as rectifier by closing the bidirectional configurator relays, R_2 and R_3, and opening relays, R_1 and R_4, in order to charge the EVB using grid power.

The fuzzy logic controller input signal, error in G2V mode, e_{G2V}, and error in V2G mode, e_{V2G}, are calculated as per Eqs. (3) and (4), respectively, as follows [24]:

$$e_{G2V} = P_{btref} - P_{bt} \tag{3}$$

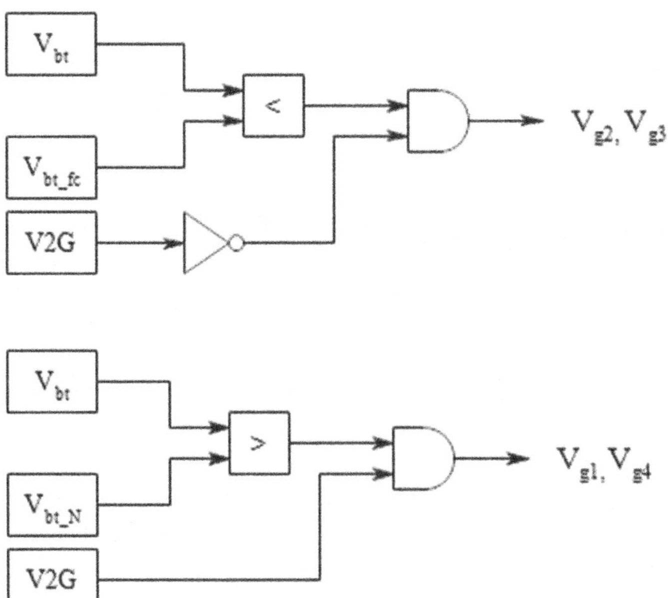

Fig. 4 Bidirectional configurator relay controller diagram

$$e_{V2G} = P_{PCCref} - P_{PCC} \tag{4}$$

$$de = e_{current} - e_{previous} \tag{5}$$

Other input of Fuzzy logic controller, which is change in error in both modes, *de* is given by Eq. (5) and the change in firing angle in both modes, dα is the output of the FLC. The FLC output dα is added with the previous instant firing angle and the obtained firing angle, and α is provided to the pulse generator to produce the firing pulses to the switches of BLCC as shown in Fig. 5.

In the fuzzification process of FLC, the two crisp inputs, error, *e*, and the change in error, *de*, are converted into fuzzy quantity using the membership functions (MFs). The triangular MFs for inputs and output are negative big (NB), negative small (NS), zero (ZR), positive small (PS) and positive big (PB) as shown in Fig. 6.

The range of error, *e*, is limited between −400 and 400 W and for change in error, *de*, it lies between −1 and 1 W, while the change in firing angle, d_alpha, ranges between −50° and 50°. The fuzzified inputs are fed to the "Mamdani"-type inference system to apply the rule base as provided in Table 1. "MIN–MAX" implication method is implemented in the FLC to locate the output, change in firing angle,

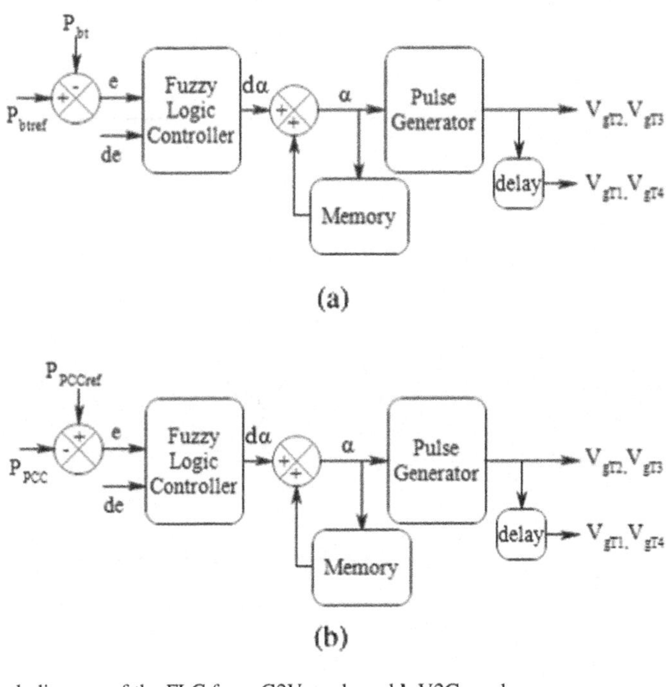

Fig. 5 Block diagram of the FLC for **a** G2V mode and **b** V2G mode

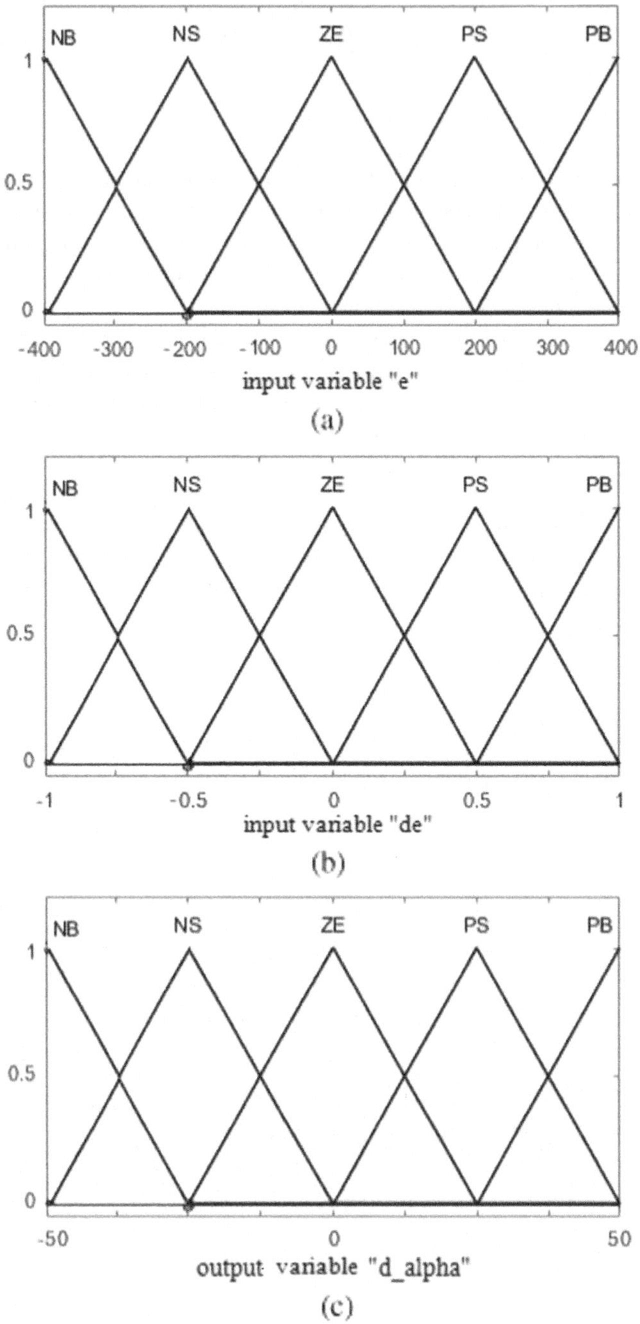

Fig. 6 MFs plots of inputs, *e* and *de* and output, dα

Table 1 Rule base table of FLC

e	*de*				
	NB	NS	ZE	PS	PB
NB	NB	NB	NB	NS	ZE
NS	NB	NB	NS	ZE	PS
ZE	NB	NS	ZE	PS	PB
PS	NS	ZE	PS	PB	PB
PB	ZE	PS	PB	PB	PB

d_alpha region based on the rule base table. From the located output region, "centroid" defuzzification method is implemented to obtain the actual change in firing angle, d_alpha from the FLC output.

4 Simulation Studies of the Proposed Charger

The proposed charger is modeled using thyristors, inductors, resistors, relays, battery and transformer available in SimPowerSystem blockset as shown in Fig. 7. The developed bidirectional line-commutated converter shown as subsystem in Fig. 7 is depicted in Fig. 8. Controller is developed using fuzzy logic controller, memory, arithmetic operator, logical operator, relational operator, pulse generator and transport delay in the Simulink library as presented in Fig. 9.

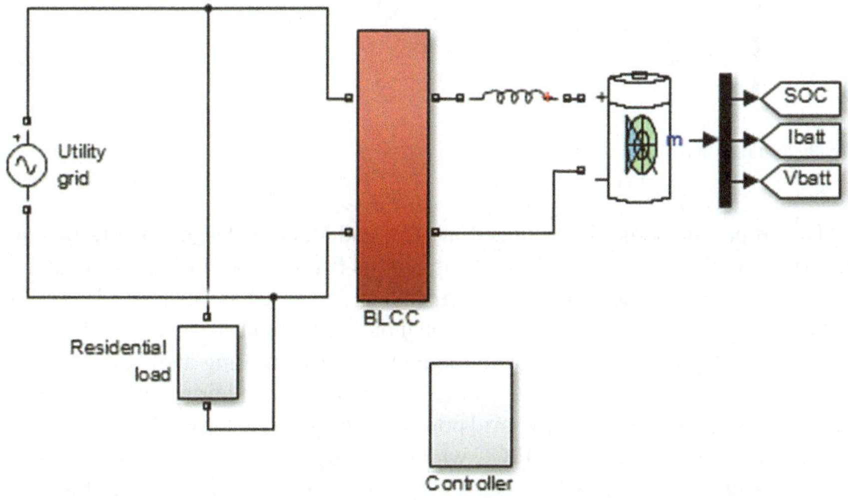

Fig. 7 Proposed charger simulation model

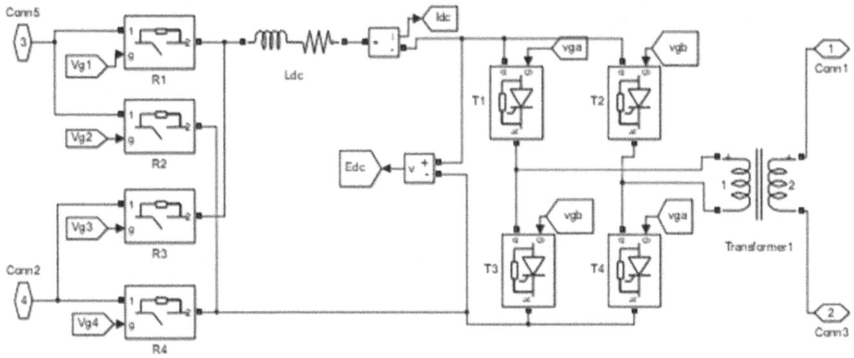

Fig. 8 Simulation diagram of BLCC

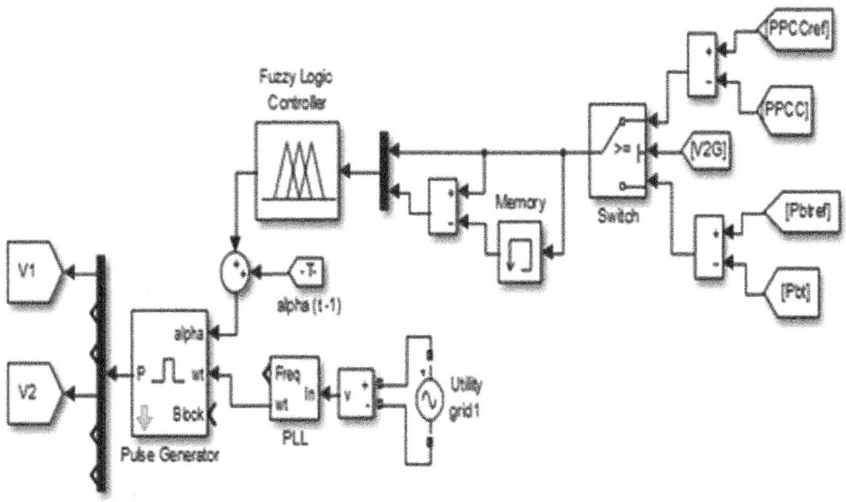

Fig. 9 Simulation diagram of proposed FLC

The proposed charger is simulated in G2V and V2G modes, and the results are furnished in this section. In G2V mode, the initial firing angle of BLCC is set as 10° and error signal and change in error signals are computed as per Eqs. (3) and (5) and fed as input to FLC. FLC processes the input signals and produces the change in firing angle, dα, which is added with the previous instant firing angle to obtain the actual firing angle α. The waveforms of error signal and actual firing angle signal are depicted in Fig. 10. In G2V mode, grid power is transferred to the EV battery through 5:1 turns ratio transformer and BLCC which operates as rectifier. The waveforms of grid, residential load, DC link and battery during G2V mode are shown in Figs. 11, 12, 13 and 14, respectively.

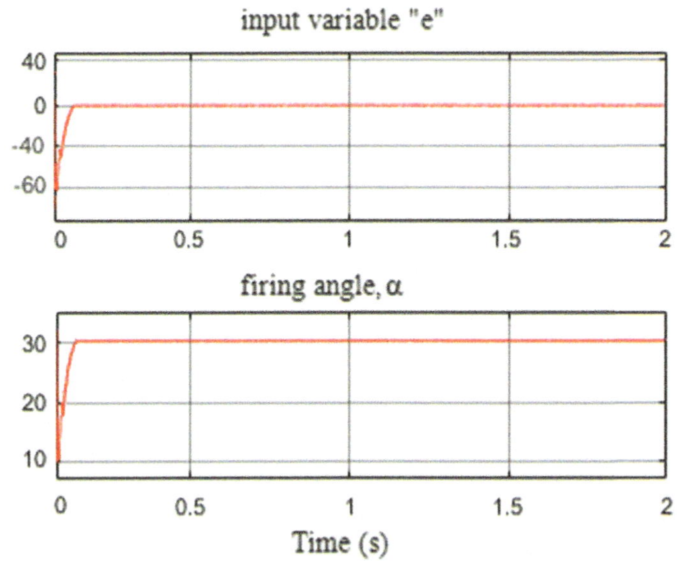

Fig. 10 Waveforms of error and firing angle signals in G2V mode

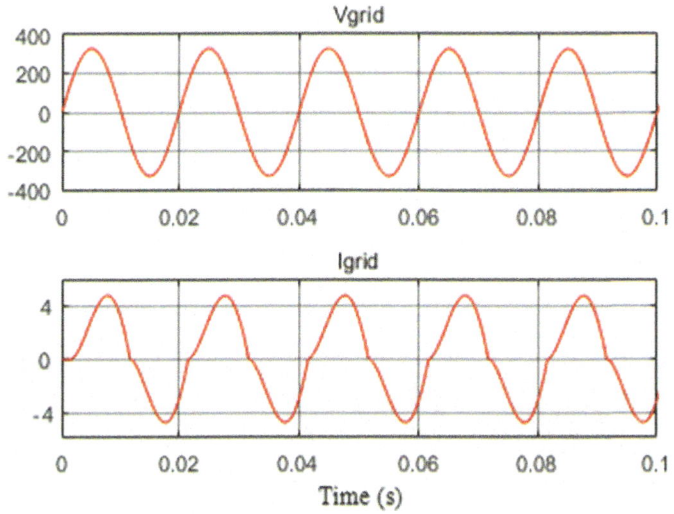

Fig. 11 Waveforms of V_{grid} and I_{grid} in G2V mode

From Figs. 11, 12 and 14, it is evident that out of 694 W grid power contributed by the grid voltage, V_{grid}, of 230 V and current, I_{grid}, of 3.11 A, 210 W is supplied to the residential load with load voltage, V_{load}, of 230 V and current, I_{load}, of 0.92 A and 395.8 W is used to charge the EVB with 26.54 V and 14.9 A. Positive DC link voltage shown in Fig. 13 ensures that the BLCC is operating in rectifier mode and

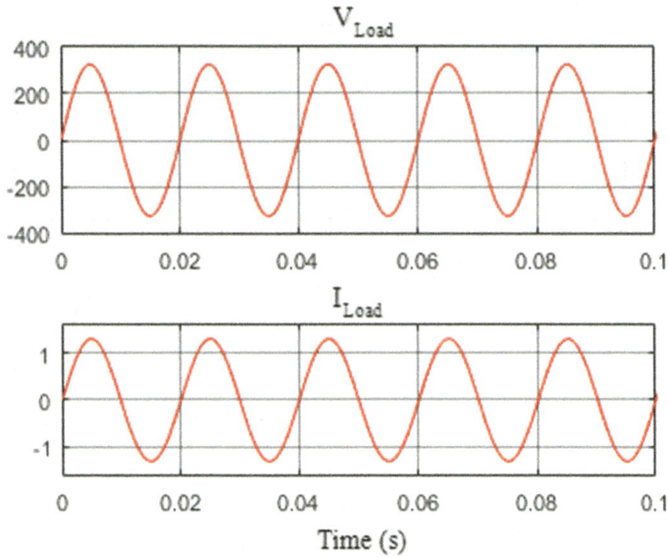

Fig. 12 Waveforms of residential load voltage and current in G2V mode

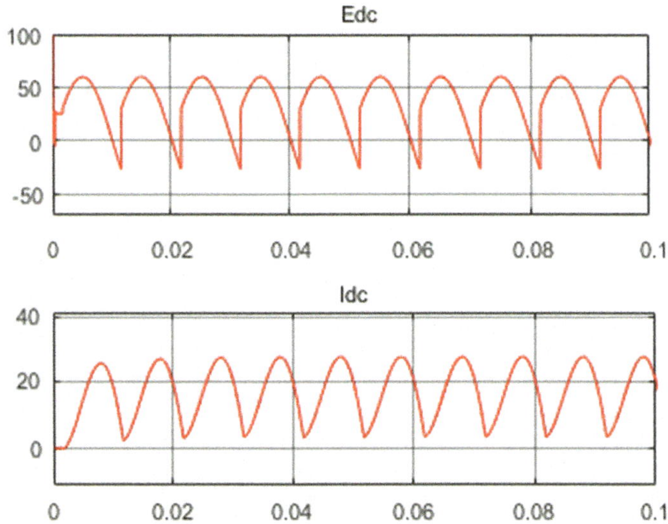

Fig. 13 Waveforms of E_{dc} and I_{dc} in G2V mode

increases in SOC and negative EVB current presented in Fig. 14 depicts the charging of EVB in G2V mode.

The waveforms of error signal and actual firing angle signal in V2G mode are depicted in Fig. 15. From Fig. 15, it is evident that the firing angle $\alpha > 90°$ in V2G

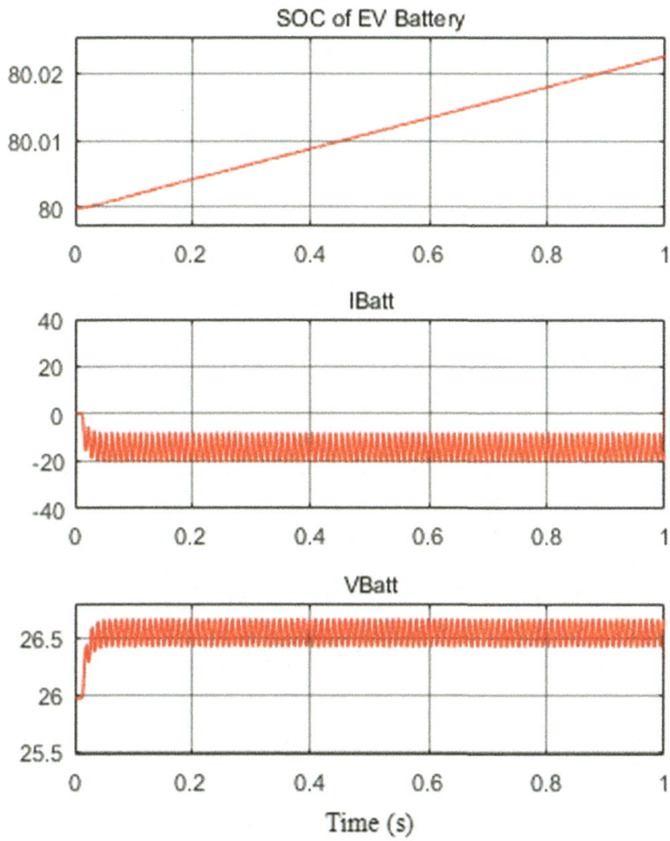

Fig. 14 Waveforms of EV battery SOC, voltage and current in G2V mode

mode. From Figs. 16, 18 and 19, it is evident that out of 320 W EV battery power contributed by the battery voltage, V_{bt}, of 25.68 V and current, I_{bt}, of 12.5 A, 210 W is supplied to the residential load with load voltage, V_{load}, of 230 V and current, I_{load}, of 0.92 A and the remaining 98 W power is fed to the grid with 230 V and 0.44 A. Decrease in SOC and positive EVB current shown in Fig. 16 depicts the discharging of EVB in V2G mode, and negative E_{dc} shown in Fig. 17 ensures that the BLCC is operating in inverter mode.

Also, dynamic response of the proposed charger is tested with mode transitions from G2V mode to V2G mode. Scheduled hours for V2G mode is considered from 1 to 2 s in simulation, and the corresponding waveforms of EV battery, DC link, residential load and grid are depicted in Figs. 20, 21, 22 and 23, respectively. From the dynamic response, it is evident that EVB gets supply from utility grid in G2V mode and discharged to support the grid in V2G mode. In G2V mode, BLCC operates in rectifier mode indicated by the positive E_{dc}, and during V2G mode, the negative E_{dc} indicates that BLCC operates in inverter mode. Power to the residential loads is

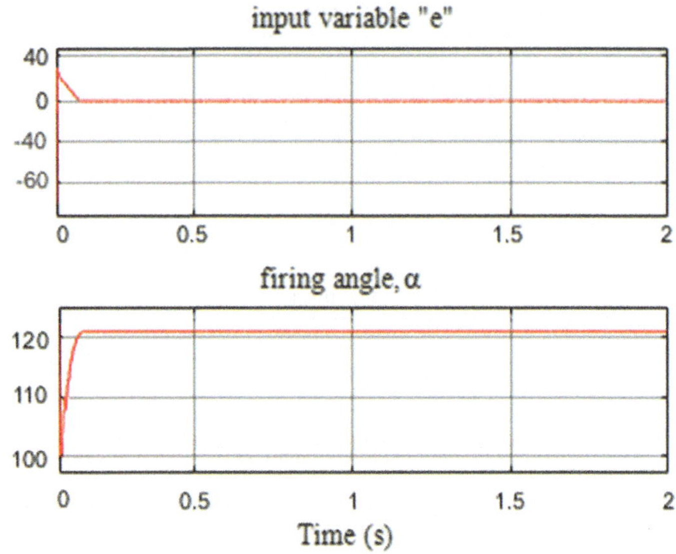

Fig. 15 Waveforms of error and firing angle signals in V2G mode

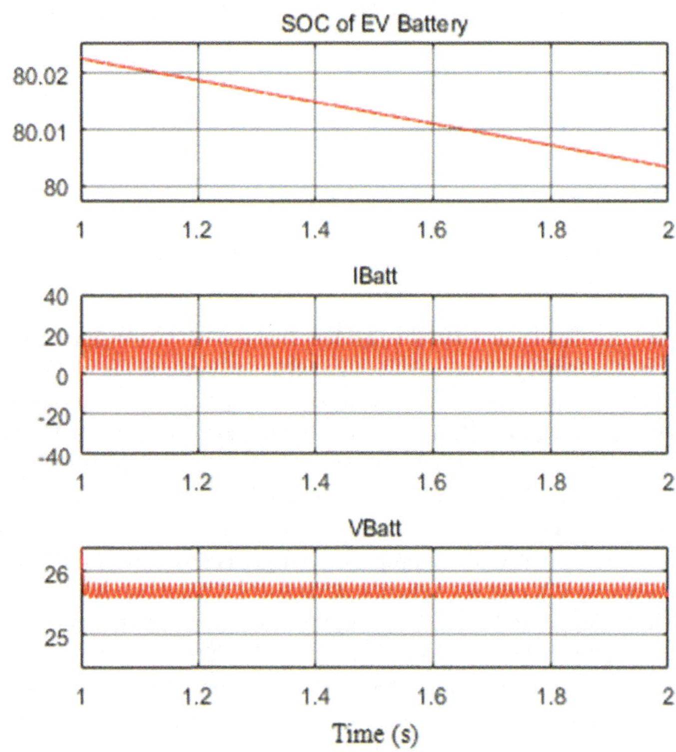

Fig. 16 EV battery SOC, V_{bt} and I_{bt} waveforms in V2G mode

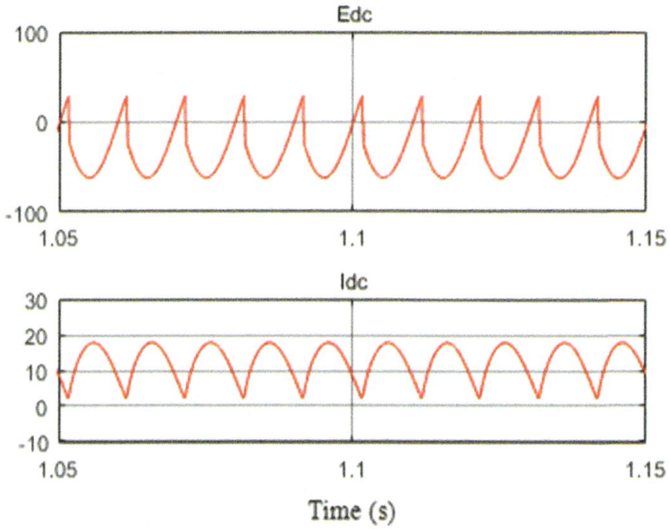

Fig. 17 Waveforms of E_{dc} and I_{dc} in V2G mode

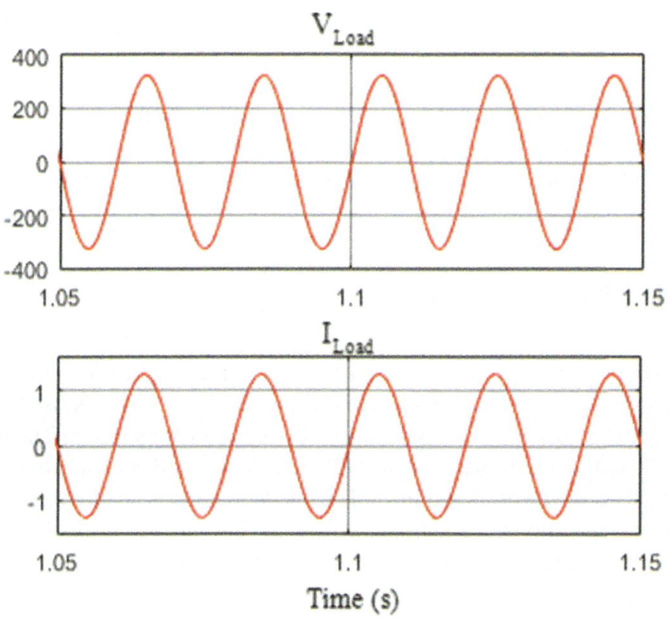

Fig. 18 Waveforms of residential load in V2G mode

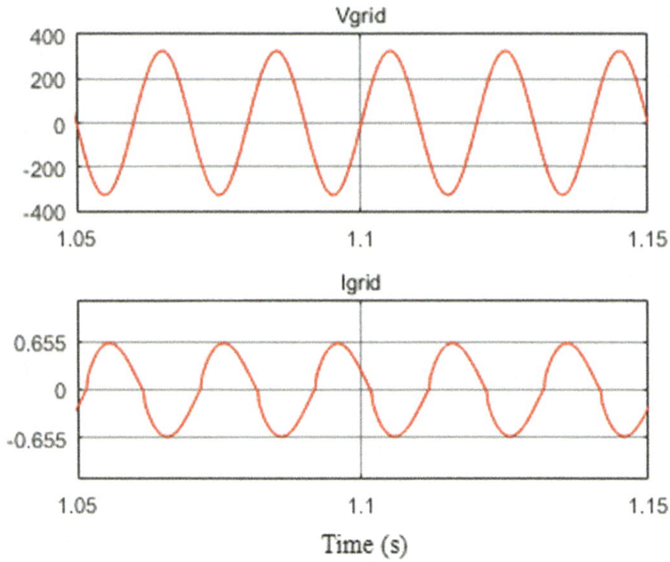

Fig. 19 Waveforms of grid parameters in V2G mode

supplied by the grid, and EVB in the G2V and V2G modes of operation, respectively, validates that the load demand on grid is reduced by the EV battery using V2G technology in the proposed charger. Also, smooth and dynamic transition from one mode to the other using FLC validates the performance of the proposed EVB charger.

5 Conclusion

This chapter presents a fuzzy logic controller for electric vehicle battery charging system in order to facilitate the G2V and V2G technology. The proposed FLC tunes the firing angle of the BLCC depending on the modes of operation. The power fed to the EVB and to the utility grid in G2V and V2G modes, respectively, is optimally controlled by the proposed FLC. The residential loads are driven by the power supplied from the utility grid and from the EVB in G2V and V2G modes, respectively. The proposed charger is modeled in MATLAB/Simulink software, and the results of the proposed charger with the minimum settling time during transition from one mode to other validate the effectiveness of the proposed fuzzy logic controller.

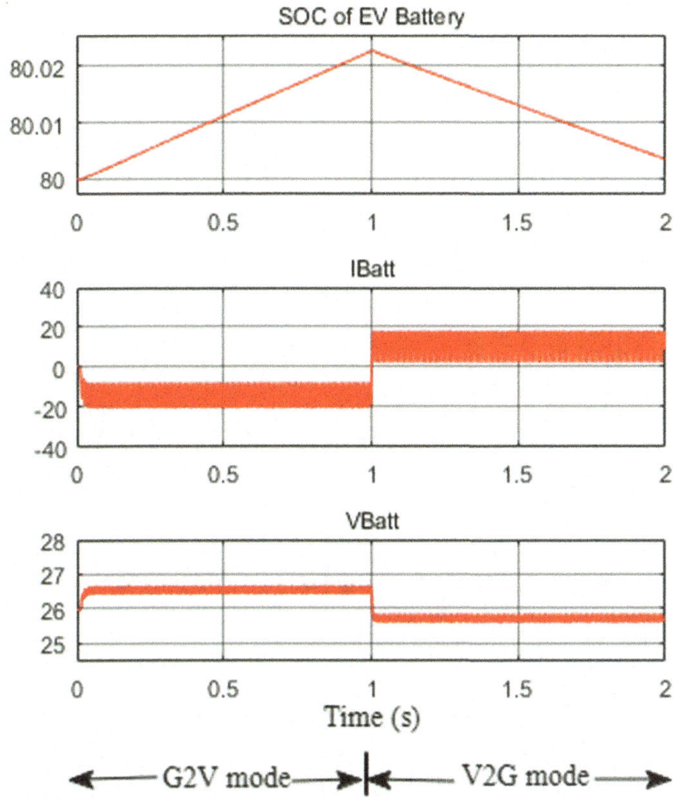

Fig. 20 EV battery waveforms with mode transition

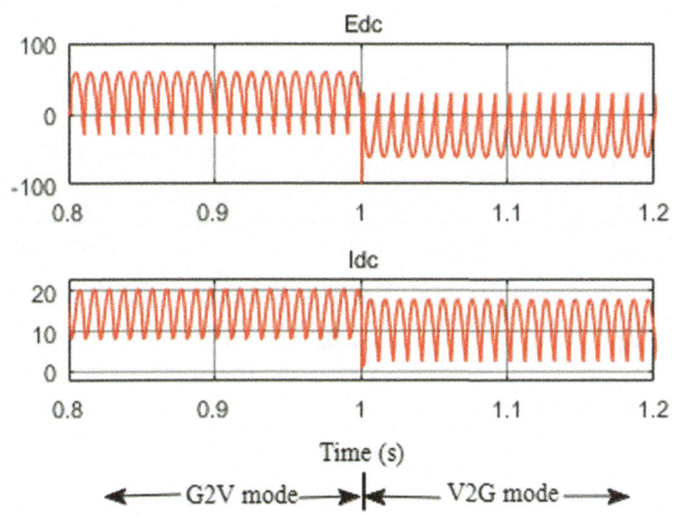

Fig. 21 Dynamic response of DC link waveforms

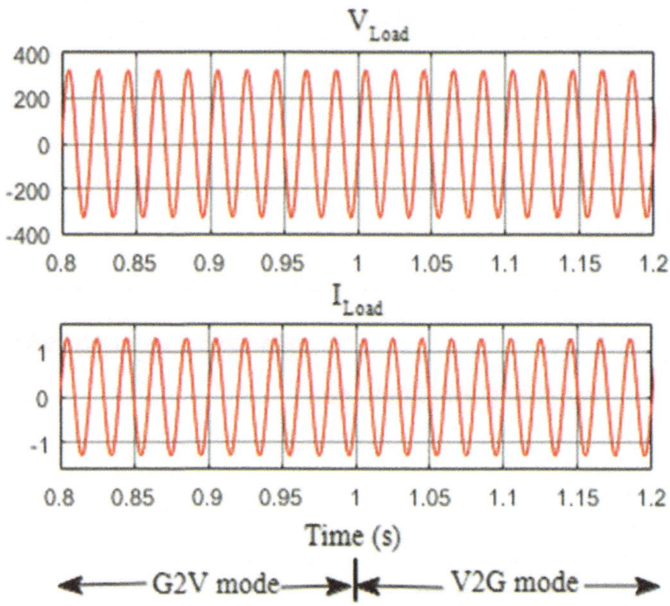

Fig. 22 Dynamic response of residential load waveforms

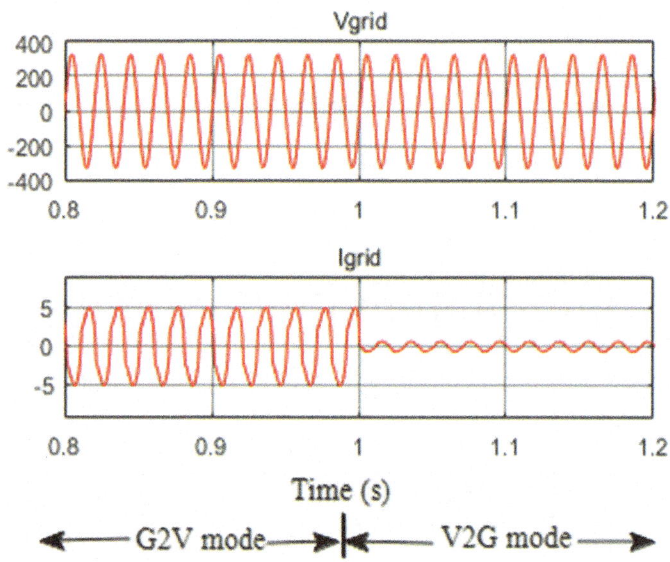

Fig. 23 Dynamic response of grid waveforms

References

1. Ashique, R.H., Salam, Z., Aziz, M.J.B.A., Bhatti, A.R.: Integrated photovoltaic-grid dc fast charging system for electric vehicle: a review of the architecture and control. Renew. Sustain. Energy Rev. **69**, 1243–1257 (2017)
2. Shi, C., Tang, Y., Khaligh, A.: A single-phase integrated onboard battery charger using propulsion system for plug-in electric vehicles. IEEE Trans. Veh. Technol. **66**(12), 10899–10910 (2017)
3. Kempton, W., Tomic, J.: Vehicle-to-grid power fundamentals: calculating capacity and net revenue. J. Power Sources **144**(1), 268–279 (2005)
4. Viswanathan, V.V., Kintner Meyer, M.: Second use of transportation batteries: maximizing the value of batteries for transportation and grid services. IEEE Trans. Veh. Technol. **60**(7), 2963–2970 (2011)
5. Marra, F., Sacchetti, D., Traeholt, C., Larsen, E.: Electric vehicle requirements for operation in smart grids. Innovative smart grid technologies (ISGT Europe). In: Proceedings of the 2nd International Conference and Exhibition on IEEE Power Energy Society, pp. 121–127, (2011).
6. Sortomme, E., El-Sharkawi, M.A.: Optimal charging strategies for unidirectional vehicle-to-grid. IEEE Trans. Smart Grid **2**(1), 131–138 (2011)
7. Su, W., Rahimi-Eichi, H., Zeng, W., Chow, M.: Survey on the electrification of transportation in a smart grid environment. IEEE Trans. Ind. Inform. **8**(1), 1–10 (2012)
8. Shariful Islam, M., Mithulananthan, N., Quoc Hung, D.: Coordinated EV charging for correlated EV and grid loads and PV output using a novel, correlated, probabilistic model. Int. J. Electr. Power Energy Syst. **104**, 335–348 (2019)
9. Gautam, D.S., Musavi, F., Edington, M., Eberle, W., Dunford, W.G.: An automotive onboard 3.3-kW battery charger for PHEV application. IEEE Trans. Veh. Technol. **61**(8), 3466–3474 (2012)
10. IEEE Trial-Use Standard Definitions for the Measurement of Electric Power Quantities Under Sinusoidal, Nonsinusoidal, Balanced, or Unbalanced Conditions. IEEE Std 1459-2000, Jan 2000
11. Xue, L., Shen, Z., Boroyevich, D., Mattavelli, P., Diaz, D.: Dual active bridge-based battery charger for plug-in hybrid electric vehicle with charging current containing low frequency ripple. IEEE Trans. Power Electron. **30**(12), 7299–7307 (2015)
12. Choi, J., Byen, B., Lee, Y., Han, D., Kho, H., Choe, G.: Design of leakage inductance in resonant DC–DC converter for electric vehicle charger. IEEE Trans. Magn. **48**(11), 4417–4420 (2012)
13. Shin, J., Lee, J.Y.: An electrolytic capacitor-less bi-directional EV on-board charger using harmonic modulation technique. IEEE Trans. Power Electron. **29**(10), 5195–5203 (2014)
14. Zhao, B., Song, Q., Liu, W., Liu, G., Zhao, Y.: Universal high-frequency-link characterization and practical fundamental-optimal strategy for dual-active-bridge DC–DC converter under PWM plus phase-shift control. IEEE Trans. Power Electron. **30**(12), 6488–6494 (2015)
15. Cho, Y.W., Cha, W.J., Kwon, J.M., Kwon, B.H.: High- efficiency bidirectional DAB inverter using a novel hybrid modulation for stand-alone power generating system with low input voltage. IEEE Trans. Power Electron. **31**(6), 4138–4147 (2016)
16. Tong, A., Hang, L., Gao, S.: Modeling and analysis of dual-active-bridge isolated bidirectional DC/DC converter to minimize RMS current with whole operating range. IEEE Trans. Power Electron. **33**(6), 5302–5316 (2017)
17. Musavi, F., Craciun, M., Gautam, D.S., Eberle, W., Dunford, W.G.: An LLC resonant DC–DC converter for wide output voltage range battery charging applications. IEEE Trans. Power Electron. **28**(12), 5437–5445 (2013)
18. Musavi, F., Craciun, M., Gautam, D.S., Eberle, W.: Control strategies for wide output voltage range LLC resonant DC–DC converters in battery chargers. IEEE Trans. Veh. Technol. **63**(3), 1117–1125 (2014)
19. Sujitha, N., Krithiga, S.: Grid tied PV-electric vehicle battery charger using bidirectional converter. Int. J. Renew. Energy Res. **9**(4), 1873–1881 (2019)

20. Yilmaz, M., Krein, P.T.: Review of the impact of vehicle-to-grid technologies on distribution systems and utility interfaces. IEEE Trans. Power Electron. **28**(12), 5673–5689 (2013)
21. Krithiga, S., Jose, D.R.B.B., Upadhya, H.R., Gounden, N.A.: Grid-tied photovoltaic array using power electronic converters with fuzzy logic controller for maximum power point tracking. Aust. J. Electr. Electron. Eng. **9**(4), 393–400 (2012)
22. Krithiga, S., Gounden, N.A.: Investigations of an improved PV system topology using multi-level boost converter and line commutated inverter with solutions to grid issues. Simul. Model. Pract. Theory **42**, 147–159 (2014)
23. Moltgen, G.: Line Commutated Thyristor Converters. Siemens Aktiengesellschaft, Pitman (1972)
24. Gounden, N.A., Sabitha, A.P., Nallandula, H., Krithiga, S.: Fuzzy logic controller with MPPT using line-commutated inverter for three-phase grid-connected photovoltaic systems. Renew. Energy **34**, 909–915 (2009)

Nature-Inspired Algorithms for Maximum Power Point Tracking in Photovoltaic Systems Under Partially Shaded Conditions

V. Vignesh Kumar and C. K. Aravind

1 Introduction

Increased power generation from photovoltaic (PV) systems has recently become more common due to several reasons such as depletion of fossil fuels and environmental concerns. Due to low conversion efficiency and higher initial investments on PV systems, maximum power point tracking (MPPT) is an essential requirement in grid-connected PV systems. The power–voltage (*P–V*) characteristic of a PV system is nonlinear and depends on solar insolation and ambient temperature. Hence, MPPT is a strenuous task and several methods have been developed and implemented [1–7]. A comprehensive review of these methods can be seen in [8–10].

Under partial-shaded conditions, multiple peaks occur in the *P–V* characteristics of PV system and the process of MPPT becomes a herculean task. The peak with highest power is called as global maximum power point (GMPP), and the remaining peaks are termed local maximum power points (LMPPs). The application of methods described in [1–7] does not guarantee convergence to GMPP and mostly gets trapped in one of the LMPPs. Generally, there are four different schemes available for extracting maximum power from partially shaded PV arrays [11]. These are modified MPPT techniques, PV array reconfiguration, PV system architectures and different topologies for power electronic interface. It can be inferred from [11] that last three schemes are expensive and involve more components and complex control when compared to modified MPPT methods. The attributes of modified

V. Vignesh Kumar (✉)
Department of Electrical and Electronics Engineering, National Institute of Technology Karnataka Surathkal, Mangalore, India
e-mail: v.vigneshkumar@gmail.com

C. K. Aravind
Department of Electrical and Electronics Engineering, Mepco Schlenk Engineering College, Sivakasi, Tamil Nadu, India
e-mail: aravindck@gmail.com

B. Vinoth Kumar et al. (eds.), *Intelligent Paradigms for Smart Grid and Renewable Energy Systems*, Algorithms for Intelligent Systems,
https://doi.org/10.1007/978-981-15-9968-2_9

MPPT methods are guaranteed convergence to global maxima, system independent and greater tracking efficiency.

A closer examination of existing MPPT methods shows that there can be three classifications; the first one corresponds to modification of classical methods such as modified P&O [12], dividing rectangle (DIRECT) search [13] and two-stage MPPT [14]. Soft computing techniques, namely fuzzy logic [15] and neural networks [16], have also been employed and can be treated as second category. The use of nature-inspired optimization techniques, such as particle swarm optimization (PSO) [17–19], genetic algorithm (GA) [20], firefly algorithm (FA) [21], artificial bee colony algorithm (ABC) [22], can be considered as the last group.

MPPT can be treated as an optimization task of maximizing PV power with PV voltage as the single variable to be identified. Since each nature-inspired algorithm varies in composition, computational complexity, hardware implementation and convergence characteristics, this chapter develops a comprehensive and unified approach towards MPPT through few commonly employed and experimentally feasible algorithms. A simple and cost-effective experimental setup is explained, and then the procedural steps of each algorithm are listed. Computed and measured MPPT curves are presented followed by performance comparison of each method. It is expected that the findings from this research work can largely help researchers working in the area of photovoltaic power generation systems.

2 Modelling and Description of PV System Under Study

2.1 Modelling of PV System

Single diode model of the PV cell which is shown in Fig. 1 is used to compute the electrical characteristics of the PV system in this work as it is simple and provides good accuracy [23].

The relationship between voltage and current of the PV module is given by Eq. (1) which is obtained by applying KCL to Fig. 1.

Fig. 1 Equivalent circuit for single diode model of PV module

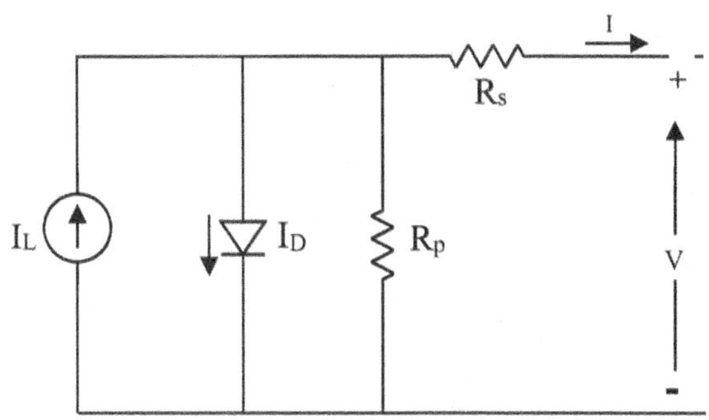

$$I = I_L - I_D - \left(\frac{V + IR_s}{R_P}\right) \quad (1)$$

where I is output current of PV module, V is output voltage of PV module, I_D is diode current, I_L is photocurrent of PV module, R_s is series resistance of PV module, R_p is parallel resistance of PV module. The diode current, I_D is mathematically represented by Eq. (2).

$$I_D = I_o\left[\exp\left(\frac{V + IR_S}{aV_t}\right) - 1\right] \quad (2)$$

where I_o is diode saturation current, a is diode ideality constant and V_t is nominal thermal voltage. The prominent equations needed to obtain the unknown parameters of Eq. (1) can be found in [23]. Equation (1) is solved numerically by the method given in [23] to obtain the current–voltage (I–V) and power–voltage (P–V) characteristics of the PV modules.

2.2 Experimental Setup

The hardware for maximum power point tracking system consists of an array of PV modules, DC–DC converter, load and one MPPT controller in the feedback loop. The PV system employed to test MPPT algorithms is shown in Fig. 2. Here, the PV power is sensed and fed to the MPPT controller. The nature-inspired algorithm embedded in the MPPT controller computes the duty ratio. Then, the pulse width modulation (PWM) block in MPPT controller generates the appropriate PWM signals to activate the DC–DC converter. The objective of MPPT algorithm is to identify the duty ratio corresponding to the GMPP.

The PV array shown in Fig. 3 has three PV modules connected in series and two such strings are connected in parallel. This type of configuration is termed as $3s2p$ configuration. For this configuration, three different shading levels as indicated in Fig. 3 are used to represent the partial-shaded conditions. Equation (1) is solved as given in [24, 25] for the PV system with three different shading configurations and

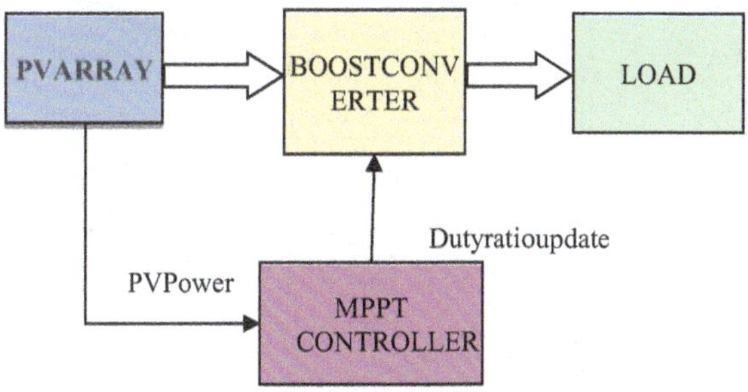

Fig. 2 Block diagram of MPPT system

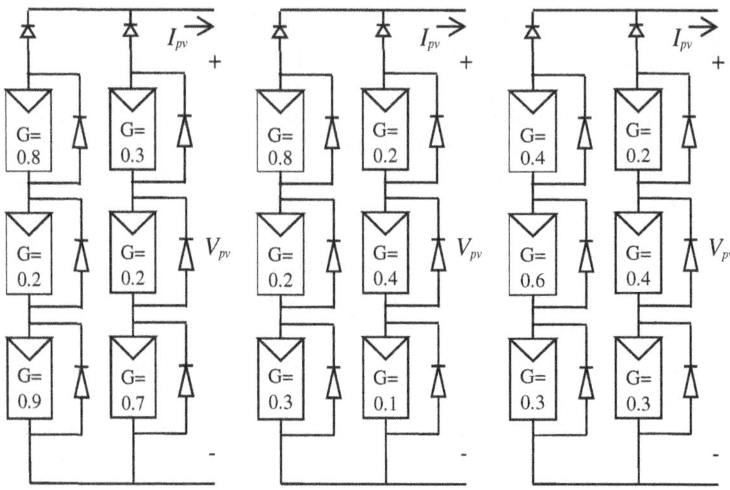

Fig. 3 Three different combination of insolation levels for 3*s*2*p* configuration (*G* represents insolation in kW/m^2)

the PV power versus duty ratio (*P*–*d*) curves under PSC are obtained. Computer program is developed in MATLAB and respective *P*–*d* plots are calculated. These curves are named as pattern 1, pattern 2 and pattern 3 and are shown in Fig. 4. As seen, pattern 1 has three power peaks with GMPP at 45.67 W and two LMPPs at 31.92 and 25.78 W. For shading pattern 2, *P*–*d* curve comprises of three power peaks with GMPP at 23.72 W and two LMPPs. *P*–*d* curve for pattern 3 also consists of three power peaks with 31.92 W as GMPP.

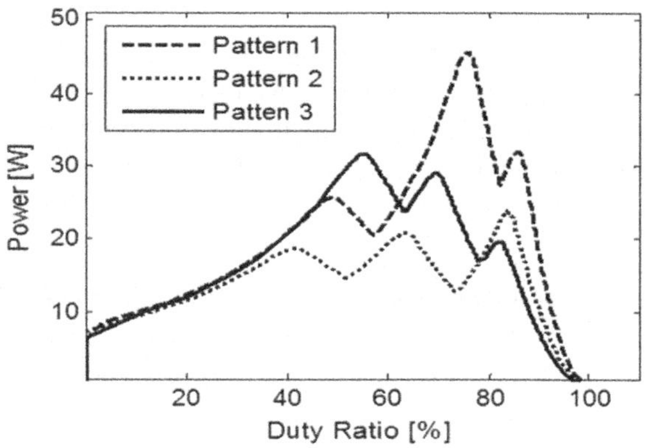

Fig. 4 Power versus duty ratio curves of 3*s*2*p* configuration

3 Formulation of the Optimization Problem

With reference to the experimental setup in Fig. 2, the MPPT can be formulated as an optimization problem of maximizing PV output power with duty ratio d of the DC–DC converter as the variable. This is given below.

$$\text{Maximize } P_{pv}(d) \tag{3}$$

$$\text{Subject to: } d_{\min} \le d \le d_{\max} \tag{4}$$

where P_{pv} is PV output power, d is duty ratio of DC–DC converter, $d_{\max}$ is maximum limit of duty ratio and $d_{\min}$ is minimum limit of duty ratio.

4 MPPT Through Nature-Inspired Algorithms

Prominent optimization algorithms employed for MPPT are described in this section. In all the optimization algorithms, the population size is taken as six and initial population is scattered between $d_{\min}$ and $d_{\max}$.

4.1 Particle Swarm Optimization (PSO)

Particle swarm optimization was introduced by Kennedy and Eberhart in 1995 and is developed by inspiring the social behaviour of bird flocking and fish schooling [26–28]. Each member in the population is called as particle. Each particle exchanges knowledge obtained from its search process with others in the population. All the particles follow the elite particle and also the best position discovered by themselves to move towards global optimum point. The operating principles of the PSO-based MPPT are given below:

1. Activate the DC–DC converter using digital controller corresponding to the position (duty ratio) of each particle and compute its fitness value (PV power) after the allowable converter settling time.
2. Compute the particle best (*pbest*) that is the best position obtained by each particle with highest fitness so far.
3. Compute the global best (*gbest*) that is particle with highest fitness value in the population.
4. Compute the velocity and position of particle i in the iteration k using Eqs. (5) and (6) given below.

$$V_i^{k+1} = w_i V_i^k + r_1 c_1 \left(pbest_i - d_i^k\right) + r_2 c_2 \left(gbest - d_i^k\right) \tag{5}$$

$$d_{pi}^{k+1} = d_{pi}^{k} + V_i^{k+1} \tag{6}$$

where k is iteration number, V_i^k is velocity of particle, i at iteration k, d_{pi}^k is position of particle i at iteration k, w is scaling factor for velocity, c_1 is cognitive constant, c_2 is social constant and r_1, r_2 are random numbers $\in (0, 1)$.

5. Terminate the program when all the particles are closer to each other; else go to step 1.

4.2 Artificial Bee Colony Algorithm (ABC)

Karaboga and Basturk have described an artificial bee colony algorithm in 2005 based on the foraging behaviour of honeybees [29–31]. In ABC algorithm, food source represents a possible solution to the optimization problem. The nectar amount of a food source corresponds to the quality of the solution represented by that food source. In order to find the best solution, the algorithm defines three classes of bees: employed bees, onlooker bees and scout bees. The employed bee searches for the food sources; the onlooker bee makes a decision to choose the food sources by sharing the information of employed bee. In ABC-based MPPT, position of food source is the duty ratio d of the DC–DC converter and the corresponding PV output power is the nectar amount. In order to reduce the convergence time, scout bee phase in conventional ABC algorithm is eliminated in this method. The procedure for MPPT through ABC algorithm is given below:

1. Calculate nectar amount (PV power) corresponding to each food position (duty ratio) and assign the bees as employed/onlooker bee based on the nectar amount.
2. Identify the new food source position through employed and onlooker bees phase
   - Employed bee phase: In this phase, the duty ratio of the DC–DC converter is updated using the following equation:

$$d_e^{k+1} = d_e^k + \varnothing\left(d_e^k - d_j^k\right) \tag{7}$$

   $e \in (1, 2..N_e)$ and j is randomly chosen index where $j \neq e$.
   In Eq. 7, d_e^k is position of employed bee at iteration k, d_h^k is position of bee with highest nectar amount, e and o are employed and onlooker bee index, respectively, N_e and N_o are number of employed and onlooker bees, respectively, $\varnothing$ is the random number $\in (-1, 1)$, k is the iteration number.
   - Calculate the probability, P_i values of the bee's positions by means of their fitness value, fit_i using Eq. (8)

$$P_i = \frac{\text{fit}_i}{\sum_{i=1}^{6} \text{fit}_i} \tag{8}$$

- Onlooker phase: Depending on the probability P_i, new positions of onlooker bees are evaluated from Eq. (9)

$$d_0^{k+1} = d_h^k + \varnothing * \left(d_o^k - d_h^k\right), \quad o \in (1, 2, ..N_0) \tag{9}$$

where d_o^k is position of onlooker bee at iteration k.

3. Terminate the algorithm if all the bees in the colony come close to each other; now operate the DC–DC converter with optimal duty ratio; else go to step-1.

4.3 *Cuckoo Search Algorithm (CSA)*

Cuckoo Search (CS) is a recently developed optimization algorithm based on the aggressive reproduction strategy of certain species of cuckoos. It was introduced by Yang and Deb in 2009 [32, 33]. Cuckoos generally lay their eggs in the other bird's (host bird) nest to hatch them, which is termed as brood parasitism. The chance of Cuckoo's egg to survive in the host bird nest depends on the quality of egg (i.e., fitness value). The sequential implementation of CSA towards MPPT is written below:

1. Evaluate the fitness value (PV output power), corresponding to each host nest's position (duty ratio).
2. Modify the host nest's current positions using Eqs. (10) and (11) and evaluate its fitness value.

$$d_{ni}^{k+1} = d_{ni}^k + s_i * \text{levy}(\lambda) * \alpha \tag{10}$$

where d_{ni}^k is position of nest i at iteration k, α is random number, $\in (-1, 1)$, s_i and levy(λ) are given as

$$s_i = d_{ni}^k - d_{nbest}^k; \quad \text{levy}(\lambda) = \left[\frac{\Gamma(1+\lambda) * \sin\frac{\pi\lambda}{2}}{\Gamma\left(\frac{1+\lambda}{2}\right) * \lambda * 2^{\left(\frac{\lambda-1}{2}\right)}}\right]^{1/\lambda} \tag{11}$$

where d_{nbest} is position of nest with best quality of eggs, λ is levy flight constant $\in (1, 3)$.

3. The probability P_i of each nest, obtained from Eq. (12) is compared with the acceptance probability P_a to identify the nests with alien eggs. The identified nests are deserted, and the new nests are created using Eq. (13).

$$P_i = (0.9 * (\text{Fit}_i/\text{Fit}_{\max})) + 0.1 \tag{12}$$

where Fit_i is fitness of nest i, $\text{Fit}_{\max}$ is fitness of nest with maximum fitness

$$d_{ni}^k = d_{\min} + \text{rand}(0, 1) * (d_{\max} - d_{\min}) \tag{13}$$

4. Repeat the steps 1 to 3 until the solutions converge to the nest position with best quality of eggs.

4.4 Ant Colony Optimization (ACO)

The ant colony optimization (ACO) is first proposed by Dorigo [34–36] in 1996. Ant colony optimization (ACO) algorithms attempt to exploit the efficiency of ant foraging behaviour by creating an abstract environment of possible paths, and simulating ants travelling along these paths. Generally, movement of ants from food location to their nest is guided by pheromone laying mechanism. The concentration of these pheromone trails helps the ant in successfully finding the shortest path between the food location and their nest.

In the MPPT process employing ACO method, ant's position is represented as duty ratio of the DC–DC converter and the PV output power respective to that duty ratio is regarded as the pheromone content. The step-by-step procedure for ACO-based MPPT is listed below:

1. The DC–DC converter is activated with respect to each ant position to compute the PV output power.
2. The movement of all other ants except the best ant is calculated by Eq. (14). The best ant is the one corresponding to highest PV output power, and it remains at same position.

$$d_{ai}^{k+1} = d_{ai}^{k} + \delta_1 \vec{a} \tag{14}$$

$$\delta_1(k) = \delta_o e^{-k}, \quad \text{where } \delta_o \text{ is taken as 10.} \tag{15}$$

   In Eqs. (14) and (15), d_{ai}^{k} is position of ant i at iteration k, δ_o is initial step size of ant movement $\delta(k)$ is step size of ant movement at iteration k, $\vec{a}$ is unit vector from ant i to the best ant's position.
3. Steps 1 and 2 are repeated until all the ants converge to maximum power point.

4.5 Genetic Algorithm (GA)

The concept of genetic algorithm (GA) was first reported by John Henry Holland [37–41]. It mimics the process of evolution of the species in the nature. It is based on the theory of natural evolution which states that only the fittest chromosomes in the population will survive and reproduce offspring for the next generation. Here duty ratio of the boost converter refers to the chromosome and the PV output power corresponds to the fitness of the chromosome. The MPPT through GA can be obtained as per the following steps:

1. Evaluate the fitness value of each chromosome in the population by activating the DC–DC converter with respective duty ratio.
2. Carry out the selection process using roulette wheel method and retain the best chromosome for next generation. This practice is called elitism which help in faster convergence to global optimum point.
3. Execute the crossover function with probability of 0.8 for the selected parents.
4. Carry out mutation with probability of 0.01 for newly generated off springs.
5. Redo Step 1 to 4 until all the chromosomes becomes identical.

4.6 Firefly Algorithm (FA)

The firefly algorithm (FA) is a population-based optimization and is introduced in 2009 [42–44]. In the FA, every firefly is assumed to attract another regardless of their sex. Since the attractiveness is proportional to the brightness, the less bright ones will move towards the brighter ones, and the brightest one moves randomly. The steps involved in FA-based MPPT are:

1. Fix the constants of the firefly algorithm, namely γ and α. In this algorithm, the position of the firefly is taken as a duty cycle of the DC–DC converter. The brightness of each firefly is taken as generated power P_{pv} of the PV system, corresponding to the position of this firefly.
2. For each duty ratio, the corresponding PV output power is calculated. The initial attractiveness of each firefly is calculated from Eq. (16)

$$\beta_i = \frac{P_{pvi}^k}{P_{pvq}^k} \tag{16}$$

   β_i is initial attractiveness of firefly i, P_{pvi} is PV power corresponding to firefly i, P_{pvq} is PV power corresponding to brightest firefly q.
3. Calculate the degree of attractiveness between the two fireflies d_i and d_j using Eq. (17)

$$\beta_{ij} = \beta_j * \exp(-\gamma\left(r_{ij}\right)^2 \tag{17}$$

   β_{ij} is degree of attractiveness between firefly i and j, γ is absorption coefficient, r_{ij} is Cartesian distance between firefly i and j.
4. The new position of firefly i is given by the following equation:

$$d_{fi}^{k+1} = d_{fi}^k + \beta_{ij}\left(d_{fi} - d_{fj}\right) + \alpha\left(r_1 - {}^1\!/_2\right), \quad i \neq q \tag{18}$$

   d_{fi}^k, d_{fj}^k are the position of firefly i and j respectively at iteration k.
5. The brightest firefly, d_q in a colony moves randomly from its position and its movement is given by:

$$d_q^{k+1} = d_q^k + \alpha\left(r_1 - \frac{1}{2}\right) \quad (19)$$

6. Terminate the program once the displacement of all fireflies in consecutive steps reaches a set minimum value. Else go to step 2.

5 Performance Evaluation

5.1 Computed and Measured Results

The $3s2p$ configuration with all the three patterns given in Fig. 4 is employed one by one each for 20 s duration. A dedicated program for all the MPPT algorithms is developed in MATLAB for carrying out simulation, and the tracking curves obtained are shown in Fig. 5. It can be observed from these tracking curves that all algorithms provide convergence to GMPP. The ripples seen in the PV output power can be attributed to the population-based strategy of these algorithms. The time taken to track the GMPP for each algorithm is given in Table 1. It can be seen from this table that FA takes least time to converge to GMPP, and consequently, the ripples in the output power prevail for smaller duration.

In order to authenticate the simulation findings, experiments were carried out on a $3s2p$ PV prototype system with two different partially shaded conditions which are obtained by placing transparent sheets of different thickness on PV modules. Two different P–V curves thus obtained for experimental work are named as pattern 4 and pattern 5 as given in Fig. 6. The LMPPs and GMPP for pattern 4 and pattern 5 are mentioned in Fig. 6. Dedicated programs employing each of the discussed methods were developed using MPLAB IDE and are dumped into digital controller. As in simulation study, the two patterns are made to exist sequentially, each one for 20 s. All the algorithms are capable of locating new GMPP successfully on the occurrence of change in shading pattern. The tracking curves employing each were recorded and reproduced in Fig. 7. The tracking time of GMPP for each algorithm is given in Table 1.

From the simulation and experimental studies, the general features of the tracking curves are observed as follows:

1. All optimization algorithms mentioned in this paper converge to GMPP successfully. Thus, all optimization algorithms are promising candidates for MPPT.
2. The convergence time for each method varies significantly. This factor decides energy generated and is therefore perfect index for comparison.
3. During tracking, large oscillations in PV output power are visibly seen. This is due to the randomness associated with each algorithm.
4. The oscillations in PV power decay as tracking progresses. This may be perceived as the learning ability of the algorithms.

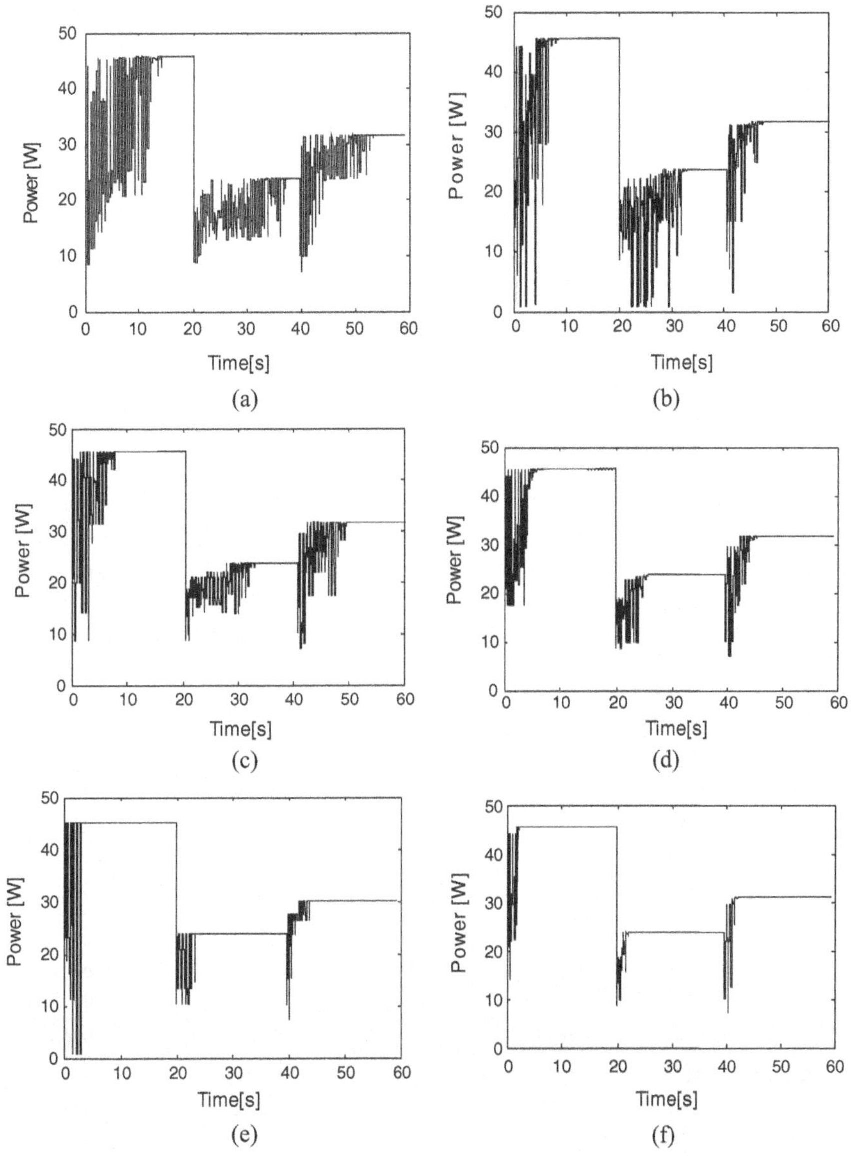

Fig. 5 MPPT curves **a** PSO **b** ABC **c** CSA **d** ACO **e** GA and **f** FA

5.2 *Performance Measures*

A closer examination of the MPPT schemes employed in this work reveals that all the proposed methods guarantee global convergence, independent of PV configuration

Table 1 Performance comparison of nature-inspired MPPT methods

Pattern	MPPT methods	Tracking speed (s)	Power tracked (Ws)	Tracking efficiency (%)	Energy tracking factor
Pattern 1 GMPP (45.67 W)	FA	3.8	45.67	100	0.955
	GA	4.2	45.18	98.92	0.937
	ACO	6.8	45.43	99.47	0.922
	CSA	8.1	45.49	99.60	0.911
	ABC	9.4	45.67	100	0.910
	PSO	14.9	45.67	100	0.867
Pattern 2 GMPP (23.72 W)	FA	3.2	23.71	100	0.975
	GA	6.5	23.65	99.5	0.957
	ACO	8.5	23.7	99.98	0.941
	CSA	9.1	23.69	99.63	0.910
	ABC	13.6	23.71	100	0.907
	PSO	14.8	23.71	100	0.901
Pattern 3 GMPP (31.92 W)	FA	3.5	31.92	100	0.967
	GA	4.6	31.79	99.38	0.941
	ACO	5.7	31.89	99.94	0.938
	CSA	13.1	31.89	99.94	0.921
	ABC	14.7	31.82	99.73	0.935
	PSO	18.7	31.79	100	0.920
Pattern 4 GMPP (25.2 W)	FA	4.2	25.14	99.76	0.952
	GA	7.1	24.96	99.04	0.941
	ACO	7.45	25.06	99.44	0.921
	CSA	9.85	25.10	99.60	0.904
	ABC	13.1	25.14	99.76	0.87
	PSO	14.25	25.14	99.76	0.852
Pattern 5 GMPP (18 W)	FA	4.95	17.97	99.83	0.96
	GA	6.25	17.90	99.44	0.936
	ACO	6.85	17.95	99.72	0.942
	CSA	10.63	17.95	99.72	0.933
	ABC	13.25	17.95	99.72	0.844
	PSO	14.05	17.97	99.83	0.876

and shading patterns. However, literature survey indicates that the performance of MPPT curves needs to be evaluated based on the following indices [45]:

Tracking Time

Tracking time is an important index and is defined as the convergence time to reach GMPP. It measures the success of any MPPT algorithm quantitatively. This value

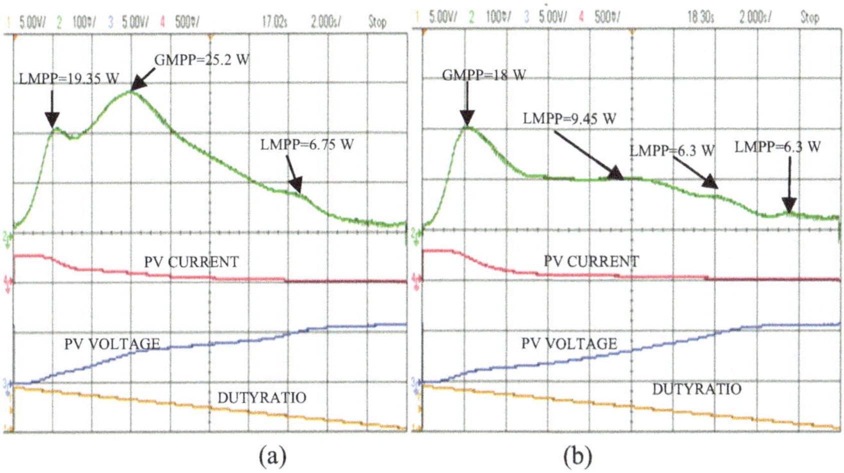

Fig. 6 Experimental PV curve for $3s2p$ configuration **a** pattern 4 **b** pattern 5 (scale: 2 s/div for time, 9 W/div for power, 4 A/div for current, 50 V/div for voltage and 100%/div for duty ratio)

should be as low as possible which will enable enhanced energy output power from the PV power generation system and will also ensure least disturbance in the PV output power.

Tracking Efficiency

Tracking efficiency reflects tracking accuracy of any MPPT method and is defined as the ratio between averaged output power obtained under steady state and maximum available power of the PV array under certain shading pattern.

Energy Tracking Factor

Energy tracking factor is calculated using the following equation for various MPPT methods:

$$E_{\text{TF}} = \frac{\text{Energy extracted by a given MPPT Scheme during } t_{\max}}{\text{maximum energy available during } t_{\max}} \tag{20}$$

where $t_{\max}$ is the time taken for convergence to GMPP by the slowest MPPT scheme considered in this work. This term helps in analysing the different performance parameters, viz. tracking time, oscillations in PV output power during tracking and the useful energy extracted during tracking period.

Ripples in the PV Output Power During Steady State

This parameter indicates the effectiveness of the convergence of the MPPT algorithm. It is expected that after successful tracking, an ideal MPPT algorithm should possess zero ripples in PV output power, voltage and current.

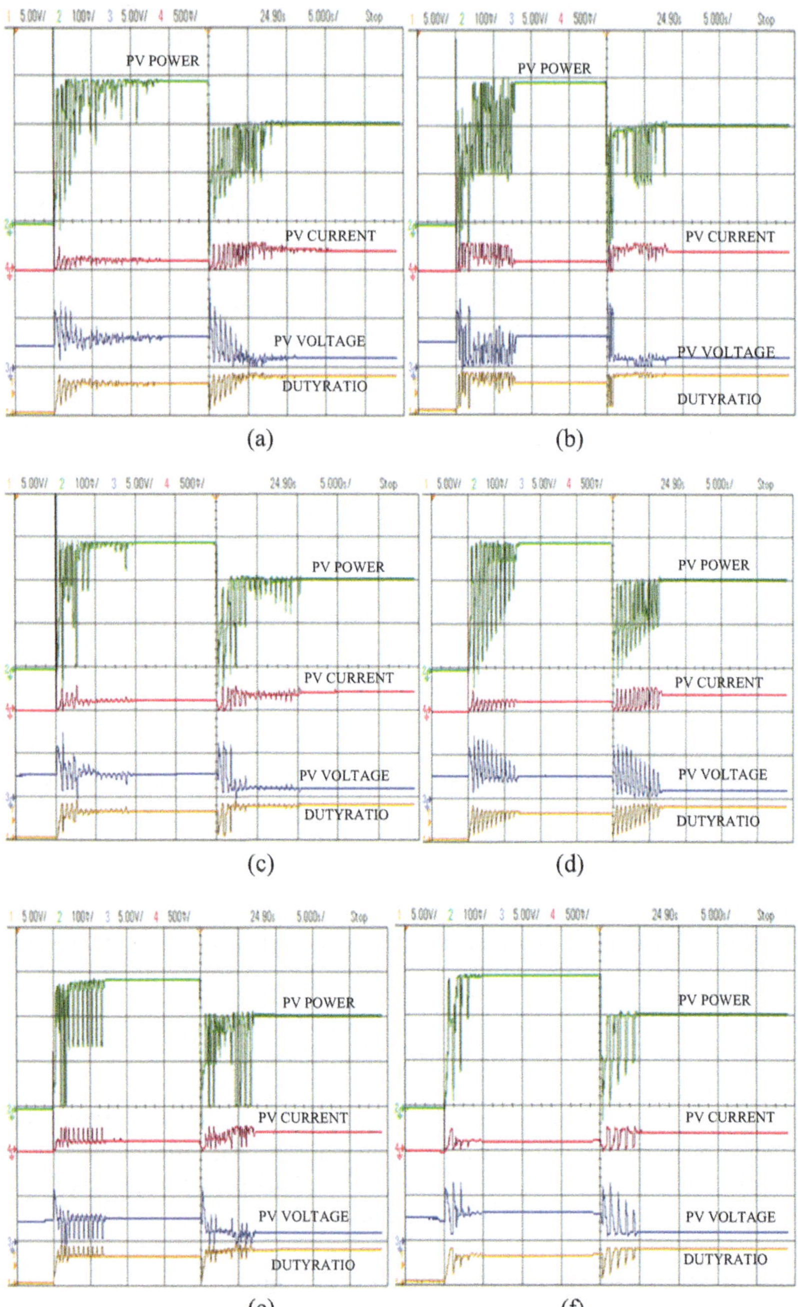

Fig. 7 Experimental tracking curves **a** PSO **b** ABC **c** CSA **d** ACO **e** GA and **f** FA (scale: 5 s/div for time, 9 W/div for power, 4 A/div for current, 50 V/div for voltage and 100%/div for duty ratio)

Dependency on PV Configuration, Array Parameters and Shading Pattern
An excellent MPPT algorithm is one which tracks the true GMPP with no prior knowledge of either type of PV configuration, array parameters or shading pattern. In other words, a perfect MPPT algorithm should be independent of PV system while tracking GMPP.

From the simulation and experimental results obtained, the performance analysis of all the MPPT methods employed in this work is carried out. The performance comparison of MPPT techniques is furnished in Table 1. Following observations are made by referring to Table 1.

1. The tracking time is least for FA method followed by GA, ACO, CSA, ABC and PSO methods.
2. The tracking efficiency of MPPT algorithms reported in this work is excellently impressive. This shows all these methods are promising candidates for MPPT.
3. The energy tracking factor of GA, ACO, CSA, ABC and PSO is lower than that of FA.
4. All the optimization algorithms employed in this work do not exhibit steady state ripples.
5. All the MPPT schemes elaborated in this work are independent of shading pattern, array parameters, while performing online MPPT.

5.3 Statistical Analysis

The effectiveness of the MPPT methods employed is further verified by the statistical analysis. The mean and standard deviation of the PV power at each iteration is plotted for the patterns 1, 2 and 3 in Fig. 8a–c, respectively. It can be seen that the mean value of all the individuals in the FA algorithm converges to GMPP faster compared to the mean of individuals in other algorithms. This further authenticates the findings of the simulation and experimental studies.

The ripples in the PV output power are plotted by taking the deviation of the power sensed from the corresponding global maximum power (GMP) after each sampling period. The plot showing ripples in the PV output power during MPPT tracking for all the methods used in this work are given in Fig. 9. It can be seen that magnitude of ripples during initial stages of transient tracking period is large and is slowly decreased as search proceeds towards GMPP. From Fig. 9, it is evident that ripples in the PV output power persists for least duration with FA method than other MPPT methods.

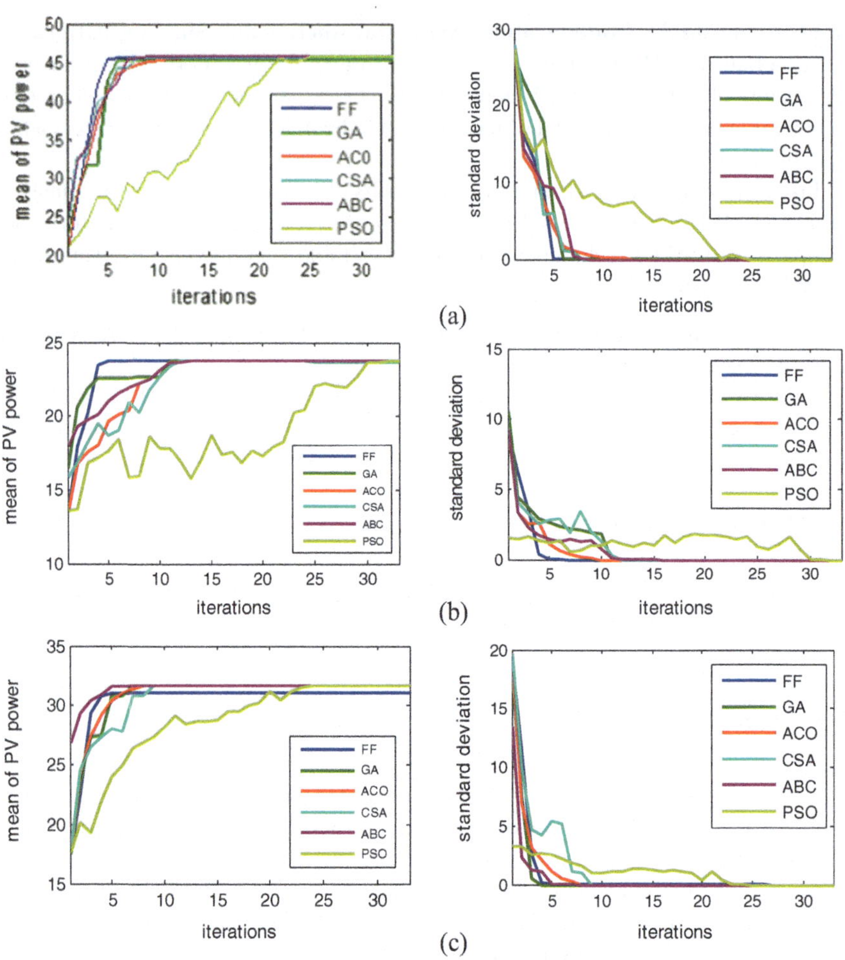

Fig. 8 Mean and standard deviation plots **a** Pattern 1 **b** Pattern 2 and **c** Pattern 3

5.4 *Discussion of Parameters of the Algorithm*

5.4.1 Particle Swarm Optimization

Velocity scaling factor w: This parameter controls the global and local search ability of the particles in the population. Generally, in the initial iterations to improve the global search this parameter is kept large and its value is reduced in the later stages for enhancing the local search.

$$w^k = w_{\max} - \left(\frac{w_{\max} - w_{\min}}{k_{\max}}\right) * k \tag{21}$$

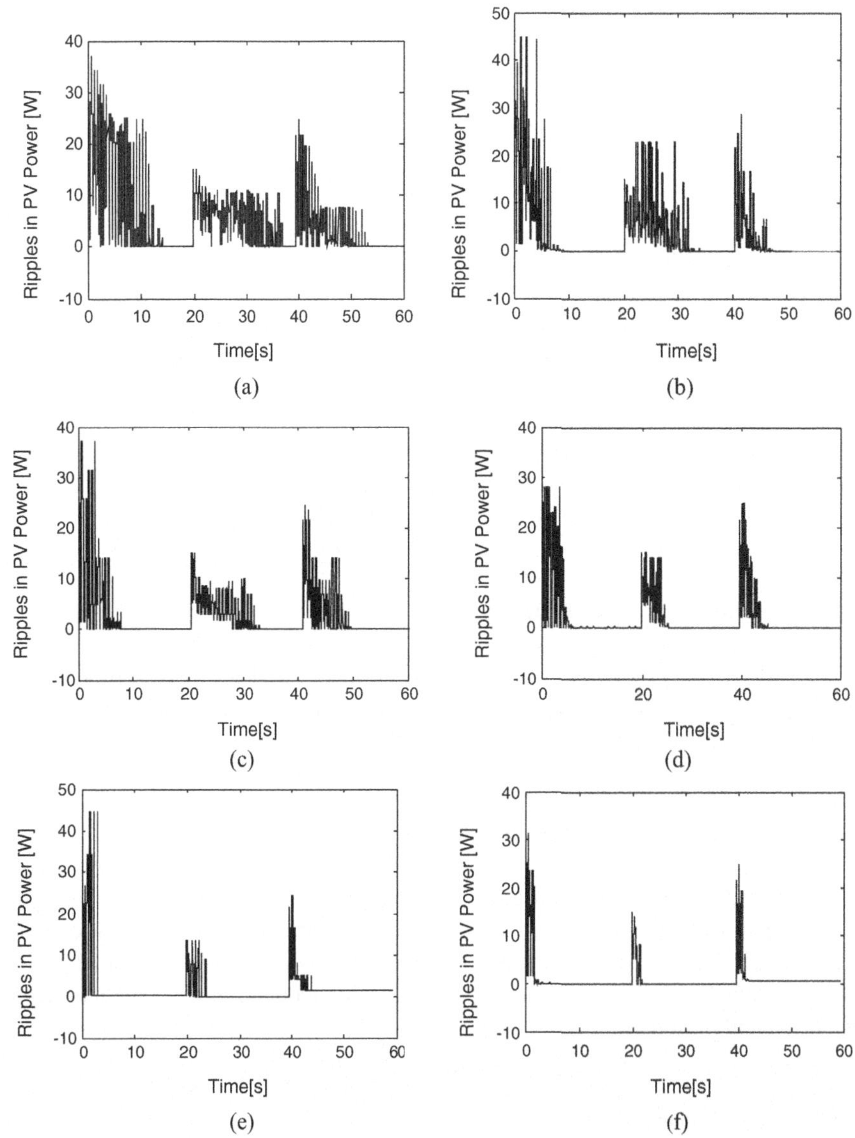

Fig. 9 Ripples during MPPT **a** PSO **b** ABC **c** CSA **d** ACO **e** GA and **f** FA

Acceleration coefficients c1 and c2: These constant multiplier terms used in velocity update formula are termed as cognitive and social learning factors, respectively. The constant c1 has a contribution towards the self-exploration (or experience) of a particle and c2 decides motion of the particles in global direction.

$$c_1^k = c_{1\max} - \left(\frac{c_{1\max} - c_{1\min}}{k_{\max}}\right) * k \tag{22}$$

$$c_2^k = c_{2\min} + \left(\frac{c_{2\max} - c_{2\min}}{k_{\max}}\right) * k \tag{23}$$

5.4.2 Artificial Bee Colony Optimization

The control parameters of the ABC algorithm are number of employed bees, onlooker bees, colony size and limit value for the bees to become scout bee. In this work, scout bee phase is eliminated, and hence, parameter limit value is not considered. The percentage of employed bees and onlooker bees is usually kept at 50% of colony size as per Karboga's ABC algorithm [31]. Thus, ABC combines the local search carried out by employed bees and the global search managed by onlooker bees, thereby providing balance between exploration and exploitation.

5.4.3 Cuckoo Search Algorithm

In CSA, there are only two parameters population size and acceptance probability, p_a [32]. Once, population size is fixed p_a essentially controls the elitism and the balance of randomization and the local search.

5.4.4 Ant Colony Optimization

The modified ACO technique used in this work has single parameter which is the initial step size of ant movement. The step size of ant with highest pheromone content is made zero. This maintains elitism in the algorithm. The step size of all other ants is exponentially decreased as they move towards the higher pheromone content position in subsequent iterations.

5.4.5 Genetic Algorithm

The parameters that characterize the search in GA are crossover probability P_c, mutation probability P_m and bit size. P_c controls the rate of crossover among the chromosomes. P_m gives the probability of modification of bits in each chromosome. The bit size of chromosome determines the resolution of the solutions obtained. Chromosomes coded with higher bits generally have more resolution than the ones with lower number of bits. P_c and P_m values are taken as given in [41].

Table 2 Parameter values of the optimization algorithms

Algorithm	Number of parameters	Parameter values
PSO	6	$w_{\max} = 1$, $w_{\max} = 0.1$, $c_{1\max} = c_{2\max} = 2$, $c_{1\min} = c_{2\min} = 0.1$
ABC	3	$N_e = 3$, $N_b = 3$, colony size = 6
CSA	1	$p_a = 0.15$
ACO	1	$\delta_0 = 10$
GA	3	$p_c = 0.8$, $p_m = 0.01$, bit size = 7
FA	3	$\beta_o = 0.9$, $\alpha = 0.95$, $\gamma = 0.002$

5.4.6 Firefly Algorithm

There are three parameters in the firefly algorithm, viz. absorption coefficient γ, initial value of attractiveness $\boldsymbol{\beta}_0$ and constant α. γ determines the variation of the attractiveness which corresponds to the variation of distance from the communicated firefly. $\boldsymbol{\beta}_0$ ranges between 0 and 1 manage the cooperative local search that aids the strongest firefly to determine the position of other fireflies. α controls the degree of randomization.

The parameters of each algorithm significantly contribute towards MPPT characteristics. The parameters of PSO and GA are well-articulated in the literature; however, parameters selection of remaining algorithms was done through repeated simulations for faster convergence. The parameters employed for various algorithms are listed in Table 2.

6 Conclusion

MPPT in PV systems is well-articulated research problem, and the contribution of this chapter lies in the formulation of MPPT as an optimization problem and application of few popular nature-inspired algorithms in a systematic manner. A simple experimental setup for closed-loop MPPT is implemented. Computed and measured results on a typical PV system under different shading patterns are obtained and presented. The MPPT characteristics are analysed qualitatively and quantitatively. It can be concluded that all the methods described in this paper are ideal candidates for MPPT in PV systems; however, firefly algorithm appears to be superior in performance.

References

1. Koutroulis, E., Kalaitzakis, K., Voulgaris, N.C.: Development of a microcontroller-based photovoltaic maximum power point tracking control system. IEEE Trans. Power Electron. **16**(1), 46–54 (2001)
2. Masoum, M.A., Dehbonei, H., Fuchs, E.F.: Theoretical and experimental analyses of photovoltaic systems with voltage and current-based maximum power point tracking. IEEE Power Eng. Rev. **22**(8), 62–62 (2002)
3. Noguchi, T., Togashi, S., Nakamoto, R.: Short-current pulse-based maximum-power-point tracking method for multiple photovoltaic-and converter module system. IEEE Trans. Ind. Electron. **49**(1), 217–223 (2002)
4. Femia, N., Petrone, G., Spagnuolo, G., Vitelli, M.: Optimization of perturb and observe maximum power point tracking method. IEEE Trans. Power Electron. **20**(4), 963–973 (2005)
5. Mei, Q., Shan, M., Liu, L., Guerrero, J.M.: A novel improved variable step-size incremental-resistance MPPT method for PV systems. IEEE Trans. Ind. Electron. **58**(6), 2427–2434 (2010)
6. Brunton, S.L., Rowley, C.W., Kulkarni, S.R., Clarkson, C.: Maximum power point tracking for photovoltaic optimization using ripple-based extremum seeking control. IEEE Trans. Power Electron. **25**(10), 2531–2540 (2010)
7. Safari, A., Mekhilef, S.: Simulation and hardware implementation of incremental conductance MPPT with direct control method using cuk converter. IEEE Trans. Ind. Electron. **58**(4), 1154–1161 (2011)
8. Eshram, T., Chapman, P.L.: Comparison of photovoltaic array maximum power point tracking techniques. IEEE Trans. Energy Convers. **22**(2), 439–450 (2007)
9. Subudhi, B., Pradhan, R.: A comparative study on maximum power point tracking techniques for photovoltaic power system. IEEE Trans. Sustain. Energy **4**(1), 89–98 (2013)
10. de Brito, M.A.G., Galotto, L., Sampio, L.P., de Azevedo e Melo, G., Canesin, V.A.: Evaluation of the main MPPT techniques for photovoltaic applications. IEEE Trans. Ind. Electron. **60**(3), 1156–1167 (2013)
11. Badram, A., Davoudi, A., Balog, R.S.: Control and circuit techniques to mitigate partial shading effects in photovoltaic arrays. IEEE J. Photovolt. **2**(4), 532–546 (2012)
12. Patel, H., Agarwal, V.: Maximum power point tracking scheme for PV systems operating under partially shaded conditions. IEEE Trans. Ind. Electron. **55**(4), 1689–1698 (2008)
13. Nguyen, T.L., Low, K.S.: A global maximum power point tracking scheme employing DIRECT search algorithm for photovoltaic systems. IEEE Trans. Ind. Electron. **57**(10), 3456–3467 (2010)
14. Kobayashi, K., Takano, I., Sawada, Y.: A study of a two stage maximum power point tracking control of a photovoltaic system under partially shaded insolation conditions. Sol. Energy Mater. Sol. Cells **90**(18/19), 2975–2988 (2006)
15. Alajmi, B.N., Ahmed, K.H., Finney, S.J., Williams, B.W.: Fuzzy- logic-control approach of a modified hill-climbing method for maximum power point in microgrid standalone photovoltaic system. IEEE Trans. Power Electron. **26**(4), 1022–1030 (2011)
16. Rai, A.K., Kaushika, N.D., Singh, B., Agarwal, N.: Simulation model of ANN based maximum power point tracking controller for solar PV system. Sol. Energy Mater. Sol. Cells **95**, 773–778 (2011)
17. Miyatake, M., Veerachary, M., Toriumi, F., Fujii, N., Ko, H.: Maximum power point tracking of multiple photovoltaic arrays: a PSO approach. IEEE Trans. Aerosp. Electron. Syst. **47**(1), 367–380 (2011)
18. Liu, Y.H., Huang, S.C., Huang, J.W., Liang, W.C.: A particle swarm optimization-based maximum power point tracking algorithm for PV systems operating under partially shaded conditions. IEEE Trans. Energy Conver. **27**(4), 1027–1035 (2012)
19. Ishaque, K., Salam, Z.: A deterministic particle swarm optimization maximum power point tracker for photovoltaic system under partial shading condition. IEEE Trans. Ind. Electron **60**(8), 3195–3206 (2013)

20. Daraban, S., Petreus, D., Morel, C.: A novel global MPPT based on genetic algorithms for photovoltaic systems under the influence of partial shading. In: 39th Annual IEEE IECON 2013, pp. 1490–1495. Vienna, 10–13 Nov 2013
21. Sundareswaran, K., Sankar, P., Palani, S.: MPPT of PV systems under partial shaded conditions through a colony of flashing fireflies. IEEE Trans. Energy. Convers. **29**(2), 463–472 (2014)
22. Sundareswaran, K., Sankar, P., Nayak, P.S.R., Simon, S.P., Palani, S.: Enhanced energy output from a PV system under partial shaded conditions through artificial bee colony. IEEE Trans. Sustain. Energy **6**(1), 198–209 (2015)
23. Villalva, M.G., Gazoli, J.R., Filho, E.R.: Comprehensive approach to modeling and simulation of photovoltaic arrays. IEEE Trans. Power Electron. **24**(5), 1198–1208 (2009)
24. Patel, H., Agarwal, V.: MATLAB-based modeling to study the effects of partial shading on PV array characteristics. IEEE Trans. Energy Convers. **23**(1), 302–310 (2008)
25. Alajmi, B.N., Ahmed, K.H., Finney, S.J., Williams, B.W.: A maximum power point tracking technique for partially shaded photovoltaic systems in microgrids. IEEE Trans. Ind. Electron. **60**(4), 1596–1606 (2013)
26. Eberhart, R., Kennedy, J.: A new optimizer using particle swarm theory. In: Proceedings of the 6th International Symposium on Micro Machine and Human Science, pp. 39–43 (1995)
27. Kennedy, J., Eberhart, R.: Particle swarm optimization. In: Proceedings of the IEEE International Conference on Neural Networks, pp. 1942–1948 (1995)
28. Kennedy, J.: The particle swarm: social adaptation of knowledge. In: Proceedings of the IEEE International Conference on Evolutionary Computation, pp. 303–308 (1997)
29. Karaboga, D.: An idea based on honey bee swarm for numerical optimization. Technical report TR06. Engineering Faculty, Computer Engineering Department, Erciyes University, Turkey, Oct 2005
30. Karaboga, D., Basturk, B.: Artificial bee colony (ABC) optimization for solving constrained optimization problems. In: Lecture notes in computer science, vol. 4529, pp. 789–798. Springer (2007)
31. Karaboga, D., Basturk, B.: On the performance of artificial bee colony (ABC) algorithm. Appl. Soft Comput. **8**(1), 687–697 (2008)
32. Yang, X.S., Deb, S.: Cuckoo search via levy flights. In: Proceedings of the NaBIC, pp. 210–214 (2009)
33. Yang, X.S., Deb, S.: Engineering optimization by cuckoo search. Int. J. Math. Model. Numer. Optim. **1**, 330–343 (2010)
34. Dorigo, M., Maniezzo, V., Colorni, A.: The ant system: optimization by a colony of cooperating agents. IEEE Trans. Syst. Man Cybern. **26**(2), 29–41 (1996)
35. Dorigo, M., Di Caro, G., Gambardella, L.M.: Ant algorithms for discrete optimization. Artif. Life **5**(2), 137–172 (1999)
36. Dorigo, M., Stutzle, T.: Ant Colony Optimization. Prentice-Hall of India Pvt. Limited, New Delhi, India (2005)
37. Holland, J.H.: Adaptation in natural and artificial systems. University of Michigan Press, Oxford, England (1975)
38. Goldberg, D.E.: Genetic algorithms in search, optimization, and machine learning. Addison-Wesley, Reading, MA (1989)
39. Beasley, D., Bull, D.R., Martin, R.R.: An overview of genetic algorithms: part I—fundamentals. Univ. Comput. **15**(2), 58–69 (1993)
40. Srinivas, M., Patnaik, L.M.: Genetic algorithms: a survey. Computer **27**, 17–26 (1994)
41. Man, K.F., Tang, K.S., Kwong, S.: Genetic algorithms: concepts and applications. IEEE Trans. Ind. Electron. **43**(5), 519–534 (1996)
42. Yang, X.S.: Nature-inspired metaheuristic algorithm. Luniver Press, Beckington, UK (2008)
43. Yang, X.S.: Firefly algorithms for multimodal optimization. Stoch. Alg. Found. Appl. (SAGA) **5792**, 169–178 (2009)
44. Yang, X.S.: Firefly algorithm, stochastic test functions and design optimization. Int. J. Bio-Inspired Comput. **2**, 78–84 (2010)

45. Jain, S.: Agarval V: Comparison of the performance of maximum power point tracking schemes applied to single-stage grid-connected photovoltaic systems. IET Electr. Power Appl. **1**(5), 753–762 (2007)

Soft Computing Techniques-Based Low Voltage Ride Through Control of Doubly Fed Induction Wind Generator

M. Maheswari, S. K. Indumathi, and A. K. Parvathy

1 Introduction

The generation of electrical energy from non-conventional energy sources has been rapidly increasing due to concern about ecological pollution, depletion of fossil fuels, and energy shortage. Technological advancement in industry and agriculture sector leads to increased power demand, and hence, renewable power generation plays a significant role in the financial growth of a country. Among diverse renewable sources, including solar, wind, hydro, and biomass, wind turbines generate the highest amount of electricity, considering the developments in power electronics during the last few decades. Among the various renewable sources, wind power is the most viable and also economically inexpensive. Several countries have provided the supportive regulations referred to as grid-codes defining the operation and grid integration of distributed generators, specifically the wind turbine generators (WTGs). Modern grid codes necessitate that a wind turbine should remain linked with the grid for a certain duration and should stream the reactive power to convey voltage support during fault conditions. The mentioned function of a wind turbine is named as 'low-voltage ride-through' (LVRT) capability.

M. Maheswari (✉) · S. K. Indumathi · A. K. Parvathy
Department of Electrical and Electronics Engineering, Hindustan Institute of Technology and Science, Chennai, Tamilnadu, India
e-mail: mmaheswari@hindustanuniv.ac.in

S. K. Indumathi
e-mail: skindumathi@hindustanuniv.ac.in

A. K. Parvathy
e-mail: akparvathy@hindustanuniv.ac.in

B. Vinoth Kumar et al. (eds.), *Intelligent Paradigms for Smart Grid and Renewable Energy Systems*, Algorithms for Intelligent Systems,
https://doi.org/10.1007/978-981-15-9968-2_10

1.1 Wind Power System Requirements and Limitations

In this section, the major grid code requirements will be discussed, and LVRT, which is the focus of this chapter, will be explained in detail.

1.1.1 International Grid Code Requirements

Grid code requirements by different countries concerning wind farm performance are specified in Fig. 1.

Due to increased penetration of wind power, wind farms must follow different grid codes to maintain grid connection and stability during various grid disturbances. If any wind farm is disconnected, then further weakening of grid stability will happen. Many countries have their own grid codes for wind farm performance regarding voltage dips and surges at the PCC. LVRT and High Voltage Ride Through (HVRT) is the capability of wind generators to bear low and high voltages at the PCC. Focus

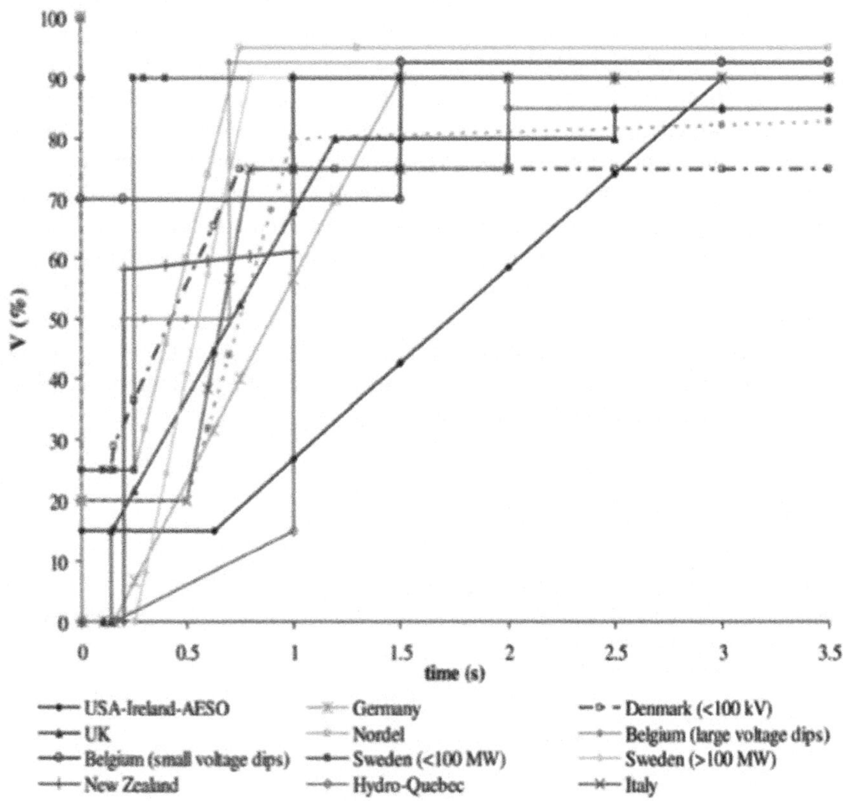

Fig. 1 LVRT requirements of different grid code (Tsili et al. 2009)

of this thesis is mainly on the analysis of symmetrical voltage sag impact on wind generator systems and implements advance strategies to alleviate their adverse effects [1].

The German grid code from E. ON Netz is used as a reference by other countries to improve their own grid codes. It deals with the voltage levels of 110, 220, and 380 kV. For an offshore wind farm nominal voltage of 155 kV is required as per German grid code. British grid code is pertinent for voltage levels of 132, 275, and 400 kV, while the Irish grid code deals with voltage levels of 110, 220, and 400 kV. Denmark, Sweden, Norway, and Finland have an interconnected power system and follow the Nordic grid code issued by Nordel. Nevertheless, Denmark is the only country that has distinct technical requisites for grid-connected wind farms with voltages below and above 100 kV. It is important to note that the lists of countries are in the process of developing grid codes.

Voltage range and control A wind farm power station should operate at a rated voltage as well as a specified operating range according to different power systems. Here, a voltage range of $\pm5\%$ is considered for many countries such as India.

Power factor requirement It is required to provide reactive power in order to maintain power factor of the wind farm. The power factor is desired to remain close to unity and wind farm to have neutral reactive power.

Real power and frequency control Wind farms should regulate their real power to achieve constant frequency in the system and to avoid transmission line overloading. In addition, frequency control must be applied by wind farms through controlling the active power level with frequency deviations.

Low-voltage ride-through In the event of grid disturbances, it is required that the wind farm should maintain its connection with the power system for a specified amount of time. This specific amount of time can vary among grid codes.

High-voltage ride through If the system voltage increases above the upper threshold limit, wind farms should maintain its connection for a certain duration.

External control of the wind farm Transmission System Operator must be capable of controlling the inclusion of wind farm to the grid or disconnection of it from the system remotely. In addition, signals corresponding to the various parameters of the wind farm such as voltage regulation must be provided by the operator to control the wind power station externally.

1.1.2 Indian Grid Code Requirements

- Power factor should be maintained in the range of 0.95 lag to 0.95 leading
- Generating units should operate at the frequency limit of 47.5–52 Hz and should deliver rated power with the frequency range of 49.5–50.5 Hz
- The abovementioned performance can be achieved with a voltage variation of $\pm5\%$

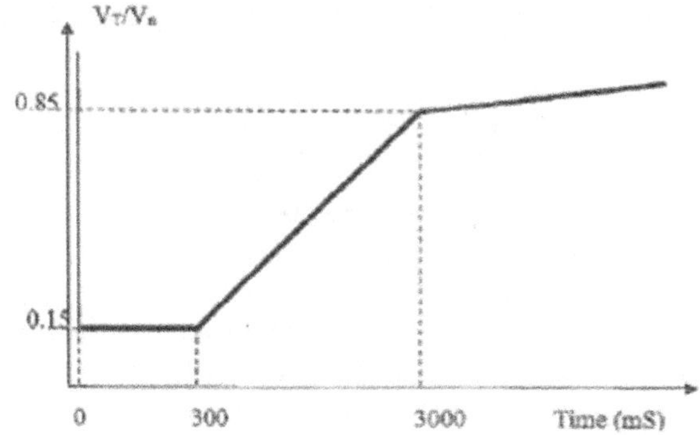

Fig. 2 FRT curve (*Source* Central Electricity Authority Regulations 2012)

- Wind-generating stations that are coupled at a voltage level of ≥66 kV should maintain its connection with the grid during voltage dip at the interconnection point on any or all the phases reaching the levels illustrated by thick lines in the curve shown in Fig. 2.
- During FRT, the WTGs should satisfy the mentioned requirements:
- Reduce the reactive power absorption.
- Provide active power in Proportion to the grid voltage dip active power need to be supplied immediately after the fault clearance.

1.2 Low-Voltage Ride Through

LVRT is a part of grid code that states that during voltage sag at the grid, wind turbines should maintain its connection with the grid for a certain duration; on the other hand, the turbines can be disconnected. This specific amount of time varies among grid codes; the severity of the fault might also vary. Different systems with various operating voltages have different fault clearing time as given in Table 1

DFIGs are the widely used type of variable-speed WECSs, and hence, this section details the LVRT of DFIGs. DFIG stator is highly influenced by the instabilities of the grid such as voltage dips. During the transient condition, a sudden decrease of

Table 1 Fault clearing time

Nominal system voltage (kV)	Fault clearing time (ms)	V_{pf}(kV)	V_f(kV)
400	100	360	60
220	160	200	33
132	160	120	19.8
110	160	96.25	16.5
66	300	60	9.9

grid voltage leads to dip in stator voltage, oscillations in stator current, and pulsations in electromagnetic torque. Since the stator and rotor are mutually coupled, any increment in stator current will cause increase in rotor current as well as a rise in DC-link voltage. Therefore, during fault conditions, DFIG performance is reduced. The transient response of the DFIG is shown in Fig. 3.

The transient state of a DFIG can be divided into three periods, such as (i) 'voltage falling', i.e. the initial period (Period 1) immediately following grid fault; (ii) 'voltage sustaining', i.e. the period during which low voltage is sustained (Period 2); and (iii) 'voltage recovery period' (Period 3) [2]. In the former days, when the penetration of wind power was low, there was much attention on protecting wind turbines, and hence, wind generators were disconnected immediately from the grid during fault condition. At present, increased wind power penetration results in huge power loss

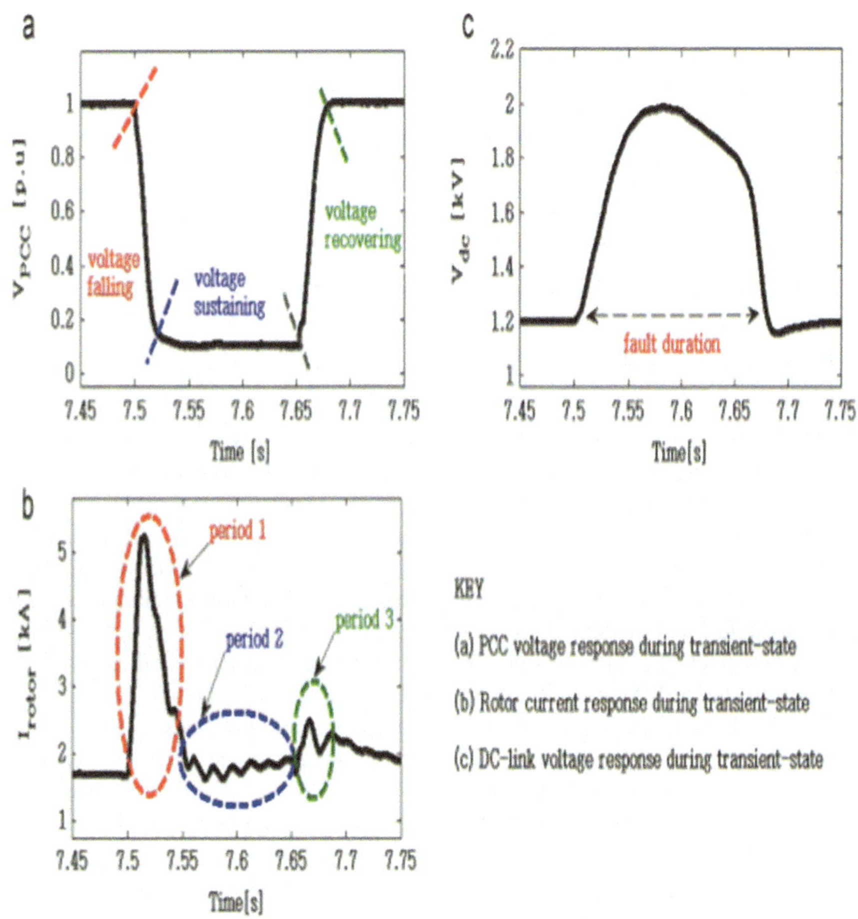

Fig. 3 Transient-state response of a DFIG wind turbine: **a** voltage at PCC; **b** transient-state rotor current; and **c** Transient-state DC-link voltage (Justo et al. 2015)

when wind turbines are disconnected. Therefore, wind turbines should maintain its connection with the grid for stable operation. Proper measures need to be followed to protect: (i) RSC from rotor inrush current and (ii) the DC-link capacitor from overvoltage. LVRT can be considered either at the system level, which consists of more wind turbines (wind farm) or at the individual turbine level. This thesis mainly focuses on the LVRT enhancement of individual wind turbines for the proper analysis of the performance parameters of DFIG.

1.2.1 LVRT Techniques for DFIG Wind Turbine

LVRT solutions can be categorized as follows based on the transient stability enhancement approach.

- Protective devices/circuits-based scheme [3–8]
- Reactive power support scheme [9]
- Modified control approach-based scheme [10–14].

This chapter is mainly focusing on Soft computing techniques based on modified control approach based scheme.

2 Modeling of DFIG and Conventional Vector Control

2.1 Structure and Basic Principle of DFIG

DFIG also termed as Wound rotor induction generators (WRIG) used invariable speed applications with few kilowatts range to several hundred Megawatts. Schematic configuration of DFIG is given in Fig. 4.

2 MW DFIG is considered for LVRT enhancement study. The parameters used in current study were obtained from the data sheet of Mitsubishi wind turbine (MWT-92). Main characteristics and equivalent model of DFIG and turbine parameters are given in Table 2 and in Table 3.

Stator magnetic field is generated by supplying three-phase power directly to the stator from the grid. Stator will be operated at a constant frequency and constant amplitude. Rotor is energized with three-phase power through back to back power electronic converter. With the help of RSC rotor circuit can be made to operate at variable voltage and variable frequency in order to meet the different speed and torque variations.VSC is used to control the power flow between rotor and grid.

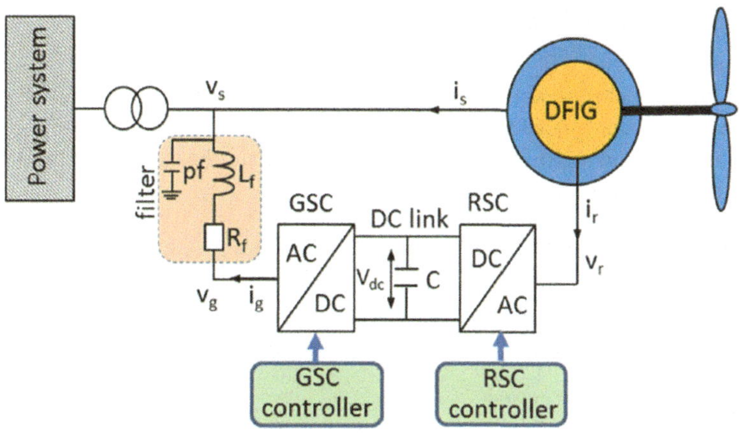

Fig. 4 Schematic configuration of DFIG

Table 2 Main characteristics and equivalent model of DFIG

Parameter	Value	Unit
Stator active power	2	MW
Rated torque	12,732	Nm
Stator voltage	690	V
Rated speed	1500	rpm
Speed range	900–2000	rpm
Pole pairs	2	–
Magnetizing inductance (L_m)	2.5e−3	H
Leakage inductance (rotor, L_{or})	87e−6	H
Leakage inductance (stator, L_{os})	87e−6	H
Resistance of rotor and stator	0.026 and 0.029	Ω

Table 3 Turbine parameters

Parameter	Value	Unit
Radius	42	m
Nominal wind speed	12.5	m/s
Variable speed ratio (minimum–maximum turbine speed)	9–18	rpm
Optimum tip speed ratio	7.2	–
Maximum power coefficient Cp_max	0.44	–
Air density *r*	1.1225	Kg/m^3

2.2 Modeling of Wind Turbine

A wind turbine converts kinetic energy of the wind into electrical energy. It can be classified as constant speed wind turbines and variable speed wind turbines. Power extraction from the wind is obtained by Aerodynamic theory. Power available in wind is obtained as,

$$\begin{aligned} P_{\text{wind}} &= \frac{1}{2}\text{(air mass per unit time)(wind velocity)2} \\ &= \frac{1}{2}(\rho A V_{\infty})(V_{\infty})^2 \\ &= \frac{1}{2}\rho A V^3 \end{aligned} \tag{1}$$

where ρ is the air density (1.225 kg/m^3), A is the rotor area in square metre, V_{∞} is the wind velocity at infinite distance from rotor. It is not possible to extract 100% efficiency from the wind. Theoritical maximum efficieny of a wind turbine is specified by the betz limit.ie.only 59.3% of the power can be extracted from wind [15].

Hence maximum extractable power from wind is given by,

$$P_{\text{max}} = \frac{8}{27}\rho A V^3 = \frac{16}{27}P_{\text{wind}} \tag{2}$$

where P_{wind} is the power contained in wind.

From Eq. (1), it is found that the power transferred to a wind turbine is directly proportional to the air density, cubic power of wind speed, and swept area. The maximum possible torque (T_{max}) developed on a turbine rotor is expressed in Equation,

$T_{\text{max}} = F_{\text{max}}.R$. Where, R is the radius of swept area. Maximum force is given by,

$$F_{\text{max}} = \frac{1}{2}\rho A V^2$$

Therefore

$$T_{\text{max}} = \frac{1}{2}\rho A V^2 R \tag{3}$$

Computing the mechanical torque in terms of wind velocity, the torque produced by the rotor can be denoted as,

$$T_t = C_t T_{\text{max}} = \frac{1}{2}\rho A V_{\infty}^2 R C_t \tag{4}$$

$$T_t = \frac{1}{2}\rho\pi R^3 V_\infty^2 C_t \tag{5}$$

tip speed ratio can be obtained as,

$$\lambda = \frac{R\Omega_t}{V_\infty} \tag{6}$$

where ρ is air density, R is the radius of the wind turbine rotor and Ω_t is the angular speed of the rotor. The tip speed ratio is the ratio between the tangential speed of the tip of a blade and the actual speed of the wind.

From Eq. (6), R is given by,

$$R = \frac{V_\infty \lambda}{\Omega_t} \tag{7}$$

Substituting R in Eq. (3)

$$T_{max} = \frac{\frac{1}{2}\rho A V_\infty^3 \lambda}{\Omega_t} \tag{8}$$

By substituting the P_{wind} in Eq. (5) and gives the Equation

$$T_{max} = \frac{\lambda P_{wind}}{\Omega_t} \tag{9}$$

Hence power available at the wind turbine shaft (P_t) is given in Eq. (10)

$$P_t = C_T T_{max} \Omega_t = C_p P_{wind} \tag{10}$$

The reduction of Eq. (10) gives the Equation

$$C_p = C_T \lambda \tag{11}$$

The torque and mechanical power developed by the turbine are expressed as in Eqs. (12) and (13) respectively.

$$T_m = \frac{1}{2}\rho A V_\infty^2 R C_t \tag{12}$$

$$P_m = \frac{1}{2} C_p \rho A V_\infty^3 \tag{13}$$

where T_m is the torque developed by wind turbine, P_m is the mechanical power developed by wind turbine, C_T is the torque coefficient, C_p is the power coefficient, F_{max} is the maximum force, T_{sh} is the shaft torque, Ω_t is the rotational frequency,

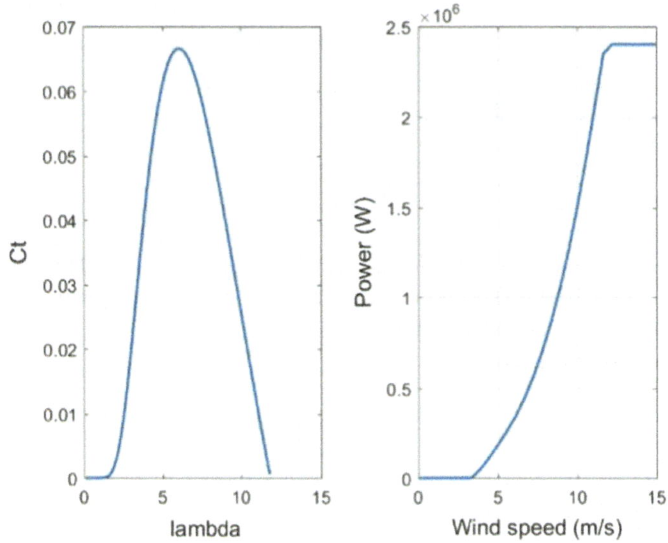

Fig. 5 Cp versus Lambda curve and power characteristics of 2 MW wind turbine

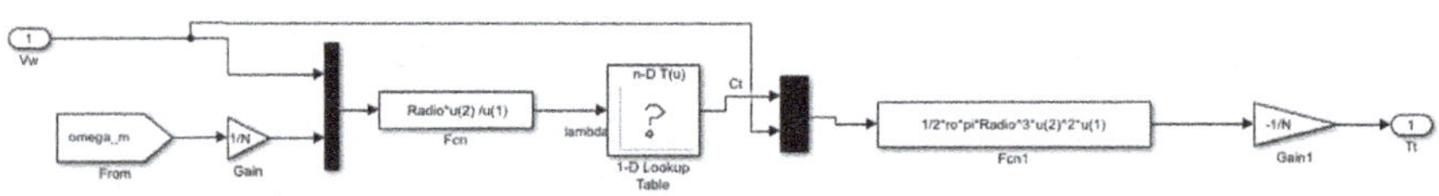

Fig. 6 Wind turbine model

Tmax is the torque at maximum efficiency, $V\infty$ is the unperturbed wind speed and Vt is the outer blade tip speed.

To model the wind turbine, Cp-λ characteristic shown in Fig. 5 is needed which can be obtained from the power curve of WEG.

A specification of the WEG chosen for the study is given in Appendix 1. Based on the above equations wind turbine is modeled in MATLAB/SIMULINK as shown in Fig. 6.

2.3 *Implementation of MPPT Control*

Indirect speed controller is considered for Maximum Power Point Tracking (MPPT). Controller circuit is given in Fig. 7. Maximum power is extracted by indirect speed control. For any speed deviation around a point in the maximum power curve, Variable speed wind turbine reaches back to its operating point.

At maximum power point opertaing condition of the turbine,

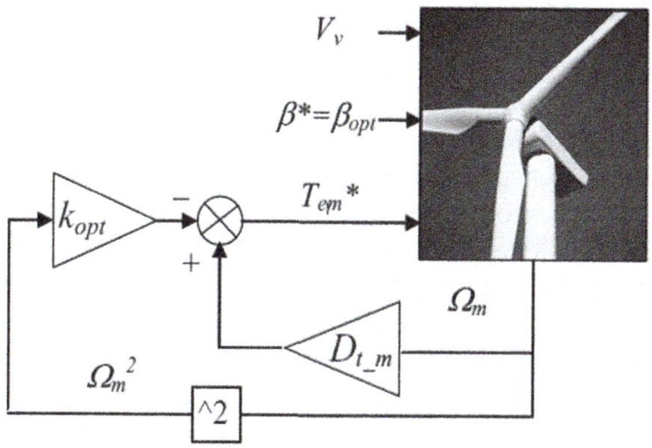

Fig. 7 Indirect speed controller for MPPT

$$\lambda_{\text{opt}} = \frac{R\Omega_{\text{t}}}{V_\infty}, C_{\text{p}} = C_{p\max} \quad \text{and} \quad C_{\text{t}} = C_{\text{topt}} \tag{14}$$

Aerodynamic torque obtained from wind turbine is given by,

$$T_{\text{t}} = \frac{1}{2}\rho\pi R^3 \frac{R^2\Omega_{\text{t}}^2 C_{\text{p}\max}}{\lambda_{\text{opt}}^2 \lambda_{\text{opt}}} \tag{15}$$

That is

$$T_{\text{t}} = \frac{1}{2}\rho\pi R^5 \frac{\Omega_{\text{t}}^2 C_{\text{p}\max}}{\lambda_{\text{opt}}^3} = K_{\text{opt}}\Omega_{\text{t}}^2 \tag{16}$$

where

$$k_{\text{opt}} = \frac{1}{2}\rho\pi \frac{R^5}{\lambda_{\text{opt}}^3 N^3} C_{\text{p\_max}} \tag{17}$$

Simulink model of indirect speed control method with MPPT control is given in Fig. 8.

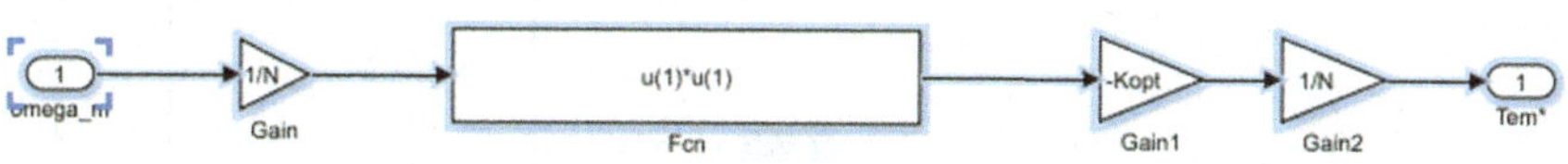

Fig. 8 Indirect speed controller simulink model

2.4 Modeling of RSC and GSC Control

DFIG comprises back-to-back converter, also known as a reversible or bidirectional converter. Voltage source converter (VSC) with two-level output is used in the present study. Hence this section mainly focuses on the model of VSC, pulse generation techniques for converter switches, and filters to remove current ripples. Finally, vector control strategy for the rotor and GSC is presented. The stator is directly linked to the grid and the supply voltage to stator will be having the constant amplitude and frequency of grid. The rotor is connected to the grid through back to back power electronic converters. Hence supply voltage of the rotor will be at different magnitude and frequency so as to obtain various speed and torque operational circumstances of the machine. This PWM converter with the suitable controller gives the mandatory rotor AC voltages to regulate the operating point of DFIG and to achieve the real and reactive power exchange to the grid via rotor.

2.4.1 Rotor Side Converter (RSC)

Rotor side converter supplies the rotor of DFIG.RSC and dv/dt filter arrangement in DFIG is shown in Fig. 9. Two-level VSC is used to feed the rotor. dv/dt filter is placed between the rotor and converter in order to protect the machine from capacitive leakage currents. Stress on the motor insulation can also be reduced.

The RSC is connected to the GSC by the DC link. It is the linkage between the GSC and RSC. Simplified model of a DC link is shown in Fig. 10. It consists of capacitor with parallel resistance.

DC link model can be derived by calculating the dc bus voltage which in turn depends on current flow through the capacitor.

$$V_{\text{bus}} = \frac{1}{C_{\text{bus}}} \int i_{\text{c}} \text{d}t \tag{18}$$

The capacitor current can be found as

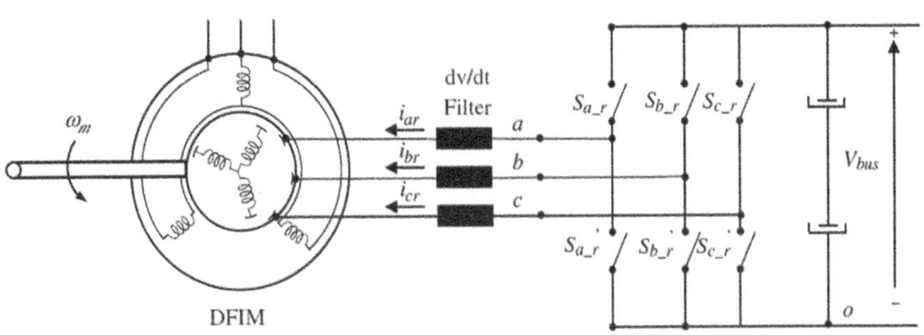

Fig. 9 Rotor side converter and dv/dt filter

Fig. 10 DC link model

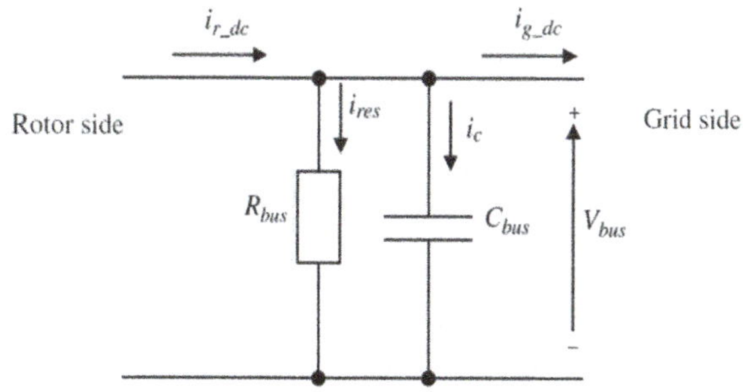

$$i_c = i_{r\_dc} - i_{g\_dc} - i_{res} \tag{19}$$

where
i_{res} = current flowing through resistance
i_{g_dc} = current from DC link to grid
i_{r_dc} = current from rotor to DC link.

2.4.2 Vector Control of RSC

The technique of vector control is used in this work for the implementation of GSC and RSC controller.DFIG vector control is achieved in a dq frame that is rotating synchronously, in which stator flux vector and *d* axis are aligned together as shown in Fig. 11. Hence, direct current of rotor is comparative to the reactive power of stator and quadrature current of rotor corresponds to active power or torque. Therefore, active as well as reactive power of stator is controlled via RSC through vector control of rotor current [16].

Rotor voltage is represented as a function of rotor current and stator flux as follows.

$$V_{dr} = R_r i_{dr} + \sigma L_r \frac{d}{dt} i_{dr} - \omega_r \sigma L_r i_{qr} + \frac{L_m}{L_s}\frac{d}{dt}\lceil \psi_s \rceil \tag{20}$$

Fig. 11. dq reference frame line up with stator flux

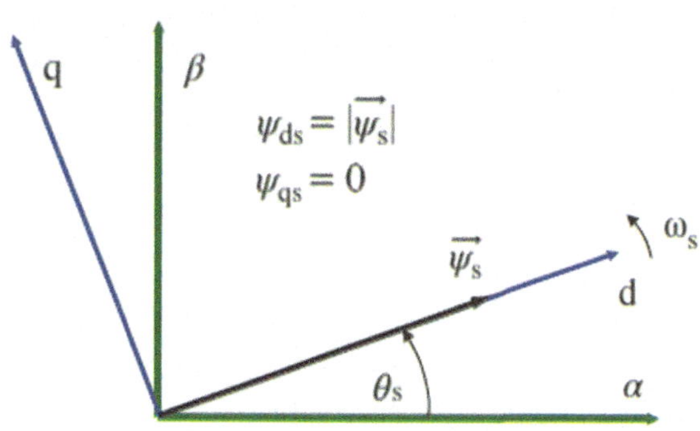

$$V_{qr} = R_r i_{qr} + \sigma L_r \frac{d}{dt} i_{qr} - \omega_r \sigma L_r i_{dr} + \omega_r \frac{L_m}{L_s} \lceil \psi_s \rceil \quad (21)$$

PI regulator is used to control the d and q axis rotor current. Cross-coupling parameters are added at the output of PI regulator. Three-phase rotor current is converted into direct and quadrature axis rotor current in dq frame. PLLis used for performing angle calculation. d-axis current is line up with the stator flux. i_{dr} is proportional to the stator reactive power. Stator flux and ω_r must be estimated. Θ_s can be calculated by obtaining the angle of space vector of stator voltage and then subtracting 90° from the estimated angle. Angle Θ_r for the transformation can be obtained as $\Theta_r = \Theta_m - \Theta_s$. In the control block diagram, all the current loops function with stator-referred rotor currents.

Immediately after the control loop for inner current has been completed, outer power and speed control loop can be included. Torque in dq reference frame is expressed as follows. Complete vector control diagram is given in Fig. 12.

$$T_{em} = \frac{3}{2} \rho \frac{L_m}{L_s} (\Psi_{qs} i_{dr} - \psi_{ds} i_{qr}) \Rightarrow T_{em} = -\frac{3}{2} \rho \frac{L_m}{L_s} |\Psi_s| i_{qr} \Rightarrow T_{em} = K_T i_{qr} \quad (22)$$

From the above equation, it is clearly specified that by controlling rotor current in quadrature axis (i_{qr}), it is possible to control torque which in turn controls the speed. Similarly, stator reactive power in terms of rotor current in d axis (i_{dr}) is derived as follows.

$$\begin{aligned} Q_s &= \frac{3}{2} (v_{qs} i_{ds} - v_{ds} i_{qs}) \Rightarrow Q_s \\ &= -\frac{3}{2} \omega_s \frac{L_m}{L_s} |\Psi_s| \left[i_{dr} - \frac{|\psi_s|}{L_m} \right] \Rightarrow Q_s = K_Q \left[i_{dr} - \frac{|\psi_s|}{L_m} \right] \end{aligned} \quad (23)$$

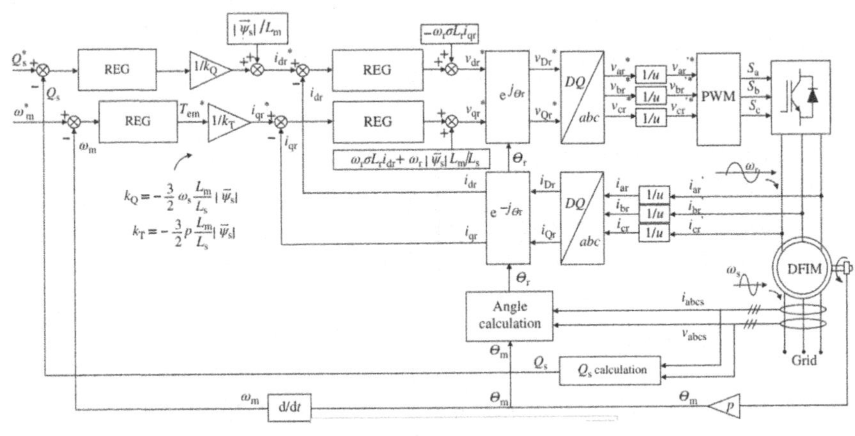

Fig. 12 Complete vector control of DFIG

Therefore, active power of stator and torque may be individually regulated through controlling the rotor current in d axis and q axis.

With inner current control, outer power and speed control loop has been added. Magnetization of DFIG machine can be controlled with reactive power control loop. Stator flux is maintained constant due to the direct connection with the grid. Stator flux is given by

$$|\Psi_{\rm s}| \cong \frac{v_{\rm s}}{\omega_{\rm s}}$$

$$|\psi_{\rm s}| = \psi_{\rm ds} = L_{\rm s} i_{\rm ds} + L_{\rm m} i_{\rm dr} \quad (24)$$

Hence stator flux can be controlled with direct axis stator and rotor current. q axis stator and rotor current cannot be used since the torque is set by quadrature axis current. The reactive power reference can be set by the wind turbine operator depending on the requirements.

2.4.3 Grid Side Converter Control

Simplified model of the GSC is given in Fig. 13. It consists of grid side converter, filter, and grid voltage. GSC is modeled with IGBT switches for the bidirectional power flow from AC to DC during rectifier mode and from DC to AC during inverter mode. Grid side filter normally consists of three inductances connected between the converter phases and grid voltage.LC filter or LCL filter can also be used during high filter requirements. Balanced and sinusoidal AC voltage is used as a grid voltage [17].

Vector control of GSC is given in Fig. 13. GSC regulates the DC link voltage and independently controls the reactive power injected in to Grid. From the Vbus and Qg references, PWM pulses are created to control the switches Sa_g, Sb_g,

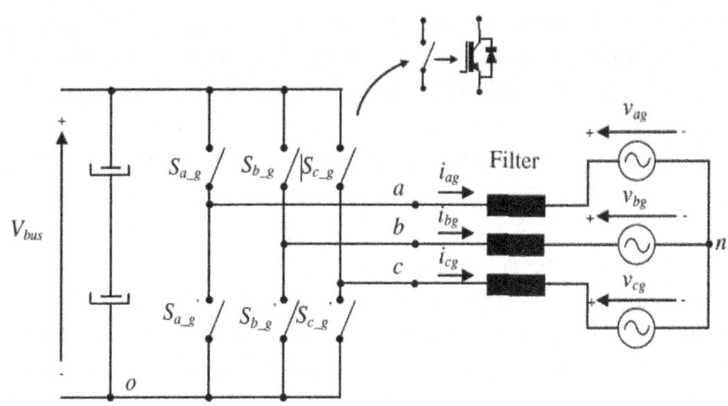

Fig. 13 Grid side converter model

and Sc_g. Thus, the modulator generates the grid side converter pulses Sa_g, Sb_g, Sc_g from the three-phase voltage references V^*_{af}, V^*_{bf} and V^*_{cf}. First, the three-phase grid current is converted to stator reference frame which is again converted into dq reference frame which gives direct and quadrature axis grid current. It must be noted that the current references (I^*_{dg}, I^*_{qg}) are decoupled from the reactive and active powers. Thus, I^*_{dg} control implies Pg control, while I^*_{qg} control implies Qg control. The dq voltage references (V^*_{df}, V^*_{qf}) are autonomously produced by the dq current (I^*_{dg}, I^*_{qg}) controllers. Hence reference three-phase voltages are created in synchronously rotating coordinates (V^*_{df}, V^*_{qf}), then transformed to stationary reference ($\alpha\beta$) coordinates ($V^*_{\alpha f}$, $V^*_{\beta f}$), and finally the abc votage references are generated.

GSC controls the DC link bus voltage and also the real and reactive power between the rotor and the grid.

3 Soft Computing Techniques Based LVRT Enhancement of DFIG

Computational intelligence based controller is employed to improve the LVRT requirement of DFIG. Conventional Vector control of RSC is modified with a fault correction parameter derived from FUZZY/BPN/ANFIS.

3.1 Description of Proposed Controller

Schematic diagram of proposed Fault confrontation controller is given in Fig. 14.

Proposed controller attenuates the system disturbance caused by the fault with optimal synchronization among two converters. Computational Intelligent controller (FUZZY/BPN/ANFIS) is used due to the requirements of efficient control in a short time duration. System should be unresponsive to measured noise quantity and to the nonexistence of information. Conventional RSC controller is modified by adding Fault confrontation controller block as shown in Fig. 14. The control of GSC remains unchanged.

Steps involved in activation of Fault detection and confrontation controller block as well as predicting the fault correction parameter are described with flowchart given in Fig. 15.

Successful LVRT can be achieved by properly controlling rotor current and DC link voltage during fault and reestablishing period. Extra energy induced must be supplied to the grid through the converter, inorder to bring back the rotor current and dc-link voltage. If the rotor current is damped quickly then there will be sudden rise in dc-link voltage. Instead, if it is reduced slowly then the rotor current reaches undesirable values. Hence rotor current correction signal must account for the respective

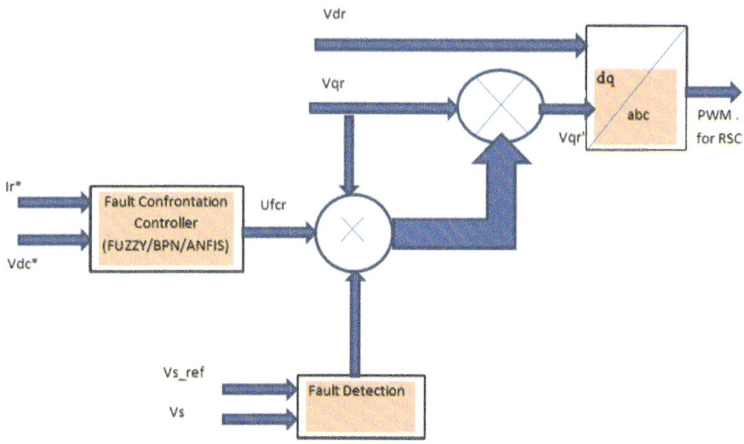

Fig. 14 Schematic diagram of fault confrontation controller

dc voltage values. V_{rq} output of the conventional RSC controller is modified by a quantity Ufcr obtained by intelligent controller (Fuzzy/BPN/ANFIS). The inputs of the controller V_{dc}* and i_r* are given by,

$$V_{dc}^{*} = \frac{V_{dc} - V_{dcss}}{V_{dc\_mav} - V_{dcss}} \tag{25}$$

$$i_r^{*} = \frac{i_r - i_{rss}}{i_{r\_mav} - i_{rss}} \tag{26}$$

where ss indicates steady-state value and mav isr maximum acceptable value, stated by the manufacturer. In proposed system, Mitsubhisi 2 MW wind turbine ratings are used. i_r is the rotor rms current. In order to ensure equal contribution to the modulation of Fault Confrontation controller output, the deviations of the dc-link voltage and rotor current are divided by their maximum tolerable deviations. Only positive deviations are considered and negative deviations are taken as zero. The concept of proposed controller can be applied in diverse ratings of machines. Controller rules remain same.

3.2 *Design of Fault Confrontation Controller with Fuzzy Logic*

Fuzzy logic is an addition to Boolean logic. Fuzzy logic provides a very appreciable flexibility for reasoning, which makes it possible to take into account inaccuracies and uncertainties. It can handle nonlinearity without accurate mathematical model [18]. Fuzzy inference system for the proposed controller is shown in Fig. 16. Mamdani

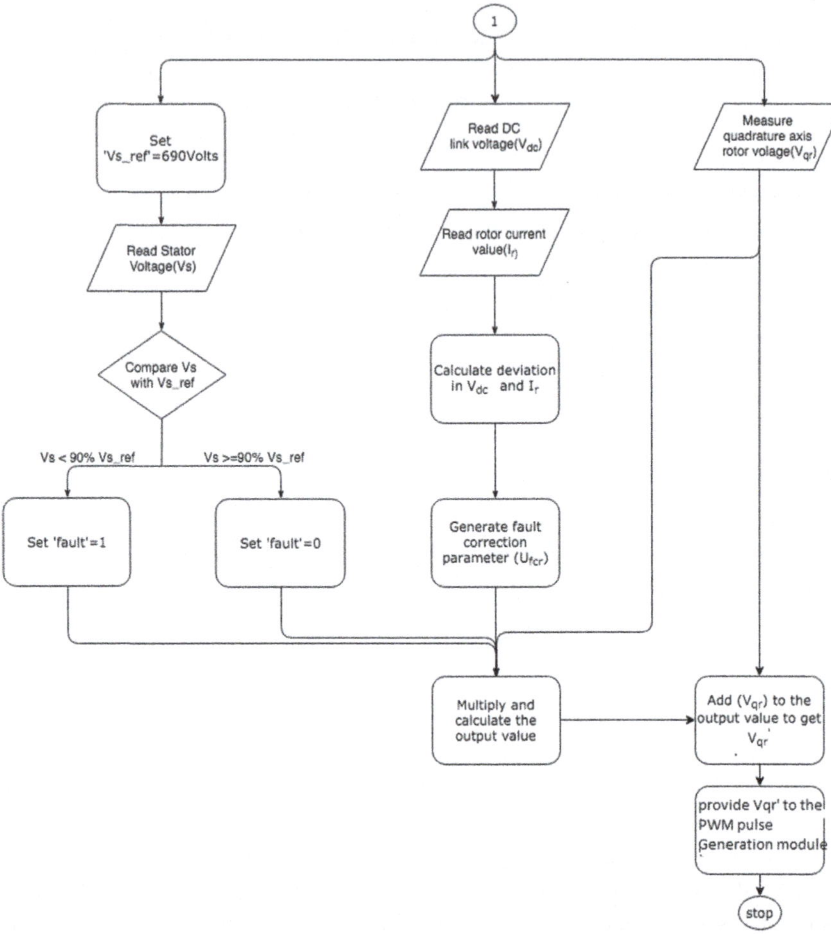

Fig. 15 Flowchart for activation of fault confrontation controller block for LVRT enhancement

Fig. 16 LVRT fuzzy inference system

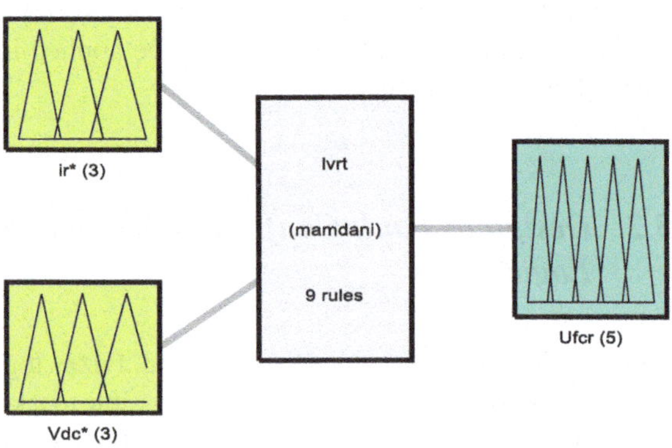

type Fuzzy inference system is used because of its widespread acceptance and also well-suited to human input.

Three subsets namely Small (S), Medium (M), and BIG (B) are used for input. In order to increase the accuracy of mapping, five levels are chosen for the output variable Ufcr. (OK, Positive Small (PS), Positive Big (PB), Negative Small (NS), and Negative Big (NB). Performance of Fuzzy Inference System is strongly dependent on the choice of membership functions. Hence in this work heuristic approach is used to select the best among the available membership functions.

Fuzzy sets are simulated with Different membership functions such as Triangular, Gbell, Gaussian, Trapezoidal, sigmoid, and Pi. Input and Output Membership Functions are given in Figs. 17, 18 and 19.

A surface graph that shows the impact of I_r* and V_{dc}* on Fault Confrontation Controller output Ufcr is shown in Fig. 20.

Performance of the FLC system is measured with the parameters root mean square error, standard deviation, minimum and maximum error and is listed in Table 4.

Performance parameters of Triangular, Gbell, Trapezoidal, Gaussian, Sigmoid, and Pi membership functions are shown in Fig. 21. From performance measures, it is clear that Triangular Membership function gives the best-optimized result for the proposed system with the root mean square error value of 0.1542. Since minimum error value is very less, it is not included in bar chart. In order to further reduce the root mean square error, standard deviation, maximum and minimum error, artificial neural network is used for predicting fault correction parameter Ufcr.

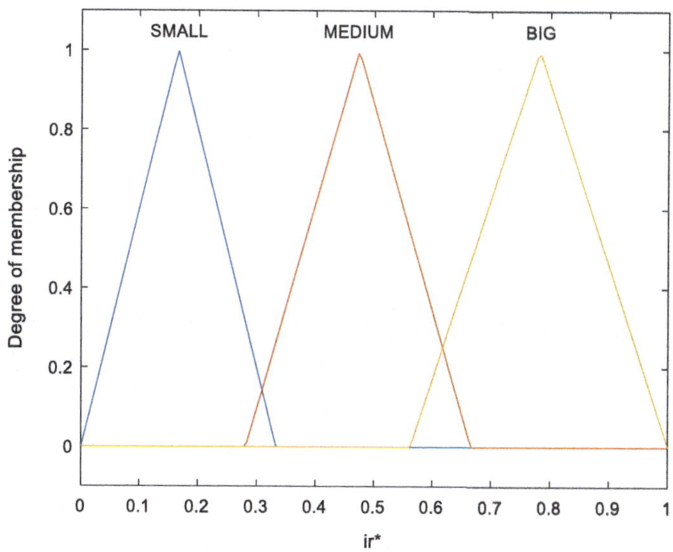

Fig. 17 Input i_r* membership functions

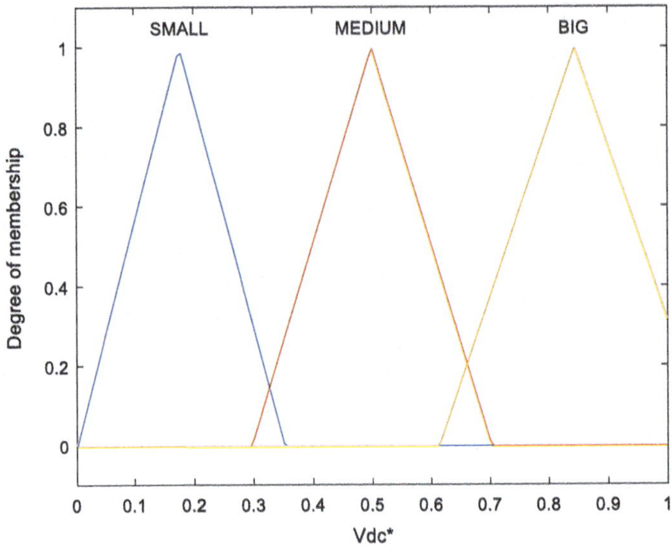

Fig. 18 Input V_{dc}* membership functions

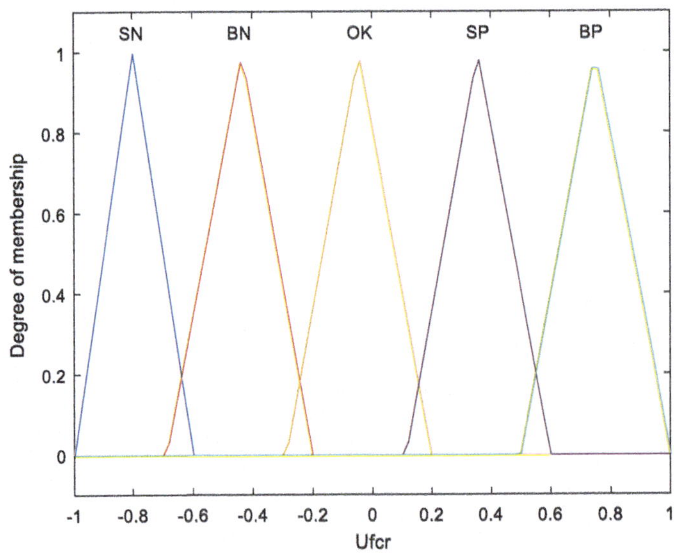

Fig. 19 Output Ufcr membership functions

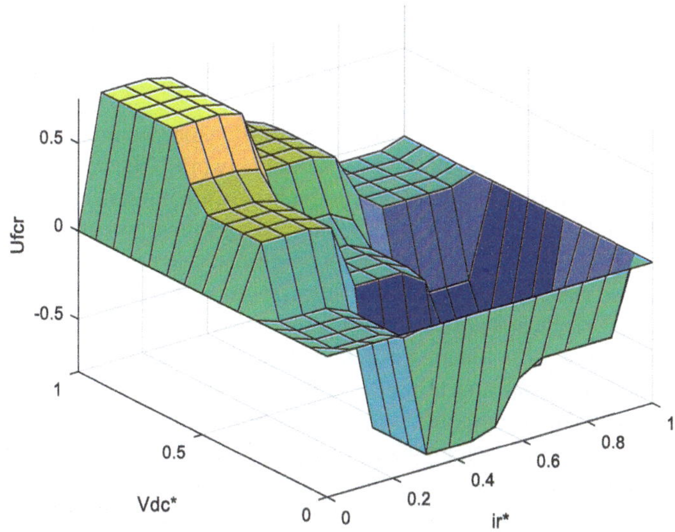

Fig. 20 Surface view of LVRT Fuzzy inference system

Table 4 Performance measure comparison with different membership function in fuzzy logic

Performance parameters	Triangular	Gbell	Trapezoidal	Gaussian	sigmoid	Pi
Root mean square error	0.1542	0.1656	0.1887	0.1641	0.4954	0.2165
Standard deviation	0.6657	0.6291	0.6447	0.6375	0.1572	0.5332
Minimum error	4.42e−05	6.76e−06	1.38e−06	2.88e−07	7.59e−05	6.86e−06
Maximum error	1	0.6514	1	0.6272	1.0548	1

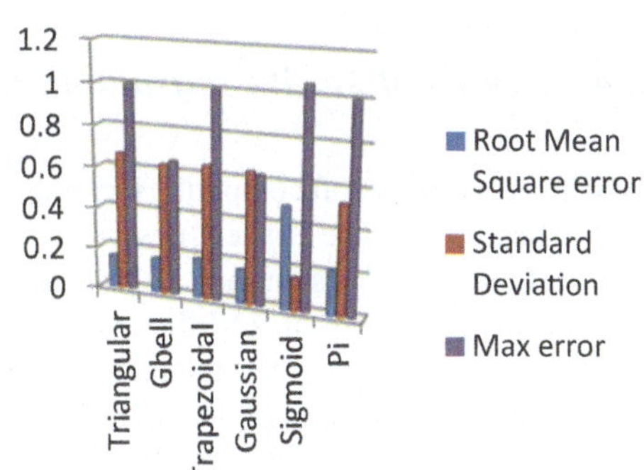

Fig. 21 Performance measures of Fuzzy logic control with different membership functions

3.3 Fault Confrontation Controller Design with Back Propagation Network

Backpropagation trains artificial neural networks. Backpropagation is a supervised learning algorithm to train a multi-layered neural network [19].

Architecture of proposed system network consists of 2 input neurons, one output neuron, and 3 hidden layers with 10:5:2 neurons. Learning process is started with the random initialization of the model with two inputs V_{dc}* and I_r*. Equations for calculating these values are expressed in Eqs. (25) and (26). After initialization, the inputs are passed via the network layer, while the actual model output is calculated by forward propagation. Loss function is defined to compare the actual output with the desired output and to determine how efficiently the neural network generates outputs that are nearly equal to the required values. In the proposed system, within the neural network, three layers between inputs and output are used to achieve more possible variations in the network's functionality. From the derivative of loss function, the error is propagated back from the end to the start. Delta rule is used for the weight updation. Learning rate 0.2 and the momentum parameter 0.3 are used to force a smooth and slow update of the weight. Among 9594 samples, 4797 rows have been taken for training the neural network and remaining for testing. The relationship between actual and desired value for trained data set and for the test data set is given in Fig. 22 and in Fig. 23.

Learning process has taken 65 iterations. Once iteration is completed, the weights are updated by a changing descent force towards a very less loss function globally. The system's performance is measured with root-mean-square error, maximum error, and minimum error and the values have been tabulated in Table 5. Root-mean-square error for the trained and test data set remains at the same value of 0.0315. Maximum and minimum error is minimum for the trained data set and for the test data set the error is maximum. But when compared to fuzzy logic system, the back-propagation network gives minimum error.

3.4 Design of Fault Confrontation Controller with ANFIS

ANFIS is a popular soft computing technique in control area which works on the basis of the fuzzy inference concept described by Takagi and Sugeno. ANFIS combines both a fuzzy logic system and neural network concepts. Fuzzy Inference System (FIS) is the main core of ANFIS, and thus it is considered as a model mapping the characteristics of an input to the membership functions of the input. Following this, the membership function of the input is mapped to rules, and consequently, the rules are mapped to a set of characteristics of the output. Finally, the characteristics of the output are mapped to the membership functions of the output, and these are in turn mapped to output with single value or an output-associated decision. ANFIS is used in modeling or control of nonlinear systems [20].

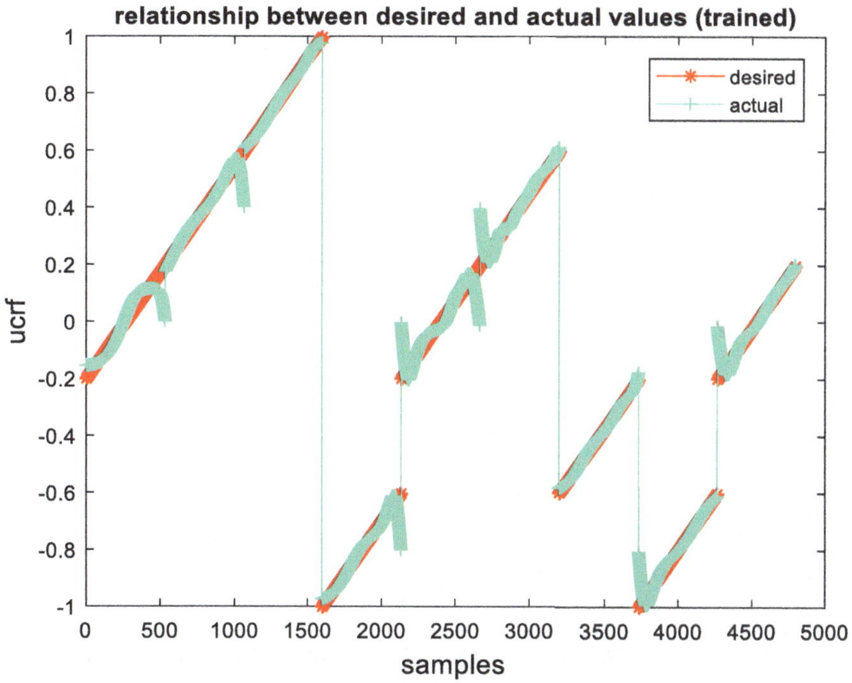

Fig. 22 Relationship between desired and actual values (Trained data set)

The proposed ANFIS architecture has two inputs consisting of five layers as shown in Fig. 24. The inputs of ANFIS, i.e., i_r* and V_{dc}* are expressed in Eqs. (25) and (26). The output of ANFIS is a fault correction signal Ufcr given for fault correction to enhance the capability of LVRT. Three cases with membership functions of triangular type, Gbell type, and PI type are used, and the corresponding performance is measured with root-mean-square error, standard deviation and minimum and maximum error. The initial and modified membership functions for triangular-type functions are shown in Fig. 25.

3.5 *ANFIS with Triangular Membership Function*

Fuzzy inference system is created with two inputs and single output. Three triangular membership functions are used for both input variables. ANFIS output is measured with root mean square value of 0.0407 and standard deviation is 0.0335. Maximum and minimum error is 0.2119 and 3.0609e−06. For the same triangular membership function, Number of variables is increased to 5, 7, and 9 and the results are tabulated as in Table 6. Triangular membership function with 7 variables gives the optimal root mean square error value of 0.0078.

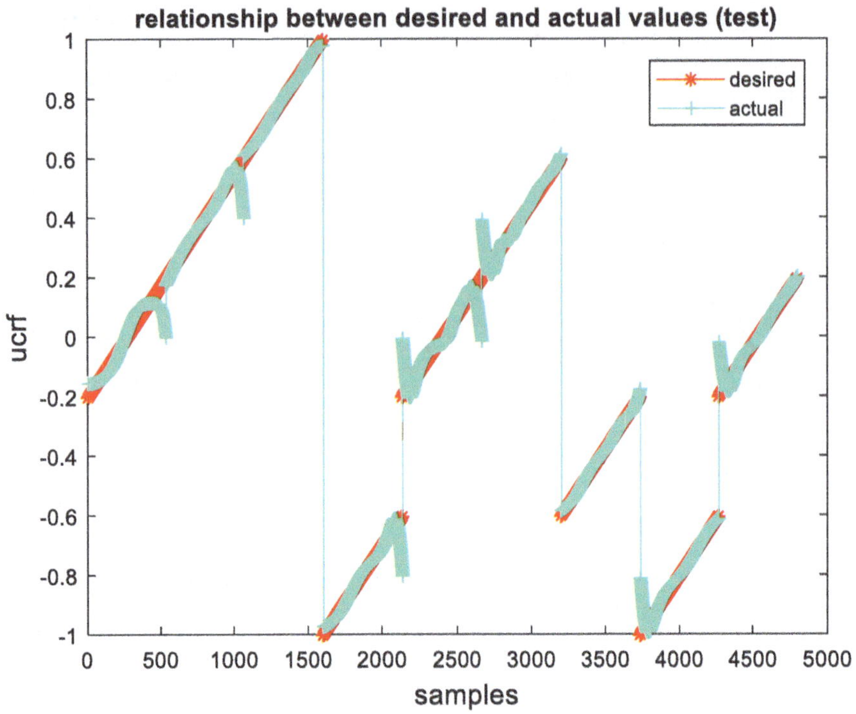

Fig. 23 Relationship between desired and actual values (Test data set)

Table 5 Performance measures with backpropagation network

Dataset/parameters	Maximum error	Minimum error	rmse	Learning parameter	Momentum parameter
Trained	0.2015	6.0318e−06	0.0316	0.2	0.3
Test	0.2042	9.3267e−07	0.0316		

4 ANFIS Controller Implementation for LVRT Enhancement

Fault confrontation controller is designed with three soft computing techniques namely fuzzy control, Backpropagation network, and ANFIS. Among which ANFIS has given the best result with minimum root mean square error value of 0.0078. Hence ANFIS technique has been adopted for predicting fault correction parameters as well as modifying the vector control of RSC. Stator voltage measurement with ANFIS controller during fault condition is shown in Fig. 26. Voltage dip of 85% is created at 3 s and cleared at 3.5 s. Voltage rebuilds to the post fault state at $t = 4.1$ s.

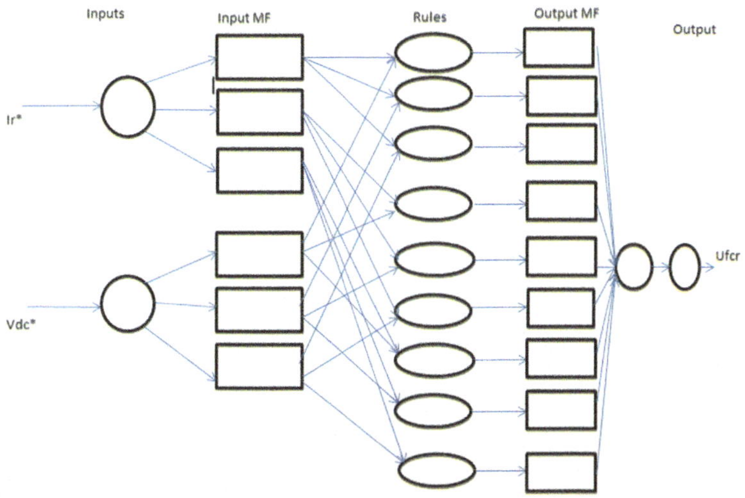

Fig. 24 ANFIS architecture with two inputs and one output

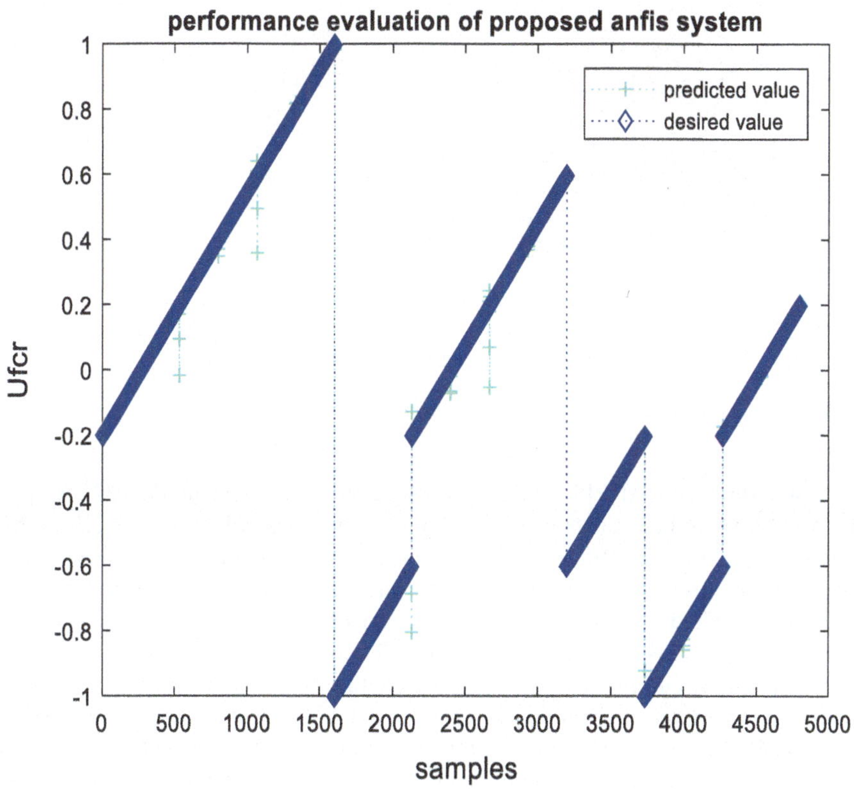

Fig. 25 Performance evaluation of ANFIS system with Triangular membership function

Table 6 ANFIS output with triangular membership function

S. No.	No. of variables	Triangular membership function			
		RMSE	ST	Max	Min
1	3	0.0407	0.0335	0.2119	3.0609e−06
2	5	0.0211	0.0200	0.2170	7.0416e−08
3	7	0.0078	0.0077	0.2484	3.3574e−08
4	9	0.0115	0.0112	0.2874	2.6523e−08

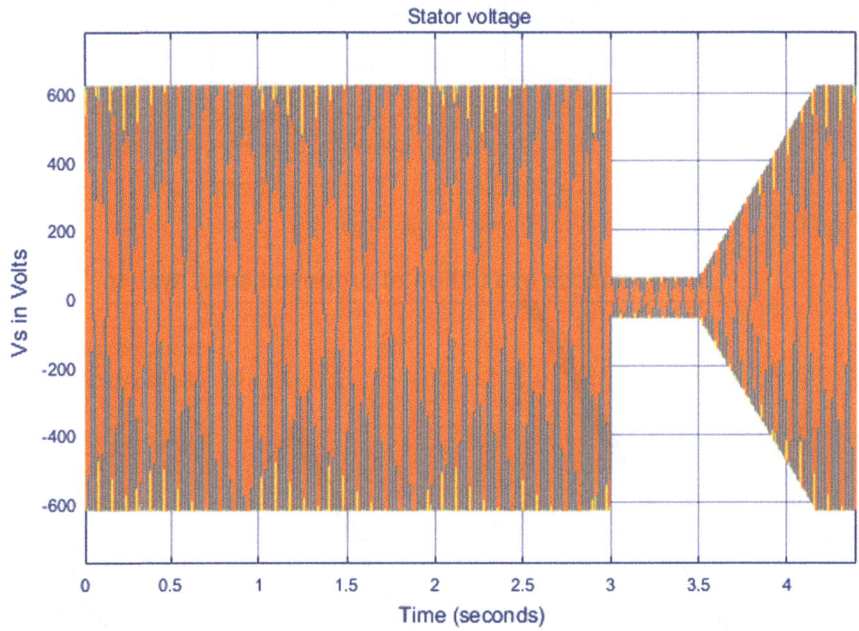

Fig. 26 Stator voltage with ANFIS modified vector control

Stator flux with ANFIS modified vector control is given in Fig. 27. As per Eq. (3.58) when there exists a voltage dip then the stator voltage is reduced to the low value very quickly. But stator flux will take large time to evolve to the low value. But with the rotor current magnitude stator flux can be quickly decayed. Crowbar protection method disconnects the DFIG rotor from RSC during a fault condition. But the proposed method maintains the DFIG connection with the converters and grid. Also, rotor current is vector controlled with rotor side converter with ANFIS fault confrontation controller which results in quick decay of stator flux and attains new steady-state and resumed back to the normal operation after fault clearance.

Rotor current during voltage dip is presented in Fig. 28. Fault is created at 3 s. Hence the rotor current increases to the value of 2000 A which is below the threshold limit which converter can withstand. But with conventional control, it has reached

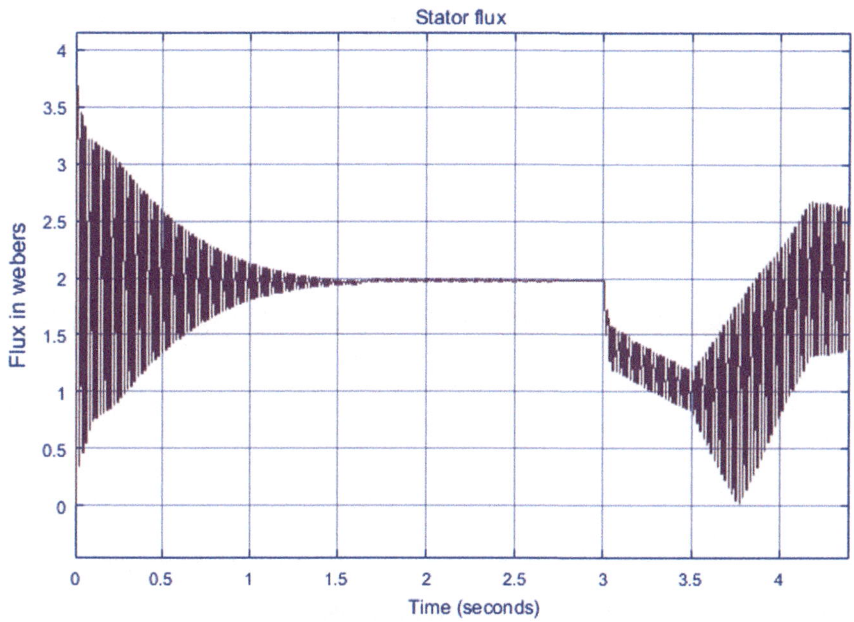

Fig. 27 Stator flux with ANFIS modified vector control

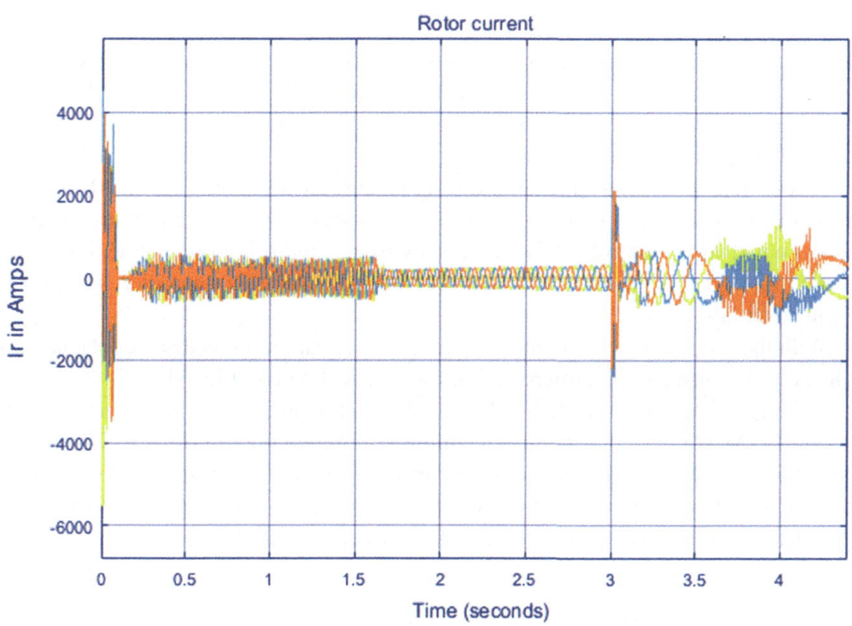

Fig. 28 Rotor current with ANFIS modified vector control

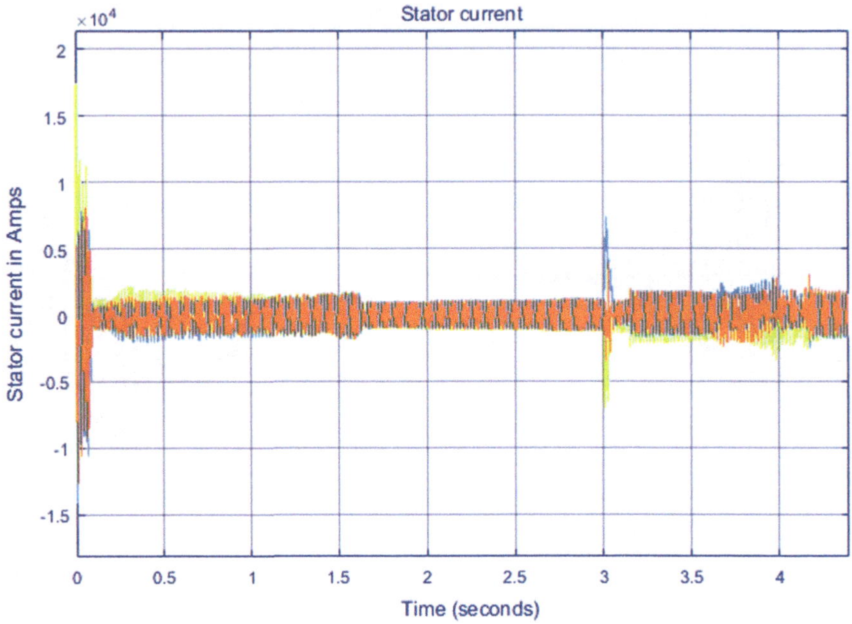

Fig. 29 Stator current with ANFIS modified vector control

the value of 4000 A. Stator current variation during transient condition is given in Fig. 29. With modified control stator current magnitude also reduced to the lower value around 7000 A at $t = 3$ s. At $t = 3.1$ s stator current reaches the low value and attains the pre-fault value around $t = 3.5$ s.

Electromagnetic torque transients during voltage dip are shown in Fig. 30. Voltage dip at $t = 3$ s will results in increase in rotor speed. Hence torque oscillations will also be present. Torque control can be achieved with the control of quadrature axis rotor current through ANFIS modified vector control of rotor side converter. Therefore, torque oscillations are reduced.

With the modified vector control technique crowbar circuit is disabled. Hence no current flows through the crowbar circuit which is shown in Fig. 31.

DC link voltage is controlled with GSC. DC link voltage during transient is given in Fig. 32. Modified vector control technique is applied only to the RSC and not to the GSC. Hence DC link voltage during transient is increased upto 1700 V which are slightly above the threshold value of 1550 V. Hence additional DC-link brake chopper can be introduced at the DC link to protect the DC link capacitor.

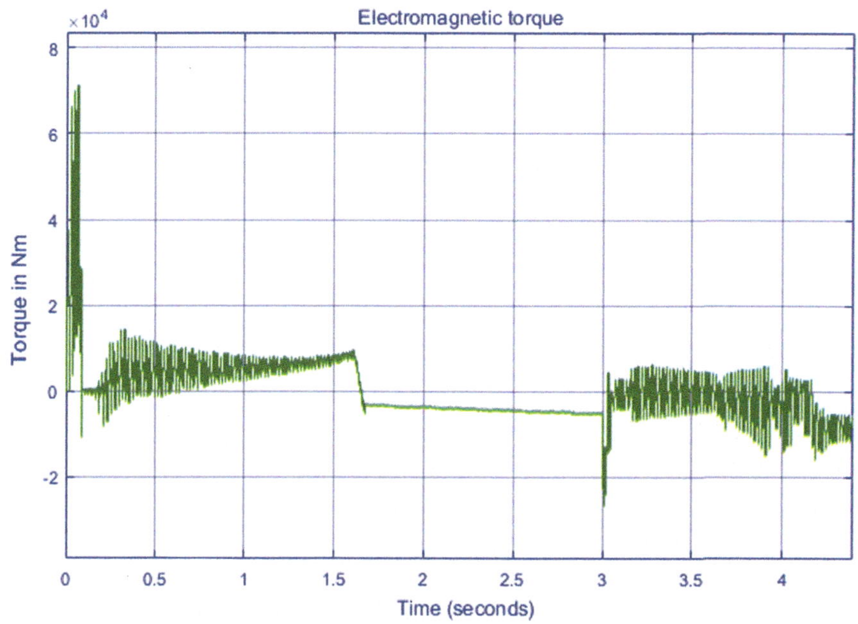

Fig. 30 Electromagnetic torque with ANFIS modified vector control

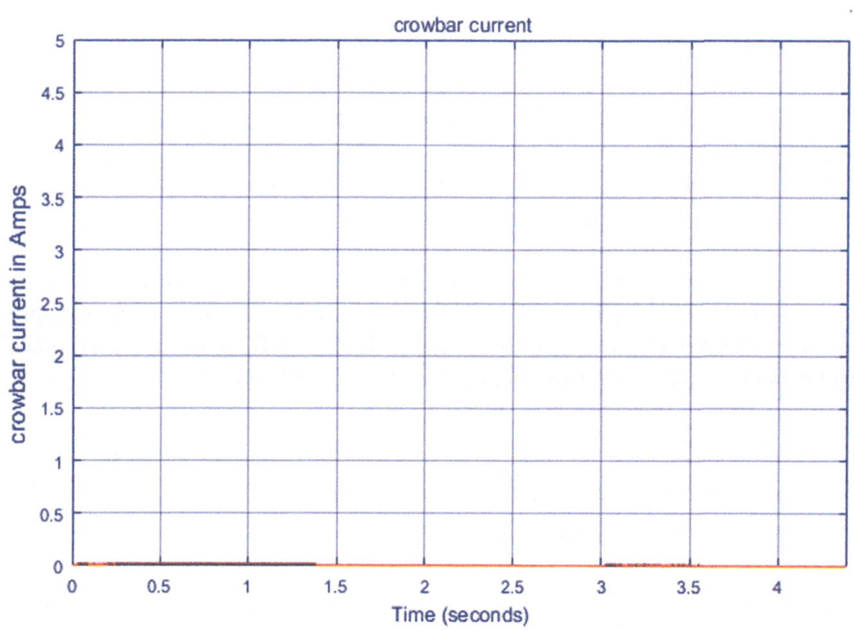

Fig. 31 Crowbar current with ANFIS modified vector control

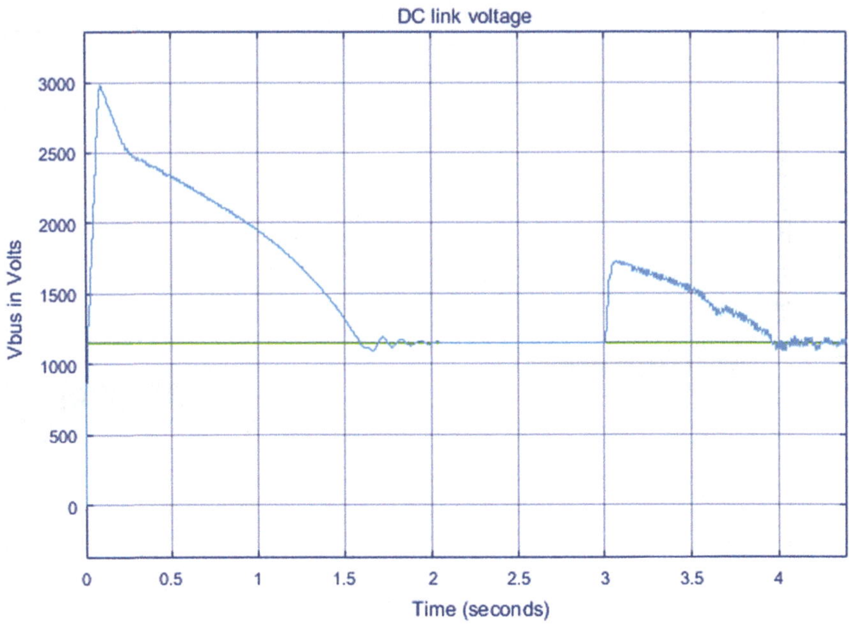

Fig. 32 DC link voltage with ANFIS modified vector control

5 Results and Discussion

Fault correction parameter prediction with three different techniques Fuzzy, BPN, and ANFIS controllers was also discussed. LVRT enhancement results with ANFIS controller are discussed. Based on performance outcomes, the most noteworthy observations and contributions are concluded.

With crowbar technique, RSC converters are protected but wind turbines are not connected as required by the new grid code. Therefore, this chapter focus on the implementation of computational intelligent controller based modified vector control which eliminates the use of crowbar and satisfies the LVRT grid code requirement by maintaining the uninterrupted operation of wind turbines.

5.1 Fault Correction Parameter Prediction with Fuzzy/BPN/ANFIS

Conventional vector control of RSC is modified with the fault correction parameter obtained from soft computing techniques. Prediction of fault correction parameter with proposed fault confrontation controller is determined with three soft computing

Table 7 Performance measure with 3 different soft computingtechniques function

S. No.	Technique	Triangular membership function			
		RMSE	SD	Max	Min
1	Fuzzy	0.1542	0.6657	1	4.427e−05
2	BPN	0.0316	0.5299	0.2025	9.3267e−07
3	ANFIS	0.0078	0.0077	0.2484	3.3574e−08

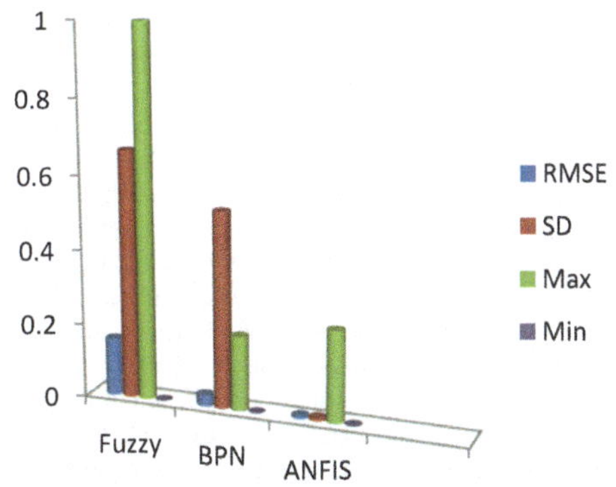

Fig. 33 Performance comparison between fuzzy/BPN/ANFIS

techniques namely fuzzy logic, backpropagation, and ANFIS. Performance parameters of three techniques namely Root Mean Square Error, Standard Deviation, Minimum and Maximum errors are measured and tabulated (Table 7). Tabulated values are plotted as shown in Fig. 33.

From results, it is inferred that backpropagation network outperforms Fuzzy logic control system. When compared to BPN, ANFIS gives the best-optimized result with minimum root mean square value of 0.0078.

6 Conclusion

The present study has been performed to enhance LVRT ability of DFIG with the Computational intelligence technique. ANFIS fault confrontation controller is selected for modifying the vector control and to improve the LVRT performance. With ANFIS modified vector control rotor inrush current, stator current, and torque oscillations are reduced during transient conditions which eliminate the use of crowbar technique and also improve the LVRT performance of DFIG. Only RSC control is modified and the GSC control remains unchanged resulting in a slight increase in DC link voltage which can be addressed with the DC link brake chopper. With the proposed modified vector control, RSC remains connected and the DFIG maintains grid connection with the grid and ensures power system stability.

References

1. Tsili, M., Papathanassiou, S.: A review of grid code technical requirements for wind farms. IET Renew. Power Gener. **3**, 308–332 (2009)
2. Justo, J.J., Mwasilu, F., Jung, J.-W.: Doubly fed induction generator based wind turbines: a comprehensive review of fault-ride through strategies. Renew. Sustain. Energy Rev. **45**, 447–467 (2015)
3. Du, K., Xiao, X., Wang, Y., Zheng, Z., Li, C.: Enhancing fault ride- through capability of DFIG-based wind turbines using inductive SFCL With coordinated control. IEEE Trans. Appl. Supercond. **29**(2), 1325–1334 (2019)
4. Firouzi, M., Gharehpetian, G.B.: LVRT performance enhancement of DFIG-based wind farms by capacitive bridge-type fault current limiter (CBFCL). IEEE Trans. Sustain. Energy. **3**(2), 242–254 (2017)
5. Guo, W., Xiao, L., Dai, S., Xu, X., Li, Y., Wang, Y.: Evaluation of the performance of BTFCLs for enhancing LVRT capability of DFIG. IEEE Trans. Power Electron. **30**(7), 3623–3637 (2015)
6. Mohammad Pour, H., Zadeh, S.G., Tohidi, S.: Symmetrical and asymmetrical low-voltage ride through of doubly-fed induction generator wind turbines using gate controlled series capacitor. IET Renew. Power Gener. **9**(7):840–846 (2015)
7. Huchel, L., Moursi, M.S.E., Zeineldin, H.H.: A parallel capacitor control strategy for enhanced FRT capability of DFIG. IEEE Trans. Sustain. Energy **6**(2), 303–312 (2015)
8. Uddin, W., Zeb, K., Tanoli, A., Haider, A.: Hardware-based hybrid scheme to improve the fault ride through capability of doubly fed induction generator under symmetrical and asymmetrical fault. IET Gener. Transm. Distrib. **12**(1), 200–206 (2017)
9. Xie, D., Zhao, Xu., Yang, L., Ostergaard, J., Xue, Y., Wong, K.P.: A comprehensive LVRT control strategy for DFIG wind turbines with enhanced reactive power support. IEEE Trans. Power Syst. **28**(3), 3302–3310 (2013)
10. Zhou, D., Blaabjerg, F.: Optimized demagnetizing control of DFIG power converter for reduced thermal stress during symmetrical grid fault. IEEE Trans. Power Electron. **33**(12), 10326–10340 (2018)
11. Hu, J., Xu, H., He, Y.: Coordinated control of DFIG's RSC and GSC nder generalized unbalanced and distorted grid voltage conditions. IEEE Trans. Industr. Electron. **60**(7), 2808–2819 (2013)
12. Liang, J., Qiao, W., Harley, R.G.: Feed-forward transient current control for low-voltage ride-through enhancement of DFIG wind turbines. IEEE Trans. Energy Convers. **25**(3), 836–843 (2010)
13. Noureldeen, O., Hamdan, I.: An efficient ANFIS crowbar protection for DFIG wind turbines during faults. In: Nineteenth International Middle East Power Systems Conference (MEPCON), pp. 263–269. ISBN: 2367-0983 (2017)
14. Zhu, R., Deng, F., Chen, Z., Liserre, M.: Enhanced control of DFIG wind turbine based on stator flux decay compensation. IEEE Trans. Energy Convers. **31**(4), 1366–1376 (2016)
15. Bhadra, S.N., Kastha, D., Banerjee, S.: Wind Electrical Systems. Oxford University Press. Chap. 1. ISBN:0195670930 (2005)
16. Abad, G., Lopez, J., Rodriguez, M.A.: Doubly Fed Induction Machine : Modeling and Control for Wind Energy Generation. Wiley, chap.2. ISBN: 9781118104965 (2011)
17. Abad, G., Iwanski, G.: Power Electronics for Renewable Energy Systems, Transportation and Industrial Applications. Wiley, Chap.10. ISBN: 9781118755525 (2014)
18. Ross, T.J.: Fuzzy Logic with Engineering Applications, 3rd edn. Wiley. ISBN: 9781119994374 (2011)
19. Freeman, J.A., Skapura, D.M.: Neural Networks: Algorithms, Applications, and Programming Techniques. Addison-Wesley. ISBN: 9780201513769 (1991)
20. Loukas, Y.L.: Adaptive neuro-fuzzy inference system: an instant and architecture-free predictor for improved QSAR studies. J. Med. Chem. **44**(17), 2772–2783 (2001)

Harmonic Current Estimation of a Non-linear Load Using Artificial Neural Network

A. Venkadesan

1 Introduction

One can say quality of the power is good only if we deliver the steady voltage/current signals within certain standard limits with the waveform shape resembling sinusoidal in nature [1]. There are many power quality problems that make the quality of the power poorer. There are numerous issues that affect the quality of power. They are voltage sag, voltage swell, short and long interruptions, voltage spike, harmonic distortion, voltage fluctuation, noise. To access the power quality, there are many power quality indices namely power factor, total harmonic distortion, VT product, unbalance factor, flicker factor, and its ranges are proposed as per the IEEE power quality standards.

In this paper, harmonic distortion is considered for the study. Harmonic distortion is nothing but the voltage and current waveform deviates from actual sinusoidal shape. It means that the waveform contains multiple fundamental frequencies. According to India, the fundamental frequency of voltage/current signal is 50 Hz. If we talk about the loads, the loads can be classified broadly classified into linear loads and non-linear loads. The linear loads are loads whose voltage/current waveforms are sinusoidal in shape. The non-linear loads are the loads whose current is not proportional to voltage. Also, its voltage/current waveform is not sinusoidal in nature. Nowadays, the use of non-linear loads is increasing day by day. The power electronic circuits are the non-linear loads. These are finding numerous applications in domestics, commercial and industrial applications. The power electronics circuits are used in switched-mode power supplies (SMPS), uninterrupted power supply (UPS), variable frequency electrical drives. It always makes the voltage/current signals distorted and makes the

A. Venkadesan (✉)
EEE Department, NIT Puducherry, Karaikal 609609, India
e-mail: Venkadesan@nitpy.ac.in

B. Vinoth Kumar et al. (eds.), *Intelligent Paradigms for Smart Grid and Renewable Energy Systems*, Algorithms for Intelligent Systems,
https://doi.org/10.1007/978-981-15-9968-2_11

waveform deviate from sinusoidal shape. These non-linear loads are the sources of harmonics. These loads will cause adverse effects like increased losses, over-heating, reduced efficiency. These effects lead to derating in the equipment [2].

These harmonics can be minimized. It can be minimized using passive filters and active filters. The passive harmonic filter can be designed using capacitors/inductors. The passive filter methods are used to reduce harmonics [3, 4]. But, nowadays, active harmonic filters (AHFs) are gaining a lot of momentum. Numerous active harmonics filtering methods are proposed in the literature. The current controlled active filter is proposed [5]. A review of active harmonic filters is proposed [6]. The adaptive control algorithm for three-phase AHF is proposed [7]. The shunt AHF is proposed for inverter fed drives [8]. Three saturation algorithms for active power filters are proposed [9]. The single-phase series active power filter is proposed [10].

Recently, artificial neural networks (ANNs) are finding applications in electrical engineering. It can map non-linear input–output data. It can offer robust performance and also offer faster operation in real-time. The capability of ANN in active harmonics filtering is demonstrated and illustrated in many works. Many works on harmonic filtering using artificial neural network based methods are reported in the literature. The comparison of neural network approach and Fourier transform approach is carried out in [11]. The literature review on neural network applications to active power filter is carried out [12]. The ANN-based predictive and adaptive controllers for shunt active power filters are proposed [13]. The Chebyshev neural network is proposed to estimate harmonics for an active power filter [14]. Harmonic monitoring of non-linear loads using neural network methods is proposed [15]. The harmonic magnitude is estimated for power electronic converter using neural network [16].

Harmonic estimation assumes importance in AHF for effective harmonic filtering. In this paper, an alternate novel approach using artificial neural network is proposed. The idea of the approach is inspired from [17]. In [17], the approach is used to extract the fundamental voltage waveform from the output of the inverter. In this paper, the same approach is applied for input current of the power electronic converter. The ANN approach depends on the type of neural architecture. The cascaded architecture is identified to provide accurate and compact model in many applications [18–20]. As per the knowledge of the author, it is first time applied for the harmonic current wave estimation of NLL. Of course, the same approach is used to estimate harmonic current for single-phase diode bridge rectifier [21]. But in this paper, it is applied for three-phase diode bridge rectifier.

2 Need for Harmonic Estimation

The AHF is popularly used for harmonic filtering. The block diagram Fig. 1 shows the concept behind the harmonic filtering. The non-linear loads make the shape of the current non-sinusoidal in nature.

The actual current is sensed and the fundamental component is extracted and subtracted from the actual current to generate harmonic reference current. Using

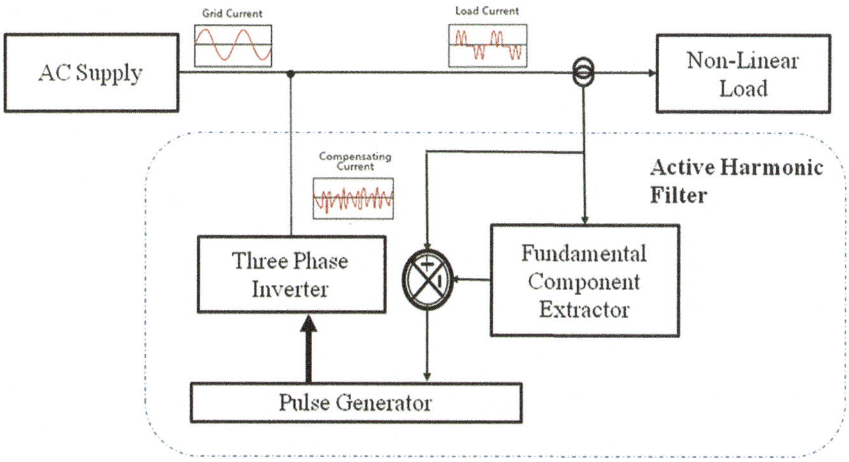

Fig. 1 Block diagram of AHF showing the need for harmonic current estimation

the reference current, the pulses are generated to switch the inverter to cancel the harmonic current in the utility grid. Various techniques namely instantaneous reactive power theory (P-Q method), Synchronous reference frame theory (SRF method), d-q method are used to generate switching pulses to the three-phase inverters. The performance of AHF to large extents depends on the accuracy of harmonic current wave estimation. Hence this necessitates the need to design an effective model/algorithm to estimate harmonic for efficient harmonic mitigation.

3 Conventional Technique for Harmonic Estimation

The Fourier series method is a popular conventional technique used for the extraction of harmonics from any periodic non-sinusoidal signals. Any periodic non-sinusoidal signal $i(t)$ with period T can be represented by a DC component and an infinite sum of harmonic signals whose angular frequencies are integral multiples of $\omega_0 = 2\pi/T$.

The d_0, e_k, g_k are called as Fourier coefficients. k = harmonic order, m_n = harmonic magnitude, φ_n = harmonic phase. In this paper, $i(t)$ is the input current signal of non-linear load. The equations are integral equations and have to be dicretized for digital implementation. It is an iterative method and takes longer time for computation.

$$i(t) = d_o + \sum_{k=1}^{\infty} [e_k \cos(k\omega_o t) + g_k \sin(k\omega_o t)] \tag{1}$$

$$d_o = \frac{1}{T}\int\limits_{-T/2}^{T/2} i(t)\mathrm{d}t \tag{2}$$

$$e_k = \frac{2}{T}\int\limits_{-T/2}^{T/2} i(t)\cos(k\omega_o t)\mathrm{d}t \quad k = 1, 2, 3, \ldots \tag{3}$$

$$g_k = \frac{2}{T}\int\limits_{-T/2}^{T/2} i(t)\sin(k\omega_o t)\mathrm{d}t \quad k = 1, 2, 3, \ldots \tag{4}$$

$$m_k = \sqrt{e_k^2 + g_k^2} \tag{5}$$

$$\phi_n = \tan^{-1}\left(\frac{e_k}{g_k}\right) \tag{6}$$

4 Artificial Neural Network Technique for Harmonic Estimation

The performance of ANN-based harmonic estimator to a large extent depends on the type of neural architecture. The cascaded architecture is proposed to design ANN-based harmonic estimator. The performance of cascaded architecture is compared with feedforward architecture for harmonic current wave estimation. A brief description of both the neural architectures are presented below.

4.1 Cascaded Neural Architecture

Figure 2 shows the schematic diagram of cascaded neural architecture. The detailed property of cascaded neural architecture is well presented in [18, 19]. The signals flow from the inputs to the outputs. The inputs to each layer are the outputs from all previous layers. The single neuron is kept in each hidden layer so that the architecture can be design automated. It is named as cascaded Architecture (CA).

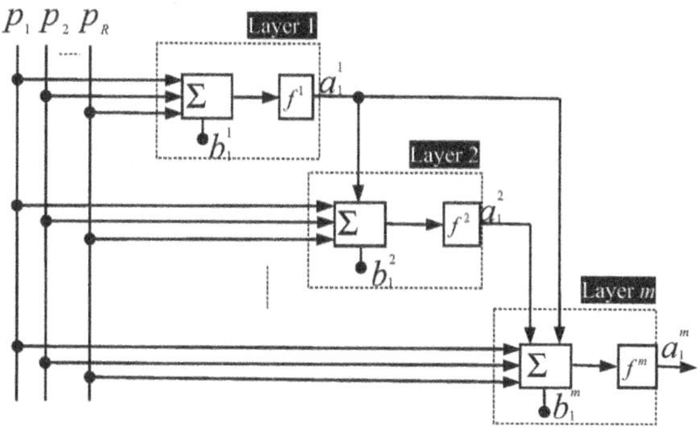

Fig. 2 Cascaded neural architecture

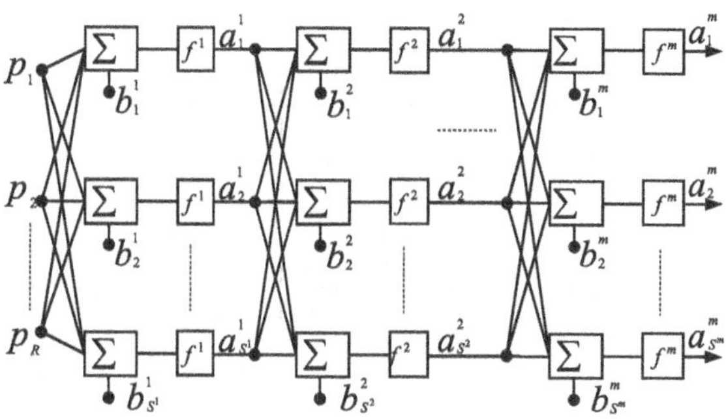

Fig. 3 Feedforward neural architecture

4.2 Feedforward Neural Architecture

Figure 3 shows the schematic diagram of feedforward neural architecture. The detailed property of the feedforward architecture is well presented in [19]. In this architecture also, the signals flow from the inputs to the outputs. The feedforward neural architecture receives inputs from only the immediate previous layer in contrast to the cascaded neural architecture. The Feedforward Architecture with Single Hidden Layer (FFASHL) and Feedforward Architecture with Multiple Hidden Layers (FFAMHL) are considered for the study.

5 Training of ANN for Harmonic Current Estimation

The harmonic current waveform estimation using ANN technique is attempted. As an example, a simple NLL namely three-phase uncontrolled rectifier with resistive inductive (RL) load is considered for study. The circuit diagram is shown in Fig. 4. The resistance and inductance values are chosen to be 10 Ω and 100 mH respectively. The three-phase input Line-Line RMS voltage is 415 V. The frequency is 50 Hz. Figure 5 shows the MATLAB Simulink diagram.

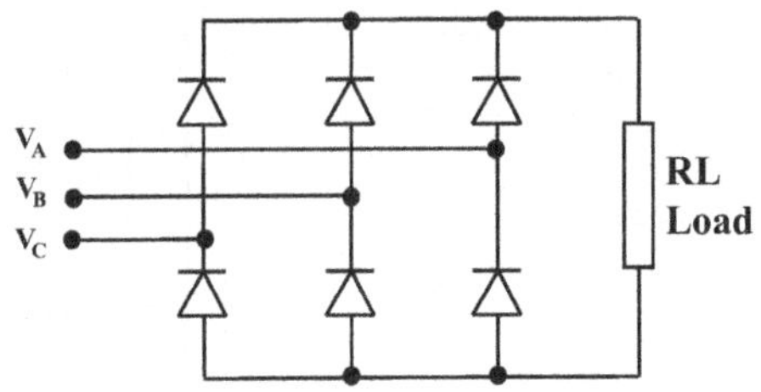

Fig. 4 3-phase uncontrolled rectifier with RL Load

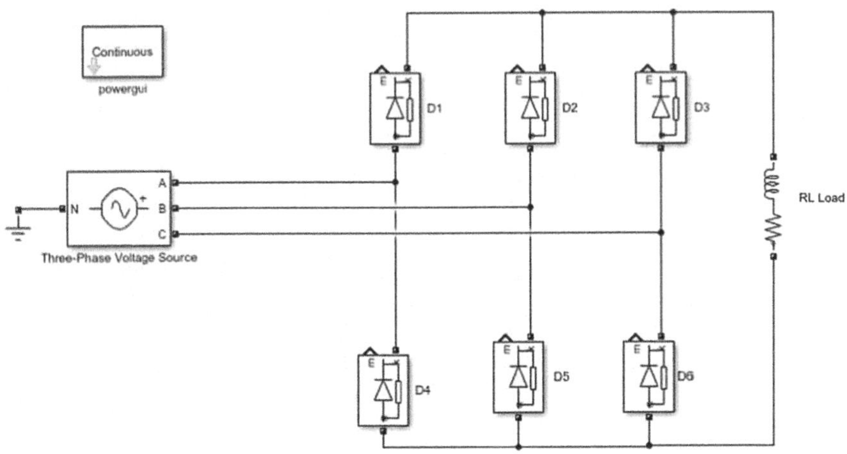

Fig. 5 MATLAB simulink diagram for 3-phase uncontrolled rectifier with RL load

To design ANN-based harmonic estimator, one thousand five hundred input–output data is collected through MATLAB simulation. The samples are collected with 40 μs of sampling time. The inputs to ANN-based harmonic estimator is chosen as actual source current and filtered source current. The output is the fundamental component of the current. The extracted fundamental component is subtracted from the actual current to generate harmonic component of the current. Figure 6 shows the schematic diagram.

The reason for choosing filtered current as one of the inputs is that ANN cannot generate continuously varying smooth sinusoidal fundamental current waveform from the square wave shaped actual source current waveform [17, 21]. The low pass filter is used to make the actual square waveform into a continuously varying waveform. The time constant of the low pass filter is chosen as 0.02 s [17, 21]. The actual input current and low pass filtered current is shown in Fig. 7 for two cycles. The actual input current contains harmonics. The harmonics are usually measured in terms of THD and it is found to be more than 5% which is not as per the IEEE standards. Hence it is to be eliminated. Therefore, active power filter is employed to filter out these harmonics. To do this effectively, accurate and fast estimation of

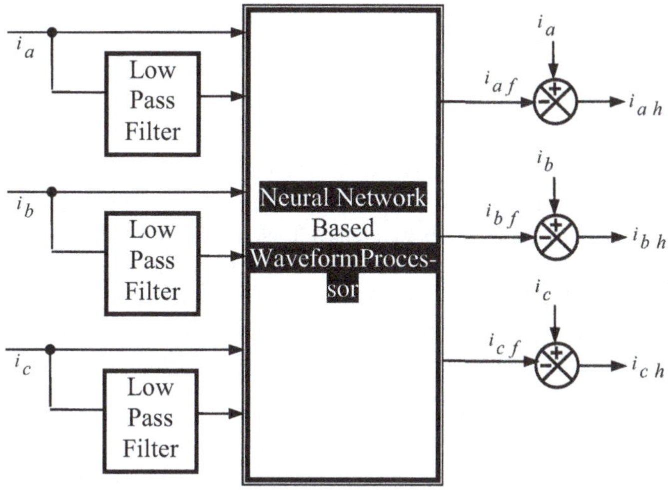

Fig. 6 ANN-based harmonic estimator

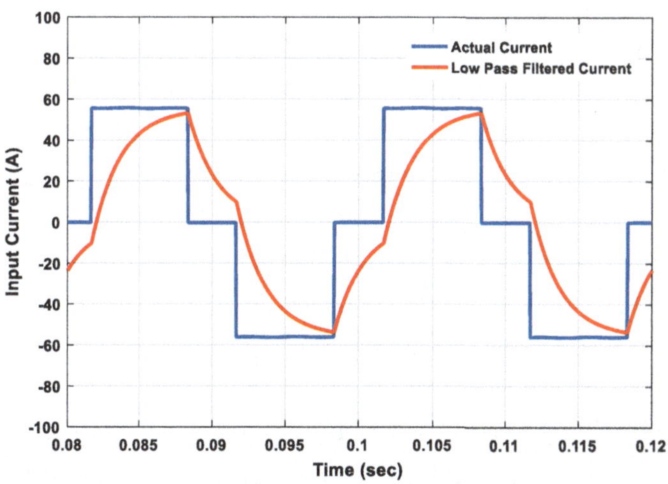

Fig. 7 Input current of the three-phase rectifier

harmonic current is needed as it is depicted earlier. This paper proposes ANN to estimate harmonic current wave.

The training algorithm employed for off-line training is the Levenberg Marquardt Algorithm (LMA). The hidden neurons use a tan-sigmoid function. The output neurons use pure-linear function. The equation for tan-sigmoid and pure-linear is presented in (1) and (2), respectively.

$$a = \frac{2}{1 + e^{-2n}} - 1 \tag{7}$$

$$a = n \tag{8}$$

The training Mean Squared Error (MSE) is chosen as 0.0000001. Using the same training data and MSE, all three neural architectures are trained. The training of ANN-based harmonic estimator is given as a flowchart in Fig. 8. In case of FFASHL based estimator, the neuron is added one by one in single hidden layer between the inputs and outputs till the target MSE is reached. The CA-based estimator is designed by adding a hidden layer with one neuron between the inputs and outputs till the target MSE is reached. The design of Feedforward architecture with multiple hidden layers is more an art than science. To make the design easy, two hidden

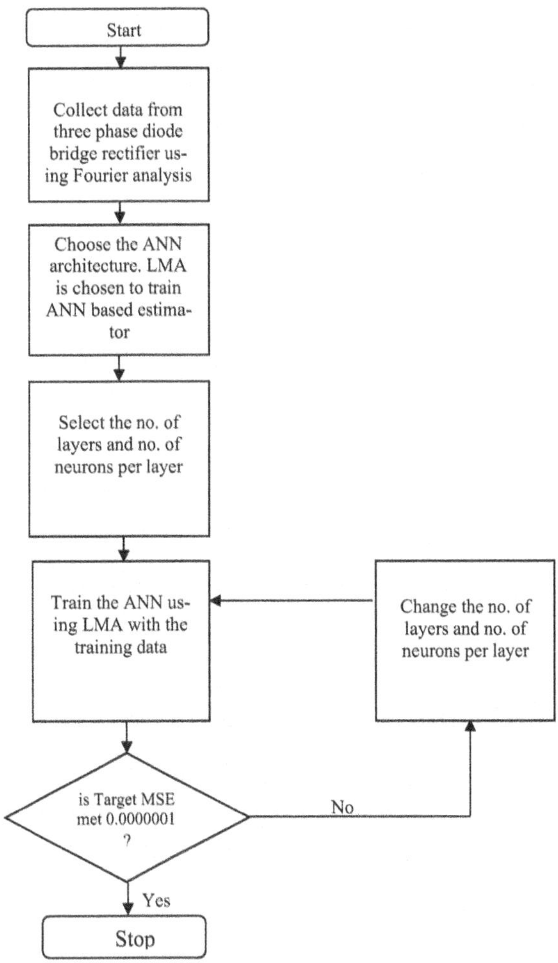

Fig. 8 Flow chart for training the ANN-based harmonic estimator

layers are considered for the study. Hence it is conveniently named as Feedforward Architecture with Two Hidden Layers (FFATHL). The neurons are added one by one equally in both the hidden layers at a time till the target MSE is reached. The MSE graph for all the architectures is shown in Figs. 9, 10 and 11 respectively. The training results are presented in Table 1. It is observed that all three architectures meet the target accuracy and the achieved training MSE of all the estimators is similar. The MATLAB 16 version is used for training ANN. On Intel I7 processor with 3.6 GHz and 8 GB RAM, FFASHL takes 1507 epochs, FFATHL takes 517 epochs, CA takes 452 epochs.

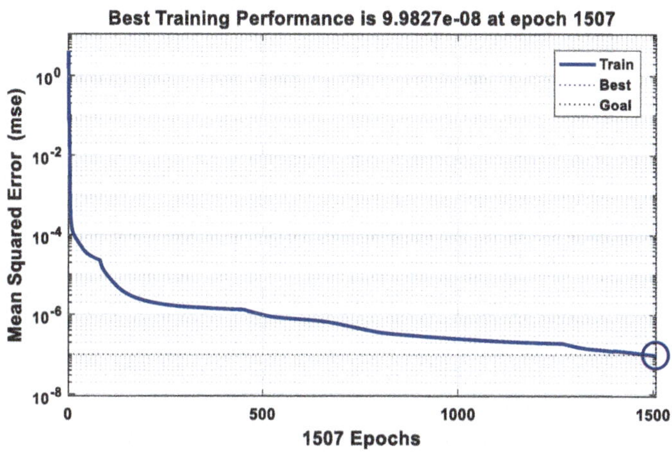

Fig. 9 Training MSE convergence graph for FFASHL-estimator

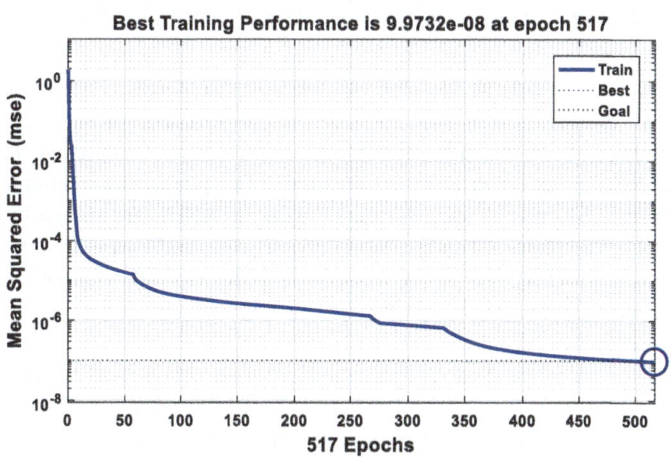

Fig. 10 Training MSE convergence graph for FFATHL-estimator

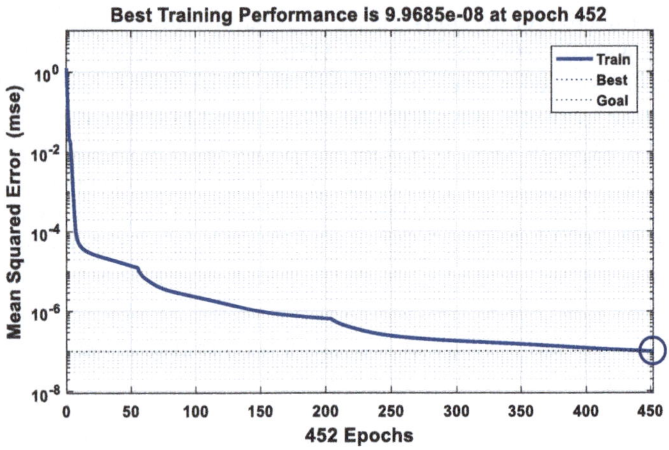

Fig. 11 Training MSE convergence graph for CA-estimator

Table 1 ANN training results for the estimation of harmonic current wave

ANN estimator	Training MSE achieved	Epochs
ANN estimator 1-Feedforward architecture with single hidden layer (FFASHL) (6-32-3)	0000000.99827	1507
ANN estimator 2-Feedforward architecture with two hidden layers (FFATHL) (6-11-11-3)	0000000.99732	517
ANN estimator 2-cascaded architecture (6-11(h*)-3) h*-hidden layer with one neuron	0000000.99685	452

6 Testing of ANN for Harmonic Current Estimation

To estimate harmonic current, the three architectures are tested. The neural network is tested with the data samples at 100 μs. The performance of both the architectures is compared in terms of test MSE and mathematical complexity. Firstly, the performance is compared in terms of accuracy. The phase a fundamental current wave and harmonic current wave estimated using FFASHL are shown in Fig. 12a, b respectively. The phase a fundamental current wave and harmonic current wave estimated using FFATHL are shown in Fig. 13a, b, respectively. The phase a fundamental current wave and harmonic current wave estimated using CA are shown in Fig. 14a, b, respectively.

The phase b fundamental current wave and harmonic current wave estimated using FFASHL are shown in Fig. 15a, b, respectively. The phase b fundamental current wave and harmonic current wave estimated using FFATHL are shown in Fig. 16a, b respectively. The phase b fundamental current wave and harmonic current wave estimated using CA are shown in Fig. 17a, b respectively.

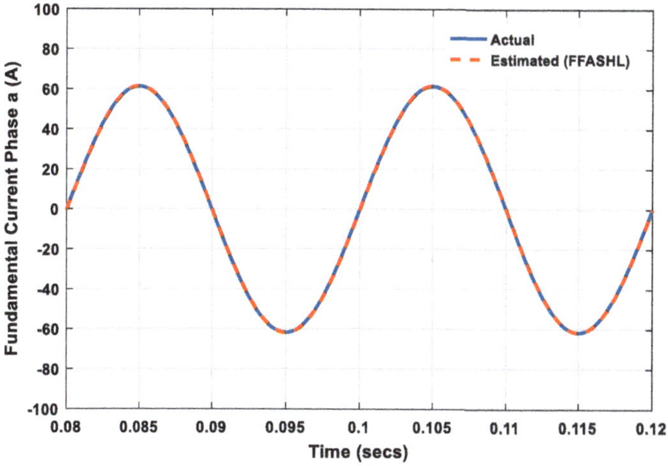

(a)

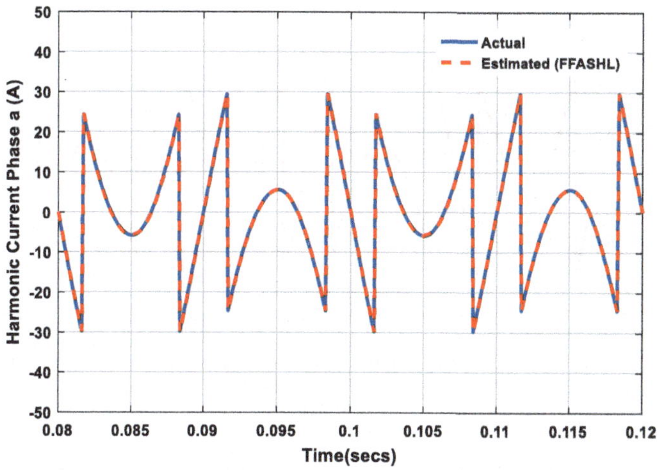

(b)

Fig. 12 Current phase a (FFASHL) **a** fundamental component, **b** harmonic component

The phase c fundamental current wave and harmonic current wave estimated using FFASHL are shown in Fig. 18a, b respectively. The phase c fundamental current wave and harmonic current wave estimated using FFATHL are shown in Fig. 19a, b respectively. The phase c fundamental current wave and harmonic current wave estimated using CA is shown in Fig. 20a, b respectively.

The estimated wave using ANN is shown along with actual wave computed using Fourier analysis. The test MSE is computed between the estimated wave and actual wave and tabulated in Table 2. All the ANN estimators compute the fundamental and harmonic current with good accuracy. But, it is noticed that ANN estimator designed using CA architecture has marginally lesser test MSE as compared to other

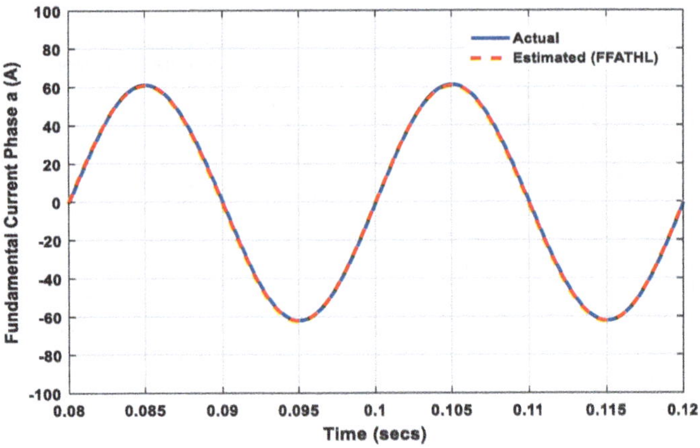

(a)

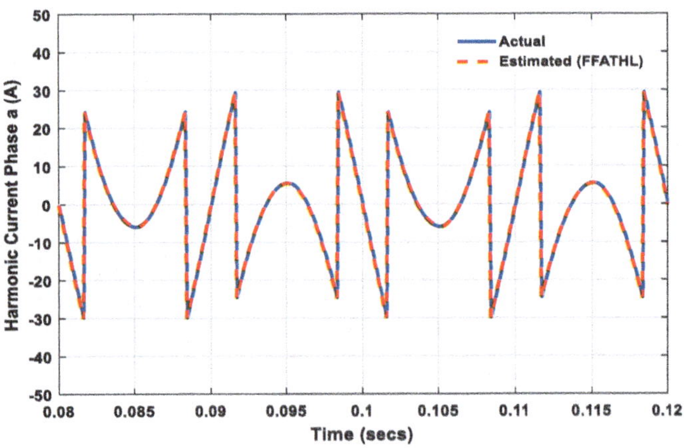

(b)

Fig. 13 Current phase a (FFATHL) **a** fundamental component, **b** harmonic component

ANN estimators. It is because of cascading of neurons and presence of massive multilayered structure.

The mathematically less complex ANN-based estimator is important for real-time implementation. Hence, it is needed to compare the ANN models in terms of mathematical complexity for harmonic current wave estimation. The mathematical computation involved in an artificial neuron with p number of inputs is represented as a block diagram in Fig. 21.

$$n = \sum_{i=1}^{R} p_i w_i + b \tag{9}$$

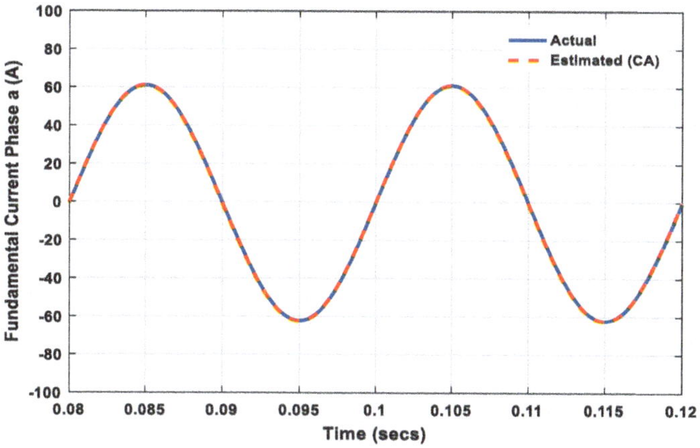

(a)

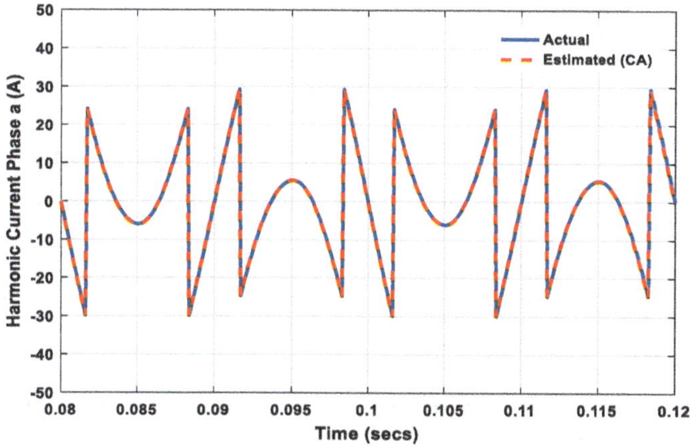

(b)

Fig. 14 Current phase a (CA) **a** fundamental component, **b** harmonic component

$$a = f(n) \tag{10}$$

$$\text{FFASHL\_}w = \text{FFATHL\_}w = \sum_{m-1}^{M} S^{m-1} S^{m} \tag{11}$$

$$CA\_w = \sum_{m=1}^{M} \sum_{q=0}^{m-1} S^{m} S^{q} \tag{12}$$

The mathematical equation involved for the same is presented in (3) and (4). The mathematical operations involved in an artificial neuron are additions, multiplications, computation of activation functions. The additions and multiplications depend

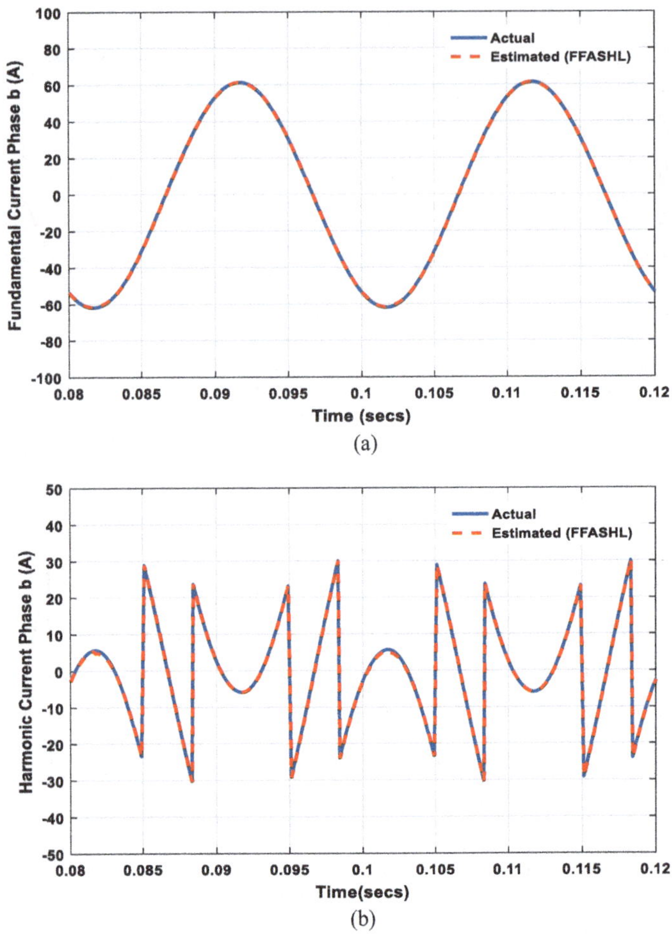

Fig. 15 Current phase b (FFASHL) **a** fundamental component, **b** harmonic component

on the total number of parameters (summation of weights and biases) of artificial neural networks. In the ANN, the weights are equal to additions. The number of additions is equal to number of multiplications. These can be computed using the formula presented in (5) and (6). Formula 3 is applicable for both FFASHL and FFATHL. S_{m} is the no. of neurons in the layer 'm' where $m = [1, 2, \ldots M]$ and $S0 = \mathrm{P}$. The number of mathematical operations involved in all the three ANN estimators is presented in Table 3. It is observed that ANN estimator 3 modeled using CA has lesser additions, multiplications, tan-sigmoid functions. Hence it is concluded that ANN Estimator 3 provides a mathematically less complex estimator and it will compute fundamental and harmonic current with faster execution time in real-time.

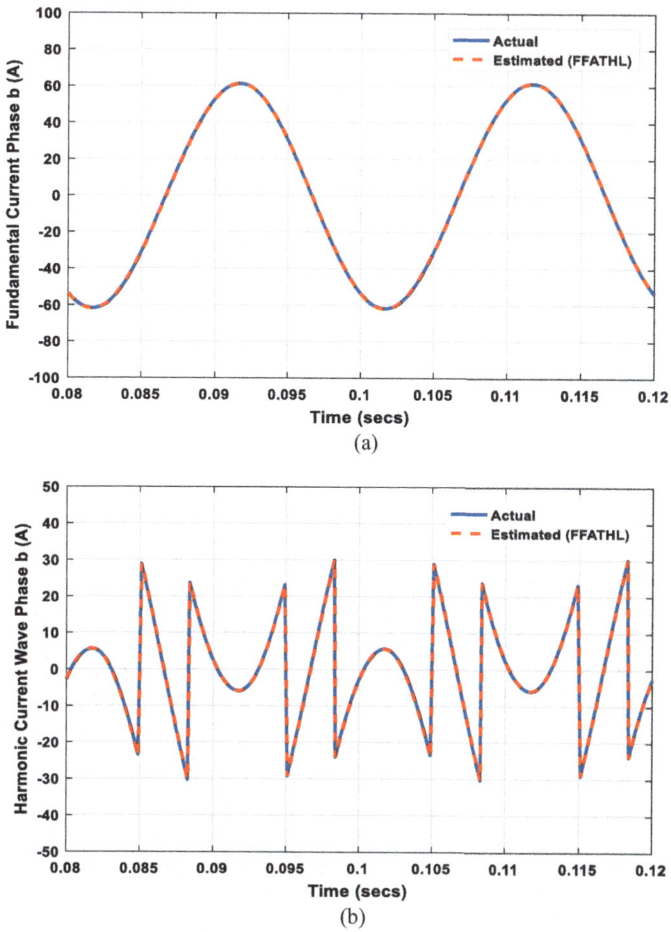

Fig. 16 Current phase b (FFATHL) **a** fundamental component, **b** harmonic component

7 Hardware Implementation Aspects of ANN Based Estimator 3

The ANN Estimator 3 designed based on Cascaded Architecture can be implemented on Field Programmable Gate Array (FPGA) as FPGA preserves parallelism and well-suited for real-time implementation of ANN estimator [18]. The most challenging issue in FPGA implementation of ANN estimator 3 is the computation of tan-sigmoid function. It involves the computation of exponential function. The exponential function can be computed using series expansion method but lesser number of terms leads to poor accuracy and higher number of terms increases the computation complexity [18]. The Look UP table method needs larger memory size for higher accuracy [18]. The 2-power-logic is proposed to compute Gaussian function which also involves

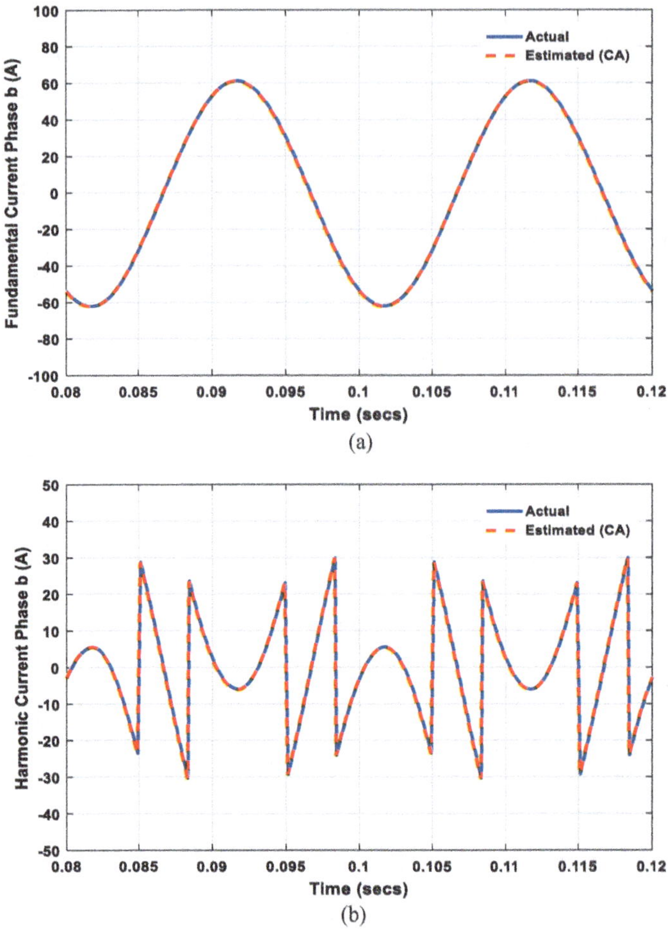

Fig. 17 Current phase b (CA) **a** fundamental component, **b** harmonic component

exponential function [22]. The base e is changed to base 2 which gives ease in computation in digital platform. The method is applied in the real-time implementation of fuzzy models. The same logic is applied in the real-time implementation of artificial neural network based space vector modulator [23]. The tansigmoid function with base 2 is presented (13). This method provides good accuracy but still it takes longer time for computation.

The simpler activation function named "Elliott function" is proposed as an alternative to tansigmoid function [18]. The equation is shown (14). The function possesses the same properties similar to tansigmoid function but it does not involve the computation of exponential function. It involves only addition and division operation. Hence the Elliott function can be used as activation function in the proposed ANN estimator. The ANN estimator can be retrained using the Elliott function and can be

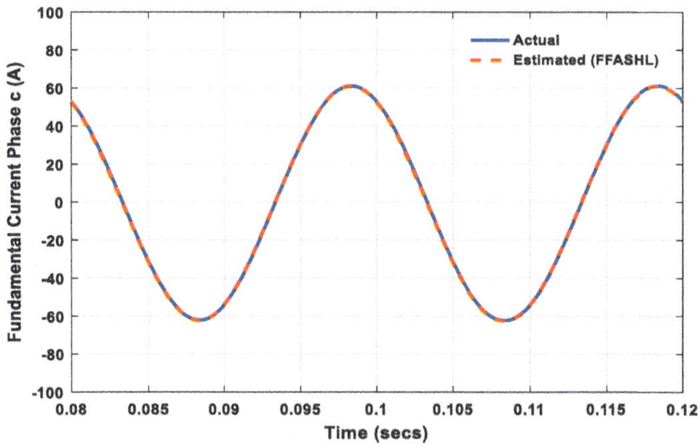

(a)

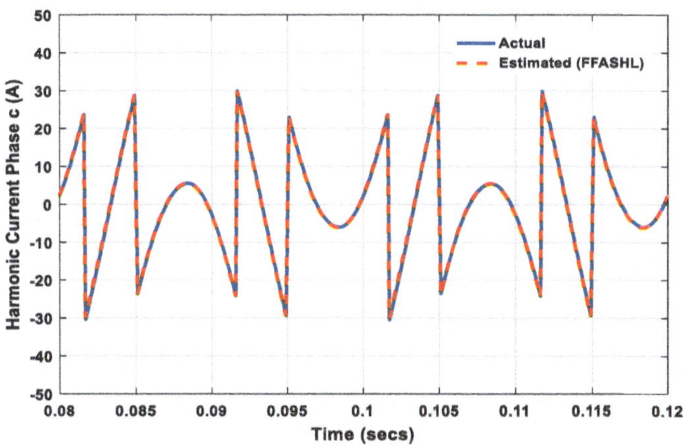

(b)

Fig. 18 Current phase c (FFASHL) **a** fundamental component, **b** harmonic component

easily implemented in real-time.

$$a = \frac{2}{1 + 2^{-2n}} - 1 \tag{13}$$

$$a = \frac{n}{1 + |n|} \tag{14}$$

The ANN estimator 3 can be implemented on FPGA with effective resource utilization using multiplexing method [24]. The ANN estimator 3 has 11 hidden layers with one neuron in each layer. The inputs to the neuron keep growing as the hidden layer increases. The maximum number of inputs to the NN estimator is 17. A single neuron with maximum of 17 inputs with non-linear activation can be

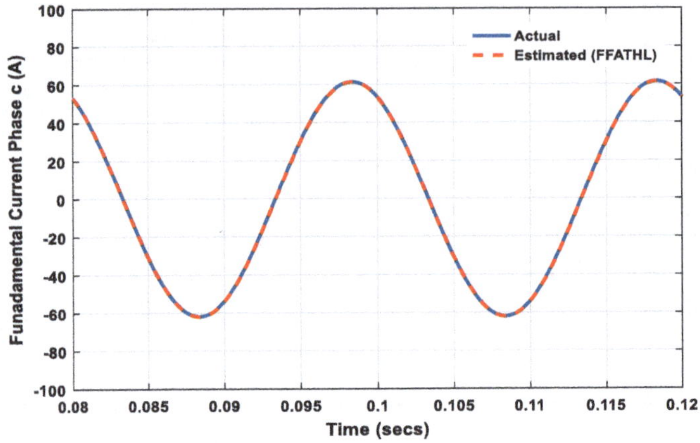

(a)

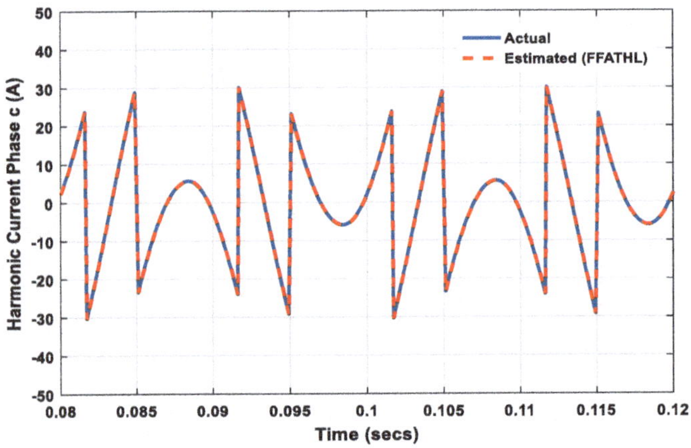

(b)

Fig. 19 Current phase c (FFATHL) **a** fundamental component, **b** harmonic component

implemented. The single neuron will act as a different neuron. The same neuron is multiplexed to realize the whole ANN architecture. The neuron multiplexing method exploits the sequential operation of the cascaded structure. The control block will place the appropriate weights and biases. When first neuron is computed, six inputs are processed into the neuron and the output is computed. When second neuron is computed, seven inputs (actual inputs and output of the first neuron) are processed into the neuron and the output is computed, and so on. The last output neuron has linear function and hence it is not processed into the non-linear activation function and the output is taken before the non-linear activation function. The three outputs are taken sequentially. The detailed operation is presented for the realization of flux estimator in motor drives [18]. The same logic can be used for the realization of

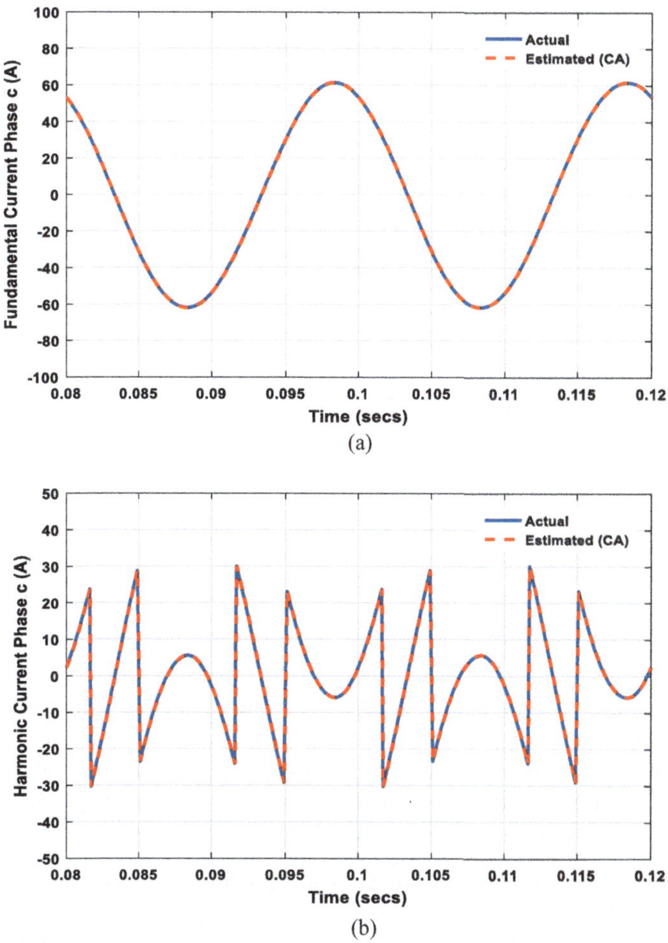

Fig. 20 Current phase c (CA) **a** fundamental component, **b** harmonic component

Table 2 TEST MSE for various ANN estimators

ANN estimator	Test MSE for harmonic current waveform (phase a)	Test MSE for harmonic current waveform (phase b)	Test MSE for harmonic current waveform (phase c)
ANN estimator 1	0.0054000	0.0274000	0.03070000
ANN estimator 2	0.0095000	0.0200000	0.00280000
ANN estimator 3	0.00026141	0.00024768	0.00053209

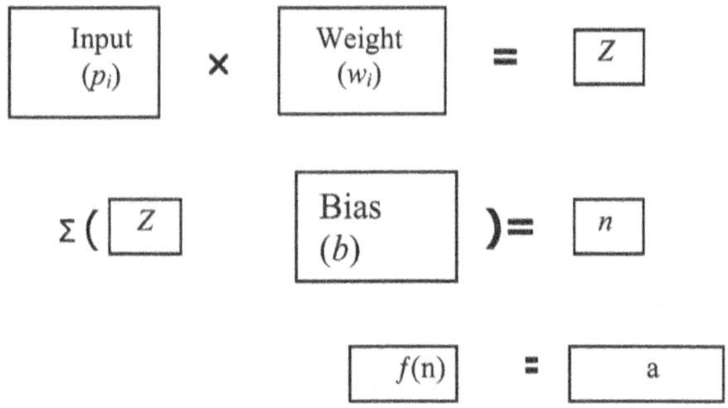

Fig. 21 Mathematical operations of an artificial neuron

Table 3 TEST MSE for various ANN estimators

ANN architectures	Additions	Multiplications	Tan-sigmoid functions
ANN estimator 1	288	288	32
ANN estimator 2	220	220	22
ANN estimator 3	172	172	11

ANN estimator 3. This method of realization includes only 17 additions, 17 multiplications, and 1 activation function. To implement whole network, 172 additions, 172 multiplications and 11 activation functions are needed which would increase the resource utilization as compared to multiplexing method. The proposed method of implementation is shown in Fig. 22. The schematic is explained as follows,

1. To initiate the fundamental component estimation process, the start signal is used.
2. An internal LAYER COUNTER is present in The LAYER CONTROL block.
3. The inputs to the first neuron and its corresponding weights & bias are placed on the internal bus. The LAYER CONTROL block initiates the read (RD) signal.
4. The ELLIOTT FUNCTION NEURON block reads the neuron inputs, weights, and bias from the bus and performs Multiplication (MUL), Add (ADD), and computation of Elliott excitation function (EF).
5. At the end of computation the output of the neuron/layer is passed back to the LAYER CONTROL block and output enable signal is asserted.
6. On receipt of the layer output, the next layer computation is initiated by LAYER CONTROL block and continues for 10 layers.
7. The output layer computation uses The LINEAR FUNCTION NEURON block.
8. The last layer uses LINEAR FUNCTION NEURON block and it's enabled by LAYER COUNTER.
9. The output is placed on the output line and the valid output signal indicates the end of fundamental component estima-tion.

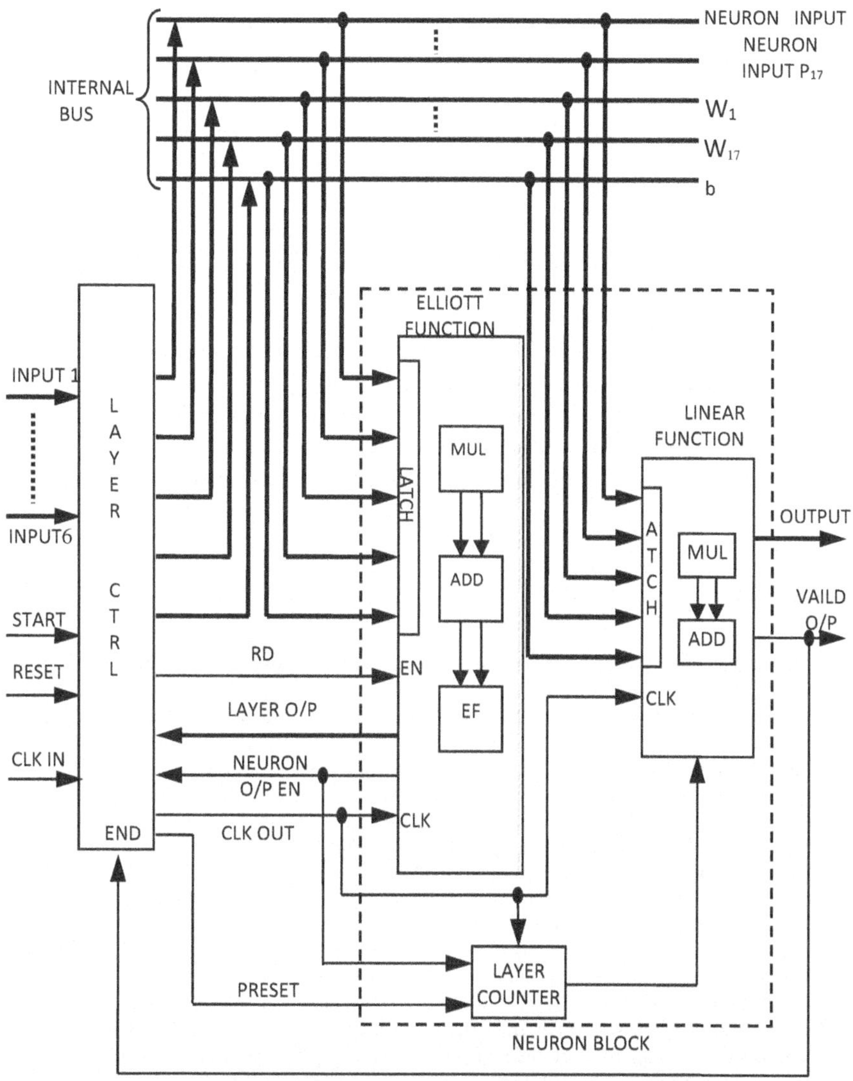

Fig. 22 Schematic diagram for implementation of proposed cascaded architecture based fundamental component estimator

10. The valid output signal is used by the LAYER CONTROL block as the end of computation process. The start signal is again asserted and the estimation is repeated for the next set of data.

8 Conclusion

The efficacy of ANN technique for HCE of non-linear load is tested in this paper. Three neural architectures namely feedforward architecture with single hidden layer, feedforward architecture with two hidden layers, and cascaded architecture are investigated for harmonic estimation. The ANN architectures are trained and tested for harmonic estimation. The models are compared in terms of accuracy and mathematical complexity for real-time implementation. The cascaded neural architecture is found to provide required accuracy and less complex model for harmonic estimation. The data-based ANN approach is specific for type of non-linear load and should be trained for that specific type of load. Once it is trained, it can estimate the harmonic current for that particular type of non-linear load. The methods to implement ANN harmonic estimator on FPGA is also comprehensively presented. Hence, it is concluded that cascaded neural estimator provides an alternate solution for harmonic estimation and is found to be suitable for active harmonic filter.

References

1. Edward Reid, W.: Power quality issues-standards and guidelines. IEEE Trans. Indus. Appl. **32**(3), 1052–1060 (1996)
2. ABB.: Control of Harmonics in Electrical Power Systems-Guidance Notes. India (2006)
3. de Campos, B.P., de Sousa, L.A.R., Ribeiro, P.F.: Mitigation of harmonic distortion with passive filters. In: International Conference on Harmonics and Quality of Power (ICHQP) (2016)
4. Yousif, S.N.A.L., Wanik, M.Z.C., Mohamed, A.: Implementation of different passive filter designs for harmonic mitigation. In: National Power and Energy Conference (2004)
5. Duke, R.M., Round, S.D.: The steady-state performance of a controlled current active filter. IEEE Trans. Power Electron. **8**(3), 140–146 (1993)
6. Kazem, H.A.: Harmonic Mitigation Techniques Applied to Power Distribution Networks. Hindawi Publishing Corporation-Advances in Power Electronics, pp. 1–10 (2013)
7. Singh, B., Solanki, J.: An implementation of an adaptive control algorithm for a three-phase shunt active filter. IEEE Trans. Indust. Electron. **56**(8) 2811–2820 (2013)
8. Ko, W.H., Jyh-Cherng, Gu.: Impact of shunt AHFon harmonic current distortion of voltage source inverter-fed drives. IEEE Trans. Industr. Electron. **52**(4), 2816–2825 (2016)
9. Ameriseet, A.: Comparison of three voltage saturation algorithms in shunt active power filters with selective harmonic control. IEEE Trans. Ind. Appl. **56**(3), 2762–2772 (2020)
10. Meloni, L.F.J., Tofoli, F.L., Rezek, A.J.J., Ribeiro, E.R.: Modeling and experimental validation of a single-phase series active power filter for harmonic voltage reduction. IEEE Access **7**, 151971–151984 (2019)
11. Chandra Sekaran, E., Anbalagan, P.: Comparison of neural network and fast fourier transform based selective harmonic extraction and total harmonic reduction for power electronic converters. Asian Power Electron. J. **2**(1), 1–9 (2008)
12. Janpong, S., Areerak, K.-L., Areerak, K.-N.: A literature survey of neural network applications for shunt active power filters. WASET **5**(12), 273–279 (2011)
13. Bhattacharya, A., Chakraborty, C.: A shunt active power filter with enhanced performance using ANN-based predictive and adaptive controllers. IEEE Trans. Industr. Electron. **8**(2), 421–428 (2011)
14. Chittora, P., Singh, A., Singh, M.: Chebyshev functional expansion based artificial neural network controller for shunt compensation. IEEE Trans. Industr. Inf. **14**(9), 3792–3800 (2018)

15. Nascimento, C.F., Oliveira Jr., A.A., Goedtel, A., Batista Dietrich, A.: Harmonic distortion monitoring for nonlinear loads using neural-network-method. Appl. Soft Comput. **13**, 475–482 (2013)
16. Venkadesan, A., Sedhuraman, K., Chandrasekaran, K., Boopathi, C.S.: Artificial neural network based harmonics estimator for a power electronics converter. Indian J. Sci. Technol. **9**(1), 1–5 (2016)
17. Zhao, J., Bose, B.K.: Neural-network-based waveform processing and Delay less filtering in power electronics and AC drives. IEEE Trans. Indust. Electron. **51**(5), 981–991 (2004)
18. Venkadesan, A., Himavathi, S., Sedhuraman, K., Muthuramalingam, A.: Design and FPGA implementation of cascade neural network based flux estimator for speed estimation in IM drives. IET-Electr. Power Appl. **11**(1), 121–131 (2017)
19. Himavathi, S., Muthuramalingam, A., Venkadesan, A., Sedhuraman, K.: Nonlinear system modelling using single neuron cascaded neural network for real-time applications. ICTACT J. Soft Comput. **2**(3), 309–318 (2012)
20. Sedhuraman, K., Himavathi, S., Muthuramalingam, A.: Performances comparison of neural architectures for on-line speed estimation in sensorless IM drives. WASET **5**(12), 1566–1573 (2011)
21. Venkadesan, A., Bhavana, G., Haneesha, D., Sedhuraman, K.: Comparison of feed forward and cascade neural network for HCE in power electronic converter. In: Proceedings of IEEE—International Conference on Innovative Research in Electrical Sciences (IICIRES-2017), E.G.S Pillay Engineering College, Nagapattinam, Tamil Nadu, India (2017)
22. Himavathi, S., Umamaheswari, B.: New membership functions for effective design and implementation of fuzzy systems. IEEE Trans. Syst. Man Cybern.—Part A: Syst. Hum. **31**(6), 717–723 (2001)
23. Muthuramalingam, A., Himavathi, S., Srinivasan, E.: Neural network implementation using FPGA: issues and application. WASET: Int. J. Electr. Comput. Eng. **2**(12), 2802–2808 (2008)
24. Himavathi, S., Anitha, D., Muthuramalingam: A feed-forward neural network implementation in FPGA using layer multiplexing for effective resource utilization. IEEE Trans. Neural Netw. **18**(3), 880–888 (2007)

Demand Response in Smart Residential Buildings

S. L. Arun and M. P. Selvan

1 Introduction

The traditional electric power system is built with different sectors such as generation, transmission, distribution, and consumption. Various types of conventional power plants like thermal, nuclear, and hydro are feeding power into the grid and enhancing a reliable operation by maintaining supply and demand balance at all times. At various stages of power systems, the voltage levels of generated electric power are stepped up or down with the help of transformers. Finally, it is distributed to end-users at required voltage levels. Hence, the power flow in the traditional system is unidirectional. Due to the rapid rise in electricity demand and depletion of fossil fuel resources, the operational challenges in the traditional power systems are increasing. Further, planning new power plants and improving the capacity of transmission lines to keep generation to demand ratio near to unity are difficult tasks to the electric utility because of environmental constraints, rapid growth in population, economic and political reasons. In order to address all these issues, the traditional electric power grid should be upgraded as smart grid.

Smart grid might be expressed as an advanced electric system which incorporates computational intelligence, two-way, cyber-secure information and communication technologies in a combined fashion across different sectors of power systems such as electric power generation, transmission, distribution and consumption in order to realize entire power network that is clean, safe, reliable, secure, efficient, sustainable

S. L. Arun (✉)
School of Electrical Engineering, Vellore Institute of Technology, Vellore, Tamil Nadu 632014, India
e-mail: arun.sl@vit.ac.in

M. P. Selvan
Department of Electrical and Electronics Engineering, National Institute of Technology, Tiruchirappalli, Tamil Nadu 620015, India
e-mail: selvanmp@nitt.edu

B. Vinoth Kumar et al. (eds.), *Intelligent Paradigms for Smart Grid and Renewable Energy Systems*, Algorithms for Intelligent Systems,
https://doi.org/10.1007/978-981-15-9968-2_12

and resilient [1]. Among various smart grid practices, Demand Side Management (DSM) is a promising activity followed by many utilities to regulate the power consumption at consumer premises in a smart grid environment [2]. The key advantage of DSM is easy to implement with the help of inexpensive systems compared to either erecting new generation units or approaching large energy storage devices [3]. DSM scheme majorly consists of efficient energy programs, energy conservation programs, demand management programs, and demand response programs [4].

The efficient energy program incorporates all permanent changes on equipment such as exchanging old incandescent lamps with either Light Emitting Diodes (LED) or Compact Fluorescent Lamps (CFLs); and enhancements on the physical properties of the system such as financing on the building shell for the incorporation of additional insulation [5]. Energy conservation is a part of an efficient energy program and it focuses on consumers and their behavioral changes to attain more efficient energy consumption.

Direct demand control and electricity pricing are two different approaches used in demand management program. In direct demand control, the consumers' appliances are directly controlled by the utility [6, 7]. In electricity pricing technique, the utilities are imposing different pricing schemes while considering the locality demand variation, generation schedule, and economic profit [8]. Simple tariff, flat rate tariff, block rate tariff, demand-based tariff, day ahead tariff [9], time of use tariff, critical peak pricing [10], and Real-Time Pricing [11] are a few energy pricing techniques followed by present utilities. In addition to this, utilities propose a flexible consumer demand limit [12] to progress the peak to average ratio of the utility. Electricity subscribers have to pay an excess amount of their total electricity consumption exceeds the utility predefined demand limit. The alterations carried out by the end-users in response to the utility implemented DSM activities is termed as demand response [13].

Demand response is expressed as modifications in energy consumption by end subscribers from their regular demand patterns as a response to variations in electricity price over time, or to incentive payments planned to induce lesser energy consumption at times of peak market prices or when system consistency is jeopardized [14]. With the help of demand response programs, end subscribers are driven to have direct communication with the utility [15, 16]. Consumers are following different techniques to alter their energy consumption pattern, for example, minimizing their energy consumption through demand reduction strategies, scheduling the operation of appliances to different time periods and using on-site standby generated power. Consumer may attain an appreciable reduction in electricity bills by implementing suitable demand response techniques. Significant price reduction is achieved by reducing consumer demand during peak price intervals. Hence, automation, monitoring, and control technologies of household appliances are essential in a smart grid environment to enrich the consumer's comfort with the minimum electricity bill.

Nowadays utilities are interested to implement real-time pricing techniques along with the time-dependent consumer demand limit to increase the revenue and reliability in utility operation. In this pricing scheme, the electricity price for different time intervals over a day varies and the utilities announce the details of the price variations just before an interval begins. Hence, the residential users schedule most of their demands during low price intervals to decrease their total electricity bill without/with compromising their comfort. Further, residential consumers prefer energy storage devices as an alternate way to manage their critical demands during high price intervals [17]. Compared to all other energy storage mechanisms, battery storage is mostly prepared by residential users due to its easy operation and affordable price [18].

Due to the rapid development and digitalization of industry processes, the electricity demand increases at faster rate. However, the growth in the generation sector is not as fast as demand development. Hence, the generation sectors concentrate more on other alternatives such as Renewable Energy Resources (RER) based power generation to fulfil the ever-increasing demand. However, the penetration of these new energy resources will significantly impact electricity price dynamics [19]. Further, utilities may be influenced by other operational challenges due to the practical difficulties associated with the power generation from RER. Apart from large scale RER power generation, end consumers are motivated by the government to set up small scale in-house RER based power generation like rooftop solar power generation and/or small wind-based power generation to decrease the grid dependency [20]. Considering the availability of RER power generation, residential users can schedule their household appliances timely and they can attain a further reduction in electricity bills [21, 22].

Depend upon the space availability at the installation site and affordable investment cost, residential consumers install their own in-house power generation. The surplus power generation beyond the own need and storage in a battery will be injected into the grid at a price set by the utility. These kinds of consumers are named as prosumers. The prosumers will use maximum of generated RER power and increase their revenue by exporting surplus power to the grid. On the other hand, the utilities are facing additional operational difficulties when many prosumers are willing to inject their surplus power into grid [23, 24]. To avoid such circumstances, a few utilities propose a time-dependent Power Export Limit (PEL) at prosumer premises [25]. The available RER power generation beyond this PEL shall be either transferred to energy storage devices like battery for future use or lost through dump load by the prosumers [26].

2 Types of Household Appliances

Nowadays, residential buildings are equipped with different household appliances. These appliances are smart enough to perform the given task easily and timely. Most of the smart appliances include advanced features like computational intelligence, remote communication with users, and flexible controls [27]. Based on the appliance's usage, they are classified into the following types: Non-Deferrable Essential Loads (NDELs), Non-Deferrable Interruptible loads (NDILs), and Deferrable loads (DLs).

2.1 Non-deferrable Essential Loads (NDELs)

NDELs are primary loads, which need to operate instantly whenever the user initializes them. All-time demand like home security systems; critical demands like fan, lights, laptop/mobile charging, desktop, and its peripherals; entertaining demands like home decorates, speakers, television and kitchen appliances such as induction stove, toaster and mixer are categorized under this type. As the operation of all NDELs in a day is basically depend upon user comfort and desire, controlling of NDELs may bother the well-being of consumers.

2.2 Non-deferrable Interruptible Loads (NDILs)

The temperature-controlled loads are categorized as NDILs. Air conditioner, refrigerator, space heater, and electric water heater are a few examples of NDILs. The operating pattern of NDILs is merely decided by the user comfort and environmental constraints. During the operation of any NDIL, the temperature is maintained at the user predefined set value but within the manufacturer's defined tolerance limit. Whenever the temperature goes beyond the tolerance limit, the NDIL moves to RUN mode and start to consume the rated power. On the other hand, NDIL continues its operation in standby mode when the temperature is within the tolerance limit.

2.3 Deferrable Loads (DLs)

The loads whose operating time can be scheduled in the users' prefixed time span are considered as DLs. Based upon the operational constraint, these DLs are further divided into two types namely: Non-Interruptible and Deferrable Loads (NIDLs) and Interruptible and Deferrable Loads (IDLs). The operation of NIDLs should be continuous till the completion of the task if they started once. Food grinder, cloth washer, and dryer are few examples of this type. However, the operation of IDLs can be either continuous or discontinuous in the user predefined time span. Well pump, electrical vehicle, dishwasher are few examples of this type.

3 Model of Household Components

In smart grid environment, demand response schemes provide considerable economic benefits to end-users. Further, the users can attain more incentives from utility when they adopt smart energy management system. These advanced system reduces the consumer electricity bill by optimally time scheduling the operation of household appliances in consideration with desire and comfort of the consumer. Further, residential consumers install energy storage devices to support their essential demand during high price intervals. In addition to these, few consumers prepare in-house renewable energy resources to reduce grid dependency. Hence, the mathematical modeling of different household components is essential for successful implementation of such energy management system.

3.1 Model of NDELs

The NDELs are usually operated immediately whenever the user initializes them. Hence, the operating pattern of these loads solely depends upon the user desire and necessity. Based on the availability of the user, the power consumption of NDELs may vary. Since the time at which the user initializes these loads in a day is highly random, all the NDELs can be gathered and assumed as a single load whose power consumption will vary dynamically.

3.2 Model of NDILs

Let us denote $\mathbb{C}$ as the set of NDILs available in a residential building and $\mathbb{Q}$ as the set of monitoring interval of non-deferrable demands (NDELs and NDILs). The status vector which represents the mode of operation (RUN/STANDBY) of a NDIL $c\left(c \in \mathbb{C} \triangleq [1, 2, \ldots, c, \ldots, C]\right)$ in each non-deferrable demand monitoring interval $q\left(q \in \mathbb{Q} \triangleq [1, 2, \ldots, Q]\right)$ is expressed as

$$O_c = \left[o_c^1, o_c^2, \ldots, o_c^q, \ldots, o_c^Q\right] \quad \forall c \in \mathbb{C} \tag{1}$$

where C represents the number of NDILs existing in the residential building and Q denotes the maximum number of non-deferrable demand intervals over a day $\left(Q = 24 \cdot \left[\frac{60}{A_{NDL}}\right]\right)$ defined by the user. Here, A_{NDL} is referred to as the time period of a non-deferrable demand interval in minute. When the user setpoint temperature (Hst_c), the manufacturer set allowable tolerance limit ($\Delta\mathrm{Htl}_c$) and the actual temperature at the end of the non-deferrable demand interval $q - 1$ are known than the operating status of NDIL c during interval q can be mathematically expressed as,

If NDIL c is a cooling load

$$o_c^q = \begin{cases} -1 & \text{if } c \text{ is not yet initialized} \\ 0 & \text{if } \mathrm{Hat}_c^{q-1} < \mathrm{Hst}_c \\ 1 & \text{if } \mathrm{Hat}_c^{q-1} > \mathrm{Hst}_c + \Delta\mathrm{Htl}_c \\ o_c^{q-1} & \text{if } \mathrm{Hst}_c \le \mathrm{Hat}_c^{q-1} \le \mathrm{Hst}_c + \Delta\mathrm{Htl}_c \end{cases} \tag{2}$$

If NDIL c is a heating load

$$o_c^q = \begin{cases} -1 & \text{if } c \text{ is not yet initialized} \\ 0 & \text{if Hat}_c^{q-1} > \text{Hst}_c \\ 1 & \text{if Hat}_c^{q-1} < \text{Hst}_c - \Delta\text{Htl}_c \\ o_c^{q-1} & \text{if Hst}_c - \Delta\text{Htl}_c \le \text{Hat}_c^{q-1} \le \text{Hst}_c \end{cases} \tag{3}$$

Considering both NDELs and NDILs, the total demand of non-deferrable demands in interval q is expressed as,

$$P_{\text{NDL}}^q = P_{\text{NDEL}}^q + P_{\text{NDIL}}^q \tag{4}$$

$$P_{\text{NDIL}}^q = \sum_{c=1}^{C} P_c^q \tag{5}$$

$$P_c^q = \begin{cases} 0 & \text{if } o_c^q = -1 \\ \text{SP}_c & \text{if } o_c^q = 0 \\ \text{RP}_c & \text{if } o_c^q = 1 \end{cases} \tag{6}$$

where, P_{NDEL}^q and P_{NDIL}^q are the aggregated power demanded by all NDELs and NDILs during interval q, respectively. SP_c and RP_c represent the stand-by power and rated power of the NDIL c, respectively.

3.3 Model of DLs

As the operation DLs can be optimally scheduled anywhere in the user pre-defined time span, they are playing a crucial role in demand response. Let us denote $\mathbb{D}$ as the set of available deferrable loads in the residential building and $\mathbb{R}$ as the set of deferrable demand intervals over a day. The scheduling vector (L_d) which represents the status (ON/OFF) of each DL $d\left(d \in \mathbb{D} \triangleq [1, 2, \ldots, D]\right)$ in each deferrable demand interval $r\left(r \in \mathbb{R} \triangleq [1, 2, \ldots, R]\right)$ can be expressed as,

$$L_d = \left[l_d^1, l_d^2, \ldots l_d^r, \ldots, l_d^R\right] \quad \forall d \in \mathbb{D} \tag{7}$$

where, D represents the existing number of DLs in the building and R represents the maximum number of deferrable demand intervals over a day $\left(R = 24 \cdot \left[\frac{60}{A_{\text{DL}}}\right]\right)$. Here, A_{DL} is the time period of a deferrable demand interval in minute defined by the user. Each component of scheduling vector l_d^r is described as,

$$l_d^r = \begin{cases} 0 \text{ if load } d \text{ is OFF} \\ 1 \text{ if load } d \text{ is ON} \end{cases} \quad \forall\ d \in \mathbb{D};\ \ r = 1, 2, \ldots, R \tag{8}$$

In order to have flexible control on DLs, the residents are expected to share the information about the initialization interval α_d (interval at which the DL d is added in the process of scheduling) and dead time interval σ_d (interval at which the task of appliance d should be finished) for each deferrable load. This sharing of information can be performed with the help of user interface module or direct settings available in the load. Any scheduling algorithm will schedule the initiated DLs only between these two-time intervals. Nowadays smart residential loads are manufactured with artificial intelligence to compute the actual number of intervals needed to finish the task τ_d (computation interval for DL d) during starting itself based on the initial conditions like available water level in water tank (if the load being a smart well pump) and weight of clothes (if the load being a smart cloth washer). The critical constraint that should be considered by a user during selection of initialization and dead time intervals is given as,

$$\tau_d \leq \sigma_d - \alpha_d \quad \forall d \in \mathbb{D} \tag{9}$$

Based on the operating nature of the work (continuous or discontinuous), the deferrable loads are further divided into two types. This classification will be distinguished by considering the preemptive status which will be set by the user. Let ϕ_d represents the deferrable load d preemptive status and its value is expressed as follows:

$$\phi_d = \begin{cases} 0 \text{ for interrptive loads (IDLs)} \\ 1 \text{ for non-interruptive (NIDLs)} \end{cases} \tag{10}$$

The total power consumption pattern of all DLs in a particular demand interval varies as the number of running DLs varies. Hence, the total power demand $\left(P_{\text{DL}}^r\right)$ by all DLs for a given deferrable demand interval r can be expressed as the function of the status of DL and the power rating of DL (RP_d).

$$P_{\text{DL}}^r = \sum_{d=1}^{D} \left(l_d^r \cdot \text{RP}_d\right) \tag{11}$$

3.4 *Model of Battery*

The battery energy storage helps the residents to reduce the electricity bill by discharging the stored energy during high price intervals. Further, charging the battery during less price intervals may reduce the electricity bill significantly. Hence, the mathematical modeling of battery is needed for optimal scheduling of battery operation. The operating mode of battery (charging mode/floating mode/discharging mode) and value of power exchange are considered to be the controllable parameters of battery. Practically, the battery is assumed to be an additional deferrable demand during charging mode whereas it is considered to be an additional resource during discharging mode.

Let us define the set of battery operating intervals as $\mathbb{U}$. The operating vector representing the operating mode of battery (charging mode $\left(S_c^u\right)$, floating mode $\left(S_f^u\right)$ and discharging mode $\left(S_d^u\right)$) during battery operating interval u, $\left(u \in \mathbb{U} \triangleq [1, 2, \ldots, U]\right)$, is defined as

$$\mathcal{S} = \left[S^1, S^2, \ldots S^u, \ldots, S^U\right] \tag{12}$$

$$S^u = \left(S_c^u, S_f^u, S_d^u\right) \tag{13}$$

where U is the maximum number of battery operating intervals in a day $\left(U = 24 \cdot \left[\frac{60}{A_{BS}}\right]\right)$. Here, A_{BS} is the time period of a battery operating interval in minute set by the user as per the suggestions received from the manufacturer for extending the battery life. Each component of the battery operating vector in interval u is expressed as,

$$S^u = \left(S_c^u, S_f^u, S_d^u\right) = \begin{cases} (1, 0, 0) & \text{if battery in charging} \\ (0, 1, 0) & \text{if battery in floating} \\ (0, 0, 1) & \text{if battery in discharging} \end{cases} \tag{14}$$

Controllable parameters of the battery can be decided by the available State of Charge (SoC). The SoC describes the available charge level of the battery which is related with its capacity. During starting of each battery operating interval u, available SoC ($\mathcal{X}_u$) is calculated by using (15).

$$\mathcal{X}_u = \left(\text{Cap}(u-1) + \zeta_{\text{Bat}}\left(\frac{P_S^u}{V_{\text{Bus}}}\right)\left(\frac{A_{\text{BS}}}{60}\right)\right) \cdot \left(\frac{1}{M\text{Cap}}\right) \tag{15}$$

where, ζ_{Bat} is the battery round trip efficiency, V_{Bus} is the voltage at the DC bus where battery is connected and MCap is the battery rated Ampere-hour capacity. The battery output power $\left(P_S^u\right)$ in interval u, is calculated using charging power $\left(P_{SC}^u\right)$ and discharging power $\left(P_{SD}^u\right)$ of the battery. It is mathematically shown in Eq. (16)

$$P_S^u = \left(1 - S_f^u\right)\left(S_c^u \cdot P_{\text{SC}}^u - S_d^u \cdot P_{\text{SD}}^u\right) \tag{16}$$

3.5 Model of RER

Nowadays, residential consumers give more attention to in-house RER generation to reduce grid dependency by meeting their own demand. However, the power generation by renewable energy resources is highly irregular and site-dependent. Among various power generation methods, the residential consumers are likely preferred to install rooftop solar PV and small wind turbine-based power generation. The amount of power generation from solar PV is highly influenced by available solar irradiation and atmospheric temperature and it is expressed as,

$$P_{\mathrm{PV}}^{j} = f_{\mathrm{PV}} P_{\mathrm{STC}} \left(\frac{G_{\mathrm{Avg}}^{j}}{G_{\mathrm{STC}}} \right) \left(1 + \left(T_{C}^{j} - T_{\mathrm{STC}} \right) \cdot C_{T} \right) \tag{17}$$

$$T_{C}^{j} = T_{A}^{j} + \left(\frac{\mathrm{NOCT} - 20}{0.8} \right) \cdot G_{A}^{j} \tag{18}$$

where, f_{PV} is the solar PV panel de-rating factor, P_{STC} is the nominal PV array power in kW under Standard Test Condition (STC), G_{Avg}^{j} is the averaged solar irradiation during RER interval $j\left(j \in \mathcal{J} \triangleq [1, 2, \ldots, J]\right)$ in kW/m^2. Here, J is maximum number of RER intervals in a day $\left(J = 24 \cdot \left[\frac{60}{A_{\mathrm{RER}}}\right]\right)$ and A_{RER} is the time period of a RER interval in minute set by the consumer as per the required accuracy. G_{STC} is the rated solar irradiation under STC (1 kW/m^2), T_{C}^{j} is the solar PV cell temperature during RER interval j in °C, T_{STC} is the rated temperature at STC in °C, C_T is the temperature coefficient of solar panel, NOCT is the normal operating cell temperature in °C and T_{Avg}^{j} is the ambient temperature averaged over a RER interval j in °C.

The power generation from wind turbine is highly dependent upon the wind speed and is computed using Eq. (19)

$$P_{W}^{j} = 0.5 \rho A_{w} (\nu^{j})^{3} C_{p} \tag{19}$$

where, P_{W}^{j} is the power generated (kW) by wind turbine in RER interval j, ρ is the air density $\left(\mathrm{kg/m^3}\right)$, A_w is the swept area (m^2), ν^j is the averaged wind velocity (m/s) during the RER interval j and C_p is the power coefficient. The aggregated power generated from RER during interval j is the sum of power generated by in-house solar PV and wind turbine during that particular interval and it is expressed as,

$$P_{\mathrm{RER}}^{j} = P_{\mathrm{PV}}^{j} + P_{W}^{j} \tag{20}$$

4 Optimal Sizing of RER

As discussed earlier, consumers can attain more economic benefits by utilizing the maximum amount of generated RER power. However, the excess power generated beyond the consumer needs can be easily managed by battery storage for future purposes or lost through dump load. Hence, random selection of RER and energy storage devices during installation may lead to more power loss through dump load. In order to avoid such power loss, the components of in-house renewable resources such as number of solar PV panels, wind turbines, and batteries have to be optimally chosen to attain a considerable reduction in capital investment cost and grid dependency. This problem is mathematically formulated as optimization problem with fitness function as minimization of the Cost of Energy (CoE) subject to various constraints. The fitness function is shown in Eq. (21).

$$\min\,(\mathrm{CoE}) = \min\left(\frac{C_{\mathrm{CC}} + C_{\mathrm{AR}} + C_{\mathrm{OM}}}{E_{\mathrm{TAR}}}\right) \tag{21}$$

where, C_{CC} is the total annual capital cost of all the elements, C_{AR} is the total annual replacement cost of all the elements, C_{OM} is the total annual operation and maintenance cost of all the elements and E_{TAR} is the expected total energy to be generated by RER over a project year. The annual capital cost C_{CC} can be computed as

$$C_C = (C_{\mathrm{CS}} + C_{\mathrm{CW}} + C_{\mathrm{CB}}) \cdot \Phi(\varepsilon, \tau) \tag{22}$$

$$\Phi(\varepsilon, \tau) = \left[\frac{\varepsilon(1+\varepsilon)^{\tau}}{(1+\varepsilon)^{\tau} - 1}\right] \tag{23}$$

where, C_{CS}, C_{CW} and C_{CB} are the initial investment costs of rooftop solar power generation, wind power generation, and battery bank, respectively. Further, the capital recovery factor Φ can be expressed as a function of annual interest rate (ε) and the project lifetime (τ) as expressed in Eq. (23). The aggregated annualized replacement cost (C_R) is expressed as

$$C_R = (C_{\mathrm{RS}} + C_{\mathrm{RW}} + C_{\mathrm{RB}}) \cdot \Phi(\varepsilon, \tau) \tag{24}$$

where, C_{RS} and C_{RW} represent the replacement costs of solar and wind power generation systems, respectively. C_{RB} represents replacement costs of battery storage systems. The aggregated annual operation and maintenance cost (C_{O}) is given in (25).

$$C_{\mathrm{O}} = C_{\mathrm{OS}} + C_{\mathrm{OW}} + C_{\mathrm{OB}} \tag{25}$$

where, C_{OS}, C_{OW}, and C_{OB} are the aggregated annual costs to be spent towards the operation and maintenance of solar and wind power generation systems and battery banks, respectively. The expected energy to be generated by in-house RER over a year can be obtained using (26).

$$E_{\mathrm{TAR}} = \sum_{j=1}^{J\cdot 365} \left(P_{\mathrm{PV}}^{j} + P_{W}^{j}\right)\left(\frac{A_{\mathrm{RER}}}{60}\right) \tag{26}$$

The objective function described in Eq. (21) is subjected to various constraints.

4.1 Capital Investment Cost Constraint

The total capital costs of in-house RER elements and battery banks should be maintained below the maximum permissible capital cost which is fixed by the user as per his/her wish.

$$\mathrm{CC} \leq \mathrm{CC}_{\max} \tag{27}$$

where CC and $\mathrm{CC}_{\max}$ are the optimal and maximum capital cost of the project, respectively.

4.2 Elements Size Constraint

The size of each element of the project such as number of PV panels (N_{PV}), wind turbines (N_{WT}) and battery banks (N_{BB}) should be within the user predefined maximum permissible limits. Generally, user will decide these maximum limits by considering the space availability and geographical constraints at the installation site.

$$\begin{aligned} 0 &\leq N_{\mathrm{PV}} \leq N_{\mathrm{PVM}} \\ 0 &\leq N_{\mathrm{WT}} \leq N_{\mathrm{WTM}} \\ 0 &\leq N_{\mathrm{BB}} \leq N_{\mathrm{BBM}} \end{aligned} \tag{28}$$

where, N_{PVM}, N_{WTM} and N_{BBM} are the consumer predefined maximum number of PV panels, wind turbines, and battery banks, respectively.

4.3 Generation and Demand Constraint

The consumer demand met by the power generated from renewable resources over a project year (AD_{RER}) must be greater than or equal to the required fraction of the aggregated yearly demand of the consumer.

$$\frac{AD_{RER}}{AD_{ACT}} \geq \Psi \tag{29}$$

where Ψ is represented as the RER energy fraction which differs from 0 to 1. AD_{ACT} is the aggregated yearly demand of the consumer.

5 Novel Residential Energy Management System

In smart grid, demand response is one of the most effective technique which will be developed in a way beneficial to both consumer and utility. As part of demand-side management scheme, utilities adopt various pricing schemes to have control over energy consumption at consumer premise. In Real-Time Pricing (RTP) scheme, the utilities impose price variations for different time intervals to maintain the system load curve almost flat. Hence, electricity price variations and penalty imposed, if used above the allowed consumer demand limit are taken as major parameters for the energy management in demand response programs.

As discussed earlier, the residential consumers are expected to schedule their operation of household appliances during low price intervals and hence the savings in electricity bills will improve further. This scheduling may be done with due consideration to the consumer's well-being, intermittence in the RER power generation, and dynamics in the behavior of consumer and utility. Manual demand response in which the user has to analyze the dynamics of the utility and operate the appliances is highly inefficient and brings more practical difficulties. In order to support the user in the most economical way, a dedicated intelligent demand response system has to be installed in the residential buildings. It requires the ability to study the dynamics in consumer behavior, renewable power generation, and utility operations. Further, these systems are expected to take decision on controlling the household appliances on its own. One such system is Novel Residential Energy Management System (NREMS). The architecture of NREMS is depicted in Fig. 1.

NREMS includes smart NDEL module, smart NDIL module, smart DL module, and smart power converter module to assist the transfer of data and control instructions between the smart computational unit and NDELs, NDILs, DLs, and power conditioning unit of RER, respectively. The NREMS obtains instant updates of utility parameters such as energy price, Consumer Demand Limit (CDL), and power export limit through the smart meter interface. Generally, utilities and smart meters which are installed in the residential building are featured with two-way communication

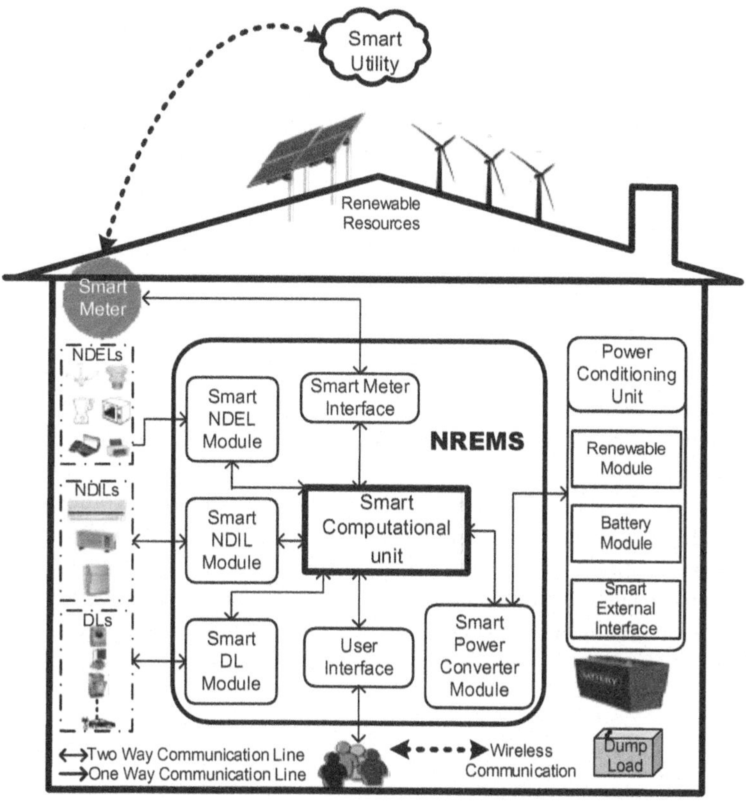

Fig. 1 Architecture of NREMS

facility to send or receive the information. Smart NDEL module combines the demand of all NDELs and displays the warning sign when the aggregated demand of NDELs crosses the consumer predefined limit. The functional parameters of NDIL such as user-defined setpoint temperature, manufacturer allowable tolerance limit, power ratings (power consumption during RUN and STAND-BY operation), and user operating status of INDLs (switch ON/OFF) are gathered by smart NDIL module. Further, it conveys the controlling instructions such as RUN or STANDBY suggested by the smart computational unit to individual NDILs. Smart DL module gets the load operational parameters like load initialization interval (σ_d), load dead time interval (τ_d), computation intervals (α_d) and preemptive status from all DLs and provides the control signals to DLs as per the instruction received from the smart computational unit.

Further, the present status of battery SoC is also computed by the battery module. The SoC information is conveyed to smart computational unit of NREMS for scheduling the battery operation. NREMS dictates the operating mode of battery and value of power exchange of battery to the power conditioning unit through smart power converter module. Power conditioning unit controls the operation of battery. In order to attain maximum benefit from RER, the operating features of smart power converter module have been extended to collect the RER power generation information and transfer to NREMS smart computational unit. Using this present and past history information, NREMS forecasts the RER power availability for future intervals. By considering the actual and forecasted renewable power, NREMS performs the scheduling of different household appliances. The user interface module which is presented in NREMS gathers the user-defined parameters such as NDELs power consumption limit and NDILs extended tolerance limit. Further, this interface module used to show the warning signs and other details like consumer electricity bill, aggregated power demand by different types of loads.

The NREMS control/schedule the operation of household appliances on basis of their operating nature. The full-time horizon of NREMS (24 h) has been separated into non-deferrable demand interval duration (A_{NDL}), deferrable demand interval duration (A_{DL}), battery scheduling interval duration (A_{BS}) and RER interval duration (A_{RER}). Generally A_{NDL} and A_{RER} are chosen to have lesser value in order to properly consider the practical dynamics in the non-deferrable demands and intermittence in power generation of RER, respectively. Effective operation of NREMS will be possible only when the intervals durations are maintained as expressed in (30). The steps associated with the working of the NREMS are represented as a flowchart in Fig. 2.

$$A_{\text{RER}} \leq A_{\text{NDL}} \leq A_{\text{BS}} \leq A_{\text{DL}} \leq A_{\text{P}} \tag{30}$$

5.1 Controlling of NDELs

Since the operating pattern of NDELs simply depends upon the requirement and comfort of the consumers, the NREMS does not have any impact on these loads. However, NREMS is featured to display a warning sign when the combined power rating of all working NDELs goes beyond the consumer set limit.

5.2 Controlling of NDILs

NREMS offers additional control on NDILs dependent upon the present status of individual NDILs and thermal dynamics of the building. The manufacturer defined tolerance limit of NDILs can be extended by a user set value. However, user may assign this value to be zero when the user gives more importance to comfort rather than electricity bill reduction. If the actual temperature of the working environment (Hat_c^q) varies in between the manufacturer tolerance limit (ΔHtl_c) and the user set extended tolerance limit (ΔHel_c), then the NREMS takes a decision to either turn

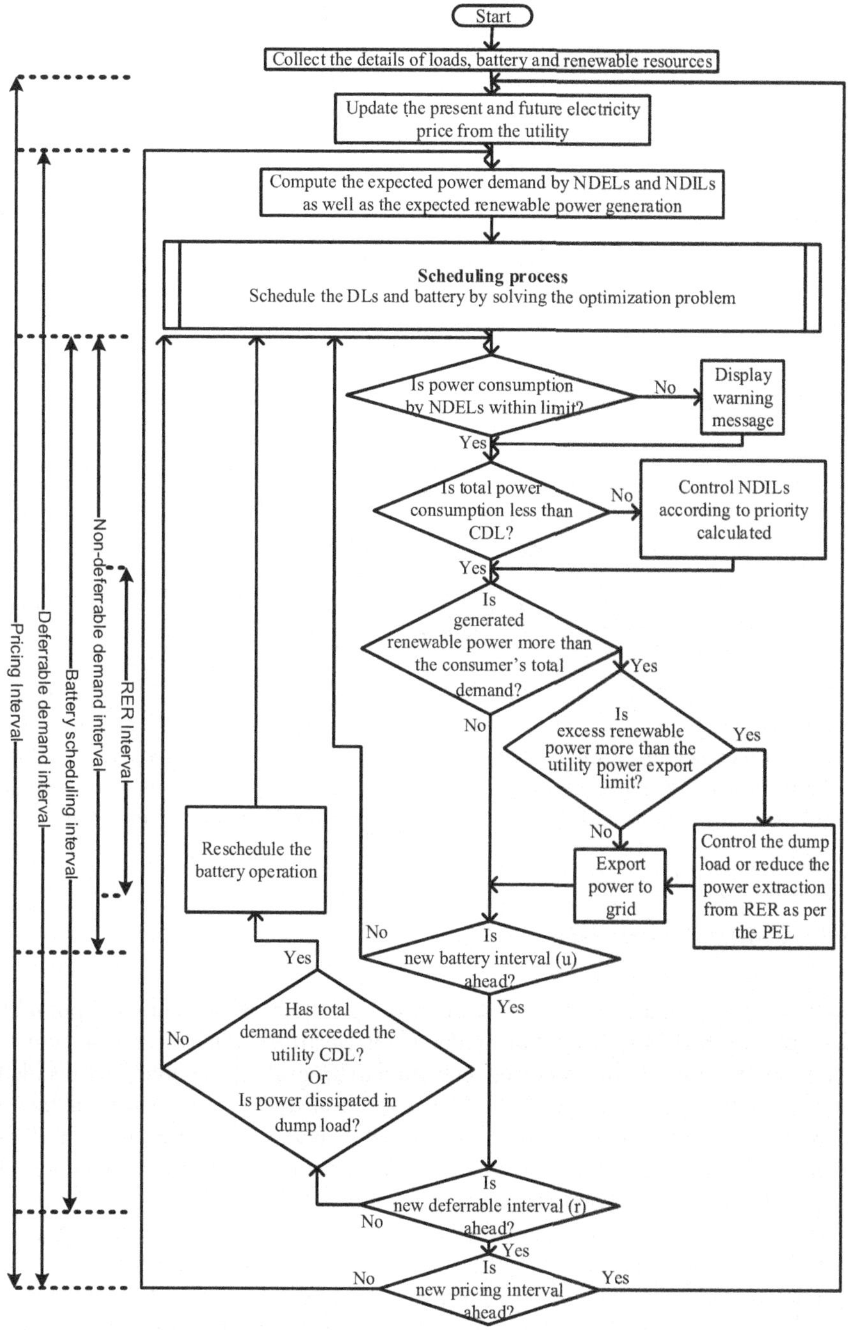

Fig. 2 Functional flowchart of NREMS

ON the NDIL or continue in stand-by mode on basis of the consumer demand limit. If the number of working NDILs increases, NREMS schedules them according to the priority of each NDIL. The priority of each NDIL c is expressed as,

If load c is a cooling NDIL,

$$\digamma_c^q = \begin{cases} 0 & \begin{cases} \text{if NDIL } c \text{ is not yet started} \\ \text{else if } \text{Hat}_c^{q-1} < \text{Hst}_c \end{cases} \\ 1 - \Delta\digamma_c^q & \text{otherwise} \end{cases} \tag{31}$$

$$\Delta\digamma_c^q = \frac{\text{Hst}_c + \Delta\text{Htl}_c + \Delta\text{Hel}_c - \text{Hat}_c^q}{\Delta\text{Htl}_c + \Delta\text{Hel}_c} \tag{32}$$

If load c is a heating NDIL,

$$\digamma_c^q = \begin{cases} 0 & \begin{cases} \text{if NDIL } c \text{ is not yet started} \\ \text{else if } \text{Hat}_c^{q-1} > \text{Hst}_c \end{cases} \\ 1 - \Delta\digamma_c^q & \text{otherwise} \end{cases} \tag{33}$$

$$\Delta\digamma_c^q = \frac{\text{Hat}_c^q - (\text{Hst}_c - \Delta\text{Htl}_c - \Delta\text{Hel}_c)}{\Delta\text{Htl}_c + \Delta\text{Hel}_c} \tag{34}$$

The operation of NDILs is immediately scheduled by NREMS if the priority of NDILs greater than or equal to 1. To avoid penalty payment, the NREMS considers the consumer demand limit and operates the remaining NDILs in the descending order of priority. When the total demand of the user exceeds the consumer demand limit, the operation of least priority NDILs is moved to forthcoming intervals.

5.3 *Controlling of DLs*

As the DLs are featured with flexible mode of operations, these loads are optimally scheduled by NREMS to avail more reduction in electricity bills. Further, prosumers can export their surplus RER power generation to the grid at profitable price decided by the utility. In order to get more incentives from utility and utilize maximum amount of generated renewable power, the DLs should be timely operated. This optimal scheduling problem is mathematically formulated with fitness function of minimization of net electricity bill payable to utility. This fitness function is subjected to various soft and hard operational constraints. In order to consider the impact of real-time modifications made by either consumer or utility, the NREMS performs the optimal scheduling problem only in between the current deferrable demand interval and highest dead time interval (r_{md}) of all DLs which are started but not finished its task. Hence, the number of intervals which is taken for the optimal scheduling process differs dynamically in every time interval.

Let $\mathcal{I}$ represents a dynamic set and contents of $\mathcal{I}$ varies dynamically according to the present operating interval (r) and the total number of initialized DLs.

$$\mathcal{I} = [r, r+1, \ldots, r_{md}] \tag{35}$$

The objective function of NREMS optimization problem can be formulated as expressed in (36).

$$\min\left(\sum_{\mathrm{i}} \Gamma_U^{\mathrm{i}}\left(E_{\text{tot}}^{\mathrm{i}}\right)\right) \quad \forall\ \mathrm{i} \in \mathcal{I} \tag{36}$$

$$\Gamma_U^{\mathrm{i}}\left(E_{\text{tot}}^{\mathrm{i}}\right) = \begin{cases} \text{if } \ P_{\text{Net}}^i > 0 & \text{if } \ 0 < P_{\text{Net}}^i < P_{\text{CDL}}^i \\ \Gamma_S^i \cdot P_{\text{Net}}^i \cdot \Delta i & \text{if } \ P_{\text{Net}}^i \geq P_{\text{CDL}}^i \\ \left(\begin{array}{c} \Gamma_S^h \cdot P_{\text{CDL}}^i \cdot \Delta i + \\ \Gamma_P^i \cdot \left(P_{\text{Net}}^i - P_{\text{CDL}}^i\right) \cdot \Delta i \end{array}\right) & \\ \text{if } P_{\text{Net}}^i \leq 0 & \text{if } \ 0 < \left|P_{\text{Net}}^i\right| < P_{\text{EGmax}}^i \\ \Gamma_B^i \cdot P_{\text{Net}}^i \cdot \Delta i & \\ \Gamma_B^i \cdot \left(-P_{\text{EGmax}}^i\right) \cdot \Delta i & \text{if } \ \left|P_{\text{Net}}^i\right| \geq P_{\text{EGmax}}^i \end{cases} \tag{37}$$

$$\Delta i = \frac{A_{\text{DL}}}{60} \tag{38}$$

$$P_{\text{Net}}^i = P_{\text{NDEL}}^i + P_{\text{NDIL}}^i + P_{\text{DL}}^i + P_S^i - P_{\text{RER}}^i \tag{39}$$

where i is a component of $\mathcal{I}$ and E_{tot}^i represents the net energy utilized by all household components during time interval i. Γ_U^{i} represents the utility price function for time interval i. Further, Γ_S^i and Γ_P^i are utility energy selling prices for users' normal and excess power consumption, respectively. Γ_B^i is the utility energy purchasing price for time interval i. P_{RER}^i and P_{EGmax}^i are the expected RER power and utility defined power export limit during interval i. The objective function defined in (36) is subjected to various soft and hard operational constraints.

5.3.1 Load Operating Constraint

The components of the deferrable demand scheduling vector (l_d^r) which represents the operating status of the DL d should be OFF in the rth deferrable demand interval if the interval r does not exist in user-defined DL time span $[\alpha_d, \sigma_d]$. Since the time span of DLs are fixed by user, this constraint is considered to be hard constraint and it is mathematically formulated as,

$$\begin{aligned} l_d^r &= 0; \quad r < \alpha_d, \ \forall d \in \mathbb{D} \\ l_d^r &= 0; \quad r > \sigma_d, \ \forall d \in \mathbb{D} \end{aligned} \tag{40}$$

5.3.2 Scheduling Interval Constraint

All the DLs must be optimally planned for the fixed number of computational intervals. The total intervals counts that DL d be planned in ON status at any deferrable demand interval r must be equal to the number of intervals needed to finish the task by load d from rth interval. This hard constraint is given in (41).

$$\sum_{k=r}^{\sigma_d} l_d^k = \Upsilon_d^k \quad \forall d \in \mathbb{D} \tag{41}$$

5.3.3 Interruption Constraint

As discussed earlier, the DLs are further divided into two types on basis of their operating nature. The NIDLs are non-preemptable loads and they are expected to run continuously till finishing the task. If the NIDL d is started to consume nominal power, the τ_d number of deferrable demand intervals should be reserved sequentially within $[\alpha_d, \sigma_d]$. On the other hand, the IDLs are preemptable loads, and hence the τ_d number of deferrable demand intervals may be reserved in any manner within $[\alpha_d, \sigma_d]$. This constraint can be expressed as a hard constraint and shown in (42).

$$\sum_{\varepsilon=0}^{\varphi-1} \prod_{\varrho=\alpha_d+\varepsilon}^{\alpha_d+\tau_d+\varepsilon-1} l_d^{\varrho} \phi_d = \phi_d \tag{42}$$

$$\varphi = \sigma_d - \alpha_d - \tau_d + 2 \tag{43}$$

5.3.4 Power Consumption Constraint

The net power consumption of the consumer during every deferrable demand interval (r) may be within the utility defined consumer demand limit (P_{CDL}^r) to avoid excess payment. However, maintain net power consumption under the consumer demand limit at all the intervals may sometimes disturb the well-being of the consumer. Hence, the power consumption constraints are formulated as soft constraint and it is mathematically represented in (44).

$$\begin{cases} P_{\mathrm{NDEL}}^r + P_{\mathrm{NDIL}}^r + P_{\mathrm{DL}}^r + P_S^r - P_{\mathrm{RER}}^r \leq P_{\mathrm{CDL}}^r \\ \frac{1}{\vartheta^{r+1}}\left(P_{\mathrm{DL}}^{r+1} + P_S^{r+1} - P_{\mathrm{RER}}^{r+1}\right) \leq P_{\mathrm{CDL}}^{r+1} \\ \vdots \\ \frac{1}{\vartheta^{r_{md}}}\left(P_{\mathrm{DL}}^{r_{md}} + P_S^{r_{md}} - P_{\mathrm{RER}}^{r_{md}}\right) \leq P_{\mathrm{CDL}}^{r_{md}} \end{cases} \tag{44}$$

where, ϑ^r is the multiplication factor for non-deferrable demands at rth deferrable demand interval. Further, the expectable power demand of NDELs for the current deferrable demand interval (r) is assumed to be that of last non-deferrable demand interval $(q-1)$. The expectable demand of NDILs for the current deferrable demand interval is anticipated by NREMS using temperature variations in the considered residential building. Further, NREMS may schedule the operation of DLs non-optimally when the impact of non-deferrable demand either not considered or considering the similar demand pattern in all upcoming deferrable demand intervals. Hence, the multiplication factor is considered to keep a part of consumer demand limit for supporting the dynamics in the non-deferrable demand. Practically, this multiplication factor changes between 0 and 1. Further, the value of this factor is decided by user based upon history data of power consumption and desire of the user.

A subscriber can pay the nominal electricity bill if the net demand of the user is kept under consumer demand limit. Whenever the net demand exceeds utility predefined limit, the users are expected to pay penalty which will significantly increase the consumer electricity bill. However, keeping the net residential demand below demand limit is not a mandatory task. The practical reasons for net power consumption exceed demand limit are listed as (1) The DLs are assumed to be uninterruptable if they are assigned to start at a deferrable demand interval (r). Unexpected changes in the operating pattern of non-deferrable demands may increase the net demand which will drive the total power consumption above the demand limit. (2) When more number of DLs are initialized and the duration between the load initialization interval and dead time intervals is short, then many DLs are scheduled to start simultaneously.

5.3.5 Battery Mode of Operation Constraint

The scheduled operating mode of the battery for any deferrable load-interval must be unique (Charging mode/floating mode/discharging mode). This battery operational constraint is mathematically expressed as a hard constraint and shown in (45).

$$S^r = S_c^r + S_f^r + S_d^r = 1 \tag{45}$$

5.3.6 Battery Boundary Limit Constraints

The operational parameters of battery such as state of charge, power exchange during charging, and discharging mode must be kept between the manufacturer defined battery boundary limits to preserve the life of battery. These boundary constraints are developed as hard constraints as presented in (46)–(48).

$$\mathcal{X}_{\min} \leq \mathcal{X}^r \leq \mathcal{X}_{\max} \tag{46}$$

$$P_{\text{SCmin}} \leq P_{\text{SC}}^{r} \leq P_{\text{SCmax}} \quad (47)$$

$$P_{\text{SDmin}} \leq P_{\text{SD}}^{r} \leq P_{\text{SDmax}} \quad (48)$$

where, $\mathcal{X}^r$ is the battery state of charge existing at the starting of deferrable demand interval r, $\mathcal{X}_{\text{max}}$ and $\mathcal{X}_{\text{min}}$ are the boundary limits for state of charge. P_{SCmax} and P_{SCmin} are the boundary power limits for battery charging. P_{SDmax} and P_{SDmin} are the boundary power limits for battery discharging.

5.3.7 Power Export Constraint

In the present scenario, the utilities are facing additional operational difficulties due to large penetration of in-house grid-connected RER. Hence, the utilities are expected to impose a Power Export Limit (PEL) at prosumers premises. Any prosumer can significantly enjoy the benefits by exporting the surplus power into grid without crossing PEL. The power generated beyond PEL shall be transferred to battery storage for future use or lost through a dump load. Further, NREMS instructs the RER power conditioning unit to add more dump load or reduce the RER power generation when the total generation (RER power generation—Residential demand) exceeds the utility set PEL. This perspective adds up a constraint expressed as prosumer grid export power $\left(P_{\text{EG}}^{r}\right)$ at any deferrable demand interval, r shall be less than or equal to the utility set power export limit $\left(P_{\text{EGmax}}^{r}\right)$.

$$P_{\text{EG}}^{r} \leq P_{\text{EGmax}}^{r} \quad (49)$$

$$P_{\text{EG}}^{r} = P_{\text{RER}}^{r} - \left(P_{\text{NINSL}}^{r} + P_{\text{INSL}}^{r} + P_{\text{SL}}^{r} + P_{B}^{r}\right) \quad (50)$$

5.4 *Controlling of Battery*

When the time period of battery operating interval and deferrable demand interval are same then the battery follows the operational instructions such as operating mode (charging mode floating mode/discharging mode) and value of power exchange as suggested by the optimal scheduling algorithm. When the time period of these two intervals differs then changing the battery operational parameters within a deferrable demand interval will significantly improve the electricity bill savings. However, this modification needs to be done according to the operational dynamics of other types of household appliances, utility defined consumer demand limit, and intermittency in RER power generation.

Let us consider that the optimal scheduling algorithm of NREMS sets the operating mode of battery as charging and fix certain value of power which needs to be transferred from grid to battery for a specific deferrable demand interval. However, the aggregated demand of NDELs and NDILs is increased suddenly due to change in number of available persons. In this scenario, if the operating mode of battery for the complete deferrable demand interval is kept as charging, the aggregated demand of the user may exceed the consumer demand limit and hence the consumer electricity bill will increase further. In order to avoid these circumstances, the battery has to be rescheduled within a deferrable demand interval. However, reconfiguring the battery operation in the time period of non-deferrable demand intervals may severely affect the life span of battery. In order to extend the battery life, NREMS modify the operation of battery only in every battery operating interval (u, where $q < u < r$). The rescheduling battery will be done on the basis of RER power generation during the battery operating interval $(u - 1)$ which exceeds the user's own demand $\left(P_{\text{EG}}^{u-1}\right)$, aggregated demand of user $\left(P_{\text{Net}}^{u-1}\right)$ which is a sum of power demanded by various types of household appliances (NDELs, NDILs, and DLs), the scheduled battery power exchange, and RER power generation during interval $(u - 1)$. The rescheduling of battery operation is mathematically shown in (51).

$$
(S_c^u, S_f^u, S_d^u) = \begin{cases} (1,0,0); \text{ if} \begin{cases} P_{\text{EG}}^{u-1} > P_{\text{EGmax}}^{u-1} \\ \mathcal{X}^u < \mathcal{X}_{\text{max}} \end{cases} \\ (0,1,0); \text{ if} \begin{cases} \begin{cases} P_{\text{EG}}^{u-1} > P_{\text{EGmax}}^{u-1} \\ \mathcal{X}^u \geq \mathcal{X}_{\text{max}} \end{cases} \\ \text{or} \\ \begin{cases} P_{\text{Net}}^{u-1} > P_{\text{CDL}}^{u-1} \\ \mathcal{X}^u \leq \mathcal{X}_{\text{min}} \end{cases} \end{cases} \\ (0,0,1); \text{ if} \begin{cases} P_{\text{Net}}^{u-1} > P_{\text{CDL}}^{u-1} \\ \mathcal{X}^u > \mathcal{X}_{\text{min}} \end{cases} \end{cases} \tag{51}
$$

6 Case Study

The case study presumes that the residential building consists of various household appliances. These appliances are categorized into NDELs, NDILs and DLs and are tabulated in Tables 1, 2, and 3 along with their power ratings and time of operation [28].

The building is supported by battery storage. Among various battery types, lead-acid batteries are commonly preferred because of their attractive features like comparatively low price, less capital investment, market availability, good performance characteristics, and better life span [29]. The specifications of battery are listed in Table 4.

The considered residential building is also equipped with in-house RER such as solar and wind power generation. Each PV panel and wind turbines are rated as

Table 1 Non-deferrable and essential loads (NDELs)

S. No.	Appliances	Rated power (kW)	Operating hours (h)	Quantity
1	Fan	0.10	00.00–06.00	4
			06.00–09.00	2
			17.00–21.00	2
			21.00–24.00	4
2	Fluorescent lamp	0.04	05.00–07.00	3
			18.00–22.00	6
3	CFL	0.02	00.00–05.00	4
			05.00–07.00	8
			18.00–22.00	8
			22.00–24.00	4
4	Television (TV)	0.25	06.00–08.00	1
			17.00–22.00	1
5	Laptop/Mobile charging	0.05	06.00–08.00	2
			17.00–19.00	2

Table 2 Non-deferrable and interruptible loads (NDILs)

S. No.	Appliances	Rated power (kW)	Operating hours (h)
1	Air conditioner (AC)	1.0	00.00–05.00
			17.00–19.00
			21.00–24.00
2	Water heater	2.0	06.00–09.00
			18.00–22.00
3	Refrigerator	0.5	00.00–24.00

Table 3 Deferrable loads (DLs)

S. No.	Load	Power (kW)	ϕ_d	Without NREMS start (h)	Time span		τ_d
					α_d	σ_d	
1	Cloth washer	0.8	1	16.00	08.00	13.00	2
2	Cloth dryer	2.2	1	18.00	13.00	19.00	1
3	Dish washer	1.5	0	08.00	08.00	12.00	1
				17.00	14.00	18.00	1
				21.00	21.00	24.00	1
4	Well pump	1.2	0	05.00	00.00	06.00	1
				17.00	09.00	18.00	1
5	Electrical, vehicle, charging	2.3	0	05.00	00.00	05.00	2
				21.00	21.00	24.00	1
6	Grinder	0.5	1	17.00	13.00	18.00	1

Table 4 Specifications of battery

S. No.	Specifications	Rating
1	Capacity (Ah)	75
2	DC voltage (V)	12
3	Efficiency during charging (%)	85
4	Efficiency during discharging (%)	95
5	State of charge boundary limit (%)	(30–90)
6	Current limit during charging	(5–20) of nominal capacity
7	Current limit during discharging	(0–20) of nominal capacity

Table 5 Specifications of solar PV panels

S. No.	Specifications	Rating
1	De-rating factor	0.8
2	STC power (kW)	0.1
3	STC irradiation (kW/m^2)	1
4	STC temperature (°C)	25
5	Temperature co-efficient	−0.0011
6	NOCT (°C)	48

0.1 kW and 1 kW, respectively. The detailed specifications of installed solar and wind power generation are tabulated in Tables 5 and 6, respectively.

Table 6 Specifications of wind turbine

S. No.	Specifications	Rating
1	Efficiency	0.4
2	Rotor diameter (m)	1.8
3	Air density (kg/m^2)	1.2
4	Rated wind speed (m/s)	12
5	Rated power (kW)	1
6	Cut-in wind speed (m/s)	3
7	Cut-out wind speed (m/s)	25

Table 7 Rating and costs of RER components and battery

S. No.	Component	Rating	Life time	CCP#	CR#	COM#
1	Solar PV panel	0.1 kW	20 year	$100	$100	$8
2	Wind turbine	1 kW	20 year	$1000	$1000	$200
3	Battery bank	12 V, 75 Ah	5 year	$125	$125	$15

CCP# Cost of capital investment; *CR#* cost of replacement; *COM#* cost for annual operation and maintenance

Table 8 Optimal sizes of RER and battery

S. No.	Resource	Quantity	Total rating
1	Solar PV panel	30 numbers	3 kW
2	Wind turbine	2 numbers	2 kW
3	Battery bank	4 in series	48 V, 75 Ah

6.1 Optimal Sizing of RER and Battery

The sizes of the renewable energy resources and batteries need to be optimally chosen in such a way that the fitness function formulated in (21) is minimized and the constraints expressed in (27)–(29) have to be satisfied. During this optimization process, all the intervals (RER interval (A_{RER}), non-deferrable demand interval (A_{NDL}), battery operating interval (A_{BS}), deferrable demand interval (A_{DL}) and pricing interval (A_{P})) are considered to be an hour i.e., time period of all intervals is 60 min. The required details for the optimal sizing of RER and battery are listed in Table 7. The other user-defined parameters such as maximum capital cost and RER energy fraction are taken as $10,000 and 0.3, respectively. The formulated optimization is solved using Genetic algorithm. The genetic algorithm parameters such as population size and iteration count are considered to be 100 and 50, respectively. The results obtained at the end of the optimal sizing are shown in Table 8.

6.2 Economic Benefit by NREMS

In order to perform the scheduling process effectively, NREMS considers various intervals with different durations. The details of the time period of intervals are listed in Table 9. Further, NREMS considers the following operational constraints of different type of household appliances: the power consumption of any non-deferrable demand during a particular non-deferrable demand interval q is constant; any DL operation is uninterruptable during a particular deferrable demand interval r, if it is planned to operate during that interval; scheduled battery operation such as operating mode and power exchange is fixed for the entire duration of battery operating interval u; the dynamics in RER parameters such as variations in solar irradiation, wind speed,

Table 9 Optimal sizes of RER and battery

S. No.	Interval	Duration	
1	RER interval	A_{RER}	1 min
2	Non-deferrable demand interval	A_{NDL}	1 min
3	Battery operating interval	A_{BS}	5 min
4	Deferrable demand interval	A_{DL}	15 min
5	Pricing interval	A_P	60 min

and temperature are assumed as constant over the entire duration of RER interval *j*; the utility price is fixed by the utility over the duration of pricing interval. The daily timeline diagram of various intervals is depicted in Fig. 3.

The utility price dynamics of a particular day is displayed in Fig. 4. Further, the consumer demand limit is taken as 5 kW and presumed to be fixed for a considered day. The additional payment for consuming beyond the demand limit is computed as 2.5 times the nominal electricity price fixed by utility. The utility assigned prosumer power export limit is taken as 0.5 kW. The utility energy purchasing price from prosumer is assumed to be the same as the nominal price.

The optimization problem associated with the optimal scheduling process of NREMS has been realized using Genetic algorithm. The observations of the case study have been gathered and compared with results obtained when NREMS is not employed. The NREMS effectively schedules the operation of DLs and battery to utilize maximum renewable power generation and reduce the dump load power wastage. The daily energy consumption cost of different household components is presented in Fig. 5. Further, the prosumer per day electricity bill is reduced from 488 cents to 368 cents which shows 24.58% decrement in daily electricity bill. It

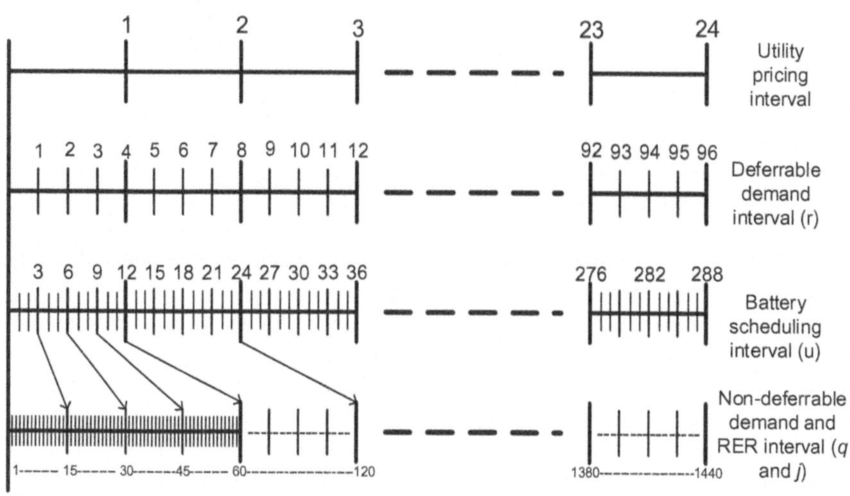

Fig. 3 Timeline diagram of NREMS

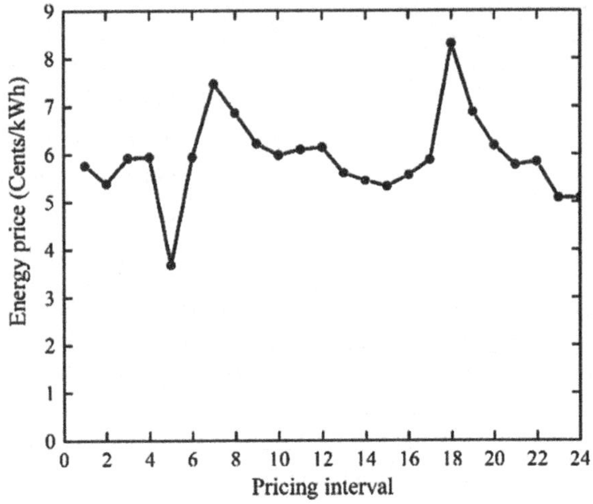

Fig. 4 Electricity price

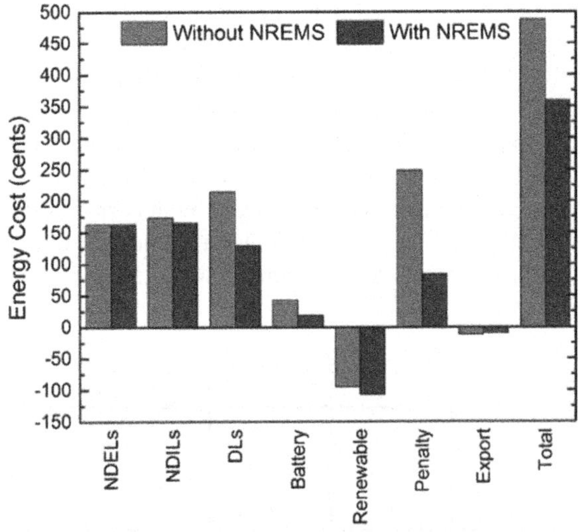

Fig. 5 Comparison of per day electricity bill without and with NREMS

can be seen from Fig. 6 that NREMS considerably reduces the peak demands of the prosumer especially during high price intervals.

The typical daily power consumption pattern of different types of components such as NDELs, NDILs, DLs, battery, and RER are averaged over a time period of deferrable demand interval and depicted in Fig. 7. The operating pattern of battery is observed as variation in state of charge and presented in Fig. 8. Further, the battery SoC during starting is considered to be within its minimum limit. However, at the end of that particular day, NREMS significantly increases the SoC level.

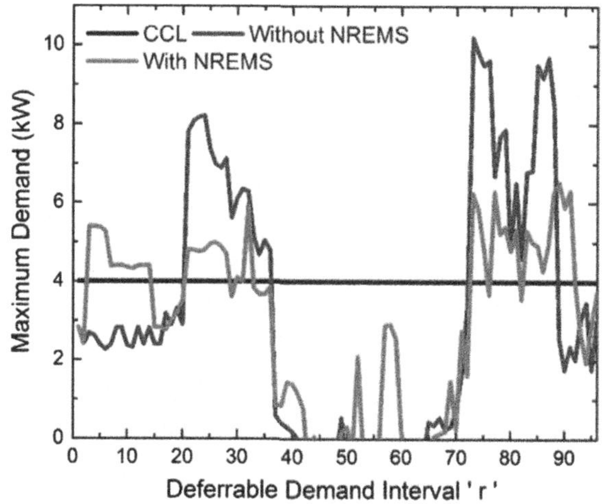

Fig. 6 Comparison of maximum demand without and with NREMS

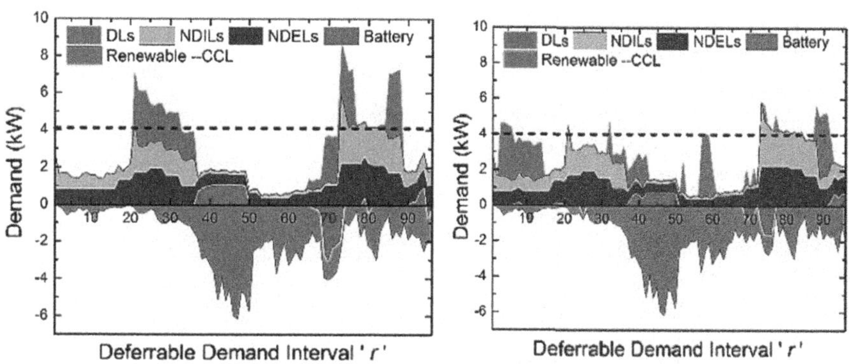

a Without NREMS

b With NREMS

Fig. 7 Average demand variation without and with NREMS

The case study on the considered residential building has been extensively conducted for a period of one year to show the effectiveness of NREMS. In addition to the variation in user behavior and utility dynamics, the sessional variations are also considered for this study. Further, time period of all the intervals (A_{RER}, A_{NDL}, A_{BS}, A_{DL}, A_{P}) are assumed to be 60 min for a better view of results. The month-wise electricity bill of the prosumer without and with employing NREMS is presented in Table 10. Further, the prosumer electricity bill per year is reduced from \$1877 to \$1501, which shows a 20% reduction in yearly electricity bill. In addition to this, NREMS reduces the annual dump load energy dissipation due to utility fixed PEL from 178 kWh to 40.86 kWh, which shows 77% of the increased utilization of generated renewable energy.

However, utilities feeding localities with meagre residential RER installations do not impose any PEL. Considering this scenario, the case study has been extended

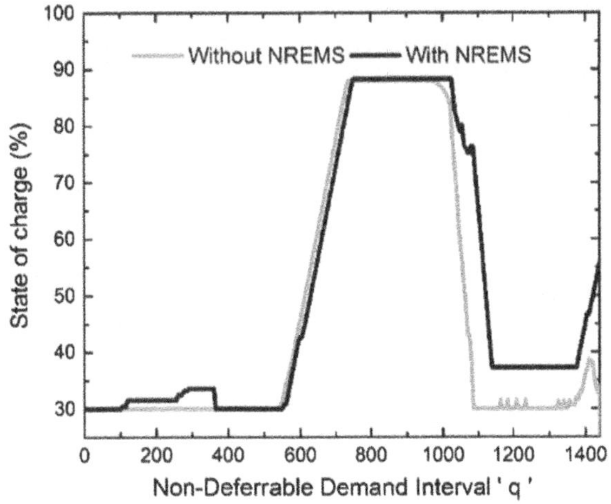

Fig. 8 Battery SoC variation without and with NREMS

Table 10 Monthly electricity bill without and with NREMS

Month	Monthly electricity bill ($)	
	Without NREMS	With NREMS
January	166	123
February	128	103
March	127	102
April	158	127
May	133	104
June	133	107
July	146	118
August	180	147
September	155	128
October	188	154
November	156	123
December	207	166

without PEL constraint. If the utility does not impose any PEL, prosumers can increase their profit by injecting their surplus generation to the grid at a utility fixed price. Further, effective scheduling of battery may improve the profit by selling the stored energy to utility during peak intervals. In addition to these benefits, the prosumer can reduce the initial investment of the project by avoiding the dump load. It is observed in this case study that NREMS reduces the prosumer per day electricity bill from 446 cents to 357 cents. This shows 19.96% decrement in per day electricity bill. The annual electricity bill of the prosumer is also reduced from $1872 to $1470 which indicates 21.47% reduction in yearly electricity bill.

7 Conclusions

In smart grid paradigm, Demand Side Management (DSM) and demand response techniques deliver more benefits, particularly in distribution systems. As part of DSM activities, utilities are adopting different pricing techniques and imposing various operational constraints such as consumer consumption limit and power export limit to end users. In order to reduce the electricity bill and attain more incentives from utilities, the end consumers are participating in demand response programs through energy management system. Novel Residential Energy Management System (NREMS) assists the end-user for active participation in utility DSM schemes to reduce the total cost payable to utility without disturbing the well-being of user. The NREMS optimally schedules the operation of deferrable loads and battery by considering power consumption dynamics of non-deferrable loads, comfort and desire of the user, utility limits, and intermittent behavior of renewable resources. The NREMS also increases the prosumers' profit in energy selling by optimally exporting the surplus renewable power generation to grid at utility desired price. Optimal sizing of renewable energy resources and battery reduces the total capital investment cost significantly and increases the effective utilization of generated renewable power. Further, the scheduling algorithm yields more benefits if the span between initialization time and dead time of deferrable loads is increased.

References

1. Fang, X., Misra, S., Xue, G., Yang, D.: Smart grid—the newand improved power grid: a survey. IEEE Commun. Surveys Tutor. **14**(4), 944–980 (2012)
2. Sheen, J.: Economic profitability analysis of demand side management program. Energy Convers. Manag. **46**(18–19), 2919–2935 (2005)
3. Palensky, P., Dietrich, D.: Demand side management: demand response, intelligent energy systems, and smart loads. IEEE Trans. Ind. Inf. **7**(3),381–388 (2011)
4. Ramanathan, B., Vittal, V.: A framework for evaluation of advanced direct load control with minimum disruption. IEEE Trans. Power Syst. **23**(4), 1681–1688 (2008)
5. Pedrasa, M.A.A., Spooner, T.D., MacGill, I.F.: Scheduling of demand side resources using binary particle swarm optimization. IEEE Trans. Power Syst. **24**(3):1173–1181 (2009)
6. Gomes, A., Antunes, C.H., Martins, A.G.: A multiple objective approach to direct load control using an interactive evolutionary algorithm. IEEE Trans. Power Syst. **22**(3), 1004–1011 (2007)
7. Ruiz, N., Cobelo, I., Oyarzabal, J.: A direct load control model for virtual power plant management. IEEE Trans. Power Syst. **24**(2), 959–966 (2009)
8. Triki, C., Violi, A.: Dynamic pricing of electricity in retail markets. 4OR **7**(1), 21–36 (2009)
9. Wu, L., Shahidehpour, M.: A hybrid model for day-ahead price forecasting. IEEE Trans. Power Syst. **25**(3), 1519–1530 (2010)
10. Herter, K.: Residential implementation of critical-peak pricing of electricity. Energy Policy **35**(4), 2121–2130 (2007)
11. Centolella, P.: The integration of price responsive demand into regional transmission organization (RTO) wholesale power markets and system operations. Energy **35**(4), 1568–1574 (2010)
12. Costanzo, G.T., Zhu, G., Anjos, M.F., Savard, G.: A system architecture for autonomous demand sideload management in smart buildings. IEEE Trans. Smart Grid **3**(4), 2157–2165 (2012)

13. Conejo, J., Morales, J.M., Baringo, L.: Real-time demand response model. IEEE Transact. Smart Grid **1**(3), 236–242 (2010)
14. Siano, P.: Demand response and smart grids—a survey. Renew. Sustain. Energy Rev. **30**, 461–478 (2014)
15. Strbac, G.: Demand side management: benefits and challenges. Energy Policy **36**(12), 4419–4426 (2008)
16. Su, C.L., Kirschen, D.: Quantifying the effect of demand response on electricity markets. IEEE Trans. Power Syst. **24**(3), 1199–1207 (2009)
17. Dell, R.M., Rand, D.A.: Energy storage—a key technology for global energy sustainability. J. Power Sour. **100**(1), 2–17 (2001)
18. Setlhaolo, D., Xia, X., Zhang, J.: Optimal scheduling of household appliances for demand response. Electr. Power Syst. Res. **116**, 24–28 (2014)
19. Aghaei, J., Alizadeh, M.I.: Demand response in smart electricity grids equipped with renewable energy sources: a review. Renew. Sustain. Energy Rev. **18**, 64–72 (2013)
20. Mesari´c, P., Krajcar, S.: Home demand side management integrated with electric vehicles and renewable energy sources. Energy Build. **108**, 1–9 (2015)
21. Saber, Y., Venayagamoorthy, G.K.: Resource scheduling under uncertainty in a smart grid with renewables and plug-in vehicles. IEEE Syst. J. **6**(1), 103–109 (2012)
22. Wu, Y., Lau, V.K.N., Tsang, D.H.K., Qian, L.P., Meng, L.: Optimal energy scheduling for residential smart grid with centralized renewable energy source. IEEE Syst. J. **8**(2), 562–576 (2014)
23. Carrasco, J.M., Franquelo, L.G., Bialasiewicz, J.T., Galvan, E., PortilloGuisado, R.C., Prats, M.A.M., Leon, J.I., Moreno-Alfonso, N.: Power–electronic systems for the grid integration of renewable energy sources: A survey. IEEE Trans. Industr. Electron. **53**(4), 1002–1016 (2006)
24. Liserre, M., Teodorescu, R., Blaabjerg, F.: Stability of photovoltaic and wind turbine grid-connected inverters for a large set of grid impedance values. IEEE Trans. Power Electron. **21**(1), 263–272 (2006)
25. Liu, N., Yu, X., Wang, C., Li, C., Ma, L., Lei, J.: Energy-sharing model with price-based demand response for microgrids of peer-to-peer prosumers. IEEE Trans. Power Syst. **32**(5), 3569–3583 (2017)
26. Varun, Prakash, R., Bhat, I.K.: Energy, economics and environmental impacts ofrenewable energy systems. Renew. Sustain. Energy Rev. **13**(9):2716–2721 (2009)
27. Arun, S.L., Selvan, M.P.: Dynamic demand response in smart buildings using an intelligent residential load management system. IET Gener. Transmiss. Distrib. **11**, 4348–4357 (2017)
28. Arun, S.L., Selvan, M.P.: Intelligent residential energy management system for dynamic demand response in smart buildings. IEEE Syst. J. **12**(2), 1329–1340 (2018)
29. Arun, S.L., Selvan, M.P.: Smart residential energy management system for demand response in buildings with energy storage devices. Front. Energy **13**, 715–730 (2019)

The manufacturer's authorised representative in the EU is Springer Nature Customer Service Centre GmbH, Europaplatz 3, 69115 Heidelberg, Germany. If you have any concerns regarding our products, please contact ProductSafety@springernature.com

Printed and bound by CPI Group (UK) Ltd, Croydon, CR0 4YY
30/09/2026
02225288-0010